Chromatin Structure and Gene Expression

Frontiers in Molecular Biology

SERIES EDITORS

B. D. Hames
Department of Biochemistry and Molecular Biology
University of Leeds, Leeds LS2 9JT, UK

D. M. Glover
Cancer Research Laboratories,
Department of Anatomy and Physiology,
University of Dundee, Dundee DD1 4HN, UK

TITLES IN THE SERIES

1. Human Retroviruses
Bryan R. Cullen

2. Steroid Hormone Action
Malcolm G. Parker

3. Mechanisms of Protein Folding
Roger H. Pain

4. Molecular Glycobiology
Minoru Fukuda and Ole Hindsgaul

5. Protein Kinases
Jim Woodgett

6. RNA-Protein Interactions
Kyoshi Nagai and Iain W. Mattaj

7. DNA-Protein: Structural Interactions
David M. J. Lilley

8. Mobile Genetic Elements
David J. Sherratt

9. Chromatin Structure and Gene Expression
Sarah C. R. Elgin

10. Cell Cycle Control
Chris Hutchison and David M. Glover

11. Molecular Immunology (Second Edn)
B. David Hames and David M. Glover

12. Eukaryotic Gene Transcription
Stephen Goodbourn

13. Molecular Biology of Parasitic Protozoa
Deborah F. Smith and Marilyn Parsons

14. Molecular Genetics of Photosynthesis
Bertil Andersson, A. Hugh Salter, and James Barber

15. Eukaryotic DNA Replication
J. Julian Blow

16. Protein Targeting
Stella M. Hurtley

17. Eukaryotic mRNA Processing
Adrian Krainer

18. Genomic Imprinting
Wolf Reik and Azim Surani

19. Oncogenes and Tumour Suppressors
Gordon Peters and Karen Vousden

Chromatin Structure and Gene Expression

EDITED BY

Sarah C. R. Elgin

Department of Biology
Washington University
St Louis, Missouri, USA

Oxford University Press, Great Clarendon Street, Oxford OX2 6DP
Oxford New York
Athens Auckland Bangkok Bogota Bombay
Buenos Aires Calcutta Cape Town Dar es Salaam
Delhi Florence Hong Kong Istanbul Karachi
Kuala Lumpur Madras Madrid Melbourne
Mexico City Nairobi Paris Singapore
Taipei Tokyo Toronto Warsaw
and associated companies in
Berlin Ibadan

Oxford is a trade mark of Oxford University Press

Published in the United States by
Oxford University Press Inc., New York

First published 1995
Reprinted 1997

A catalogue record for this book is available from the British Library

Library of Congress Cataloging in Publication Data
Chromatin structure and gene expression / edited by Sarah C. R. Elgin.
(Frontiers in molecular biology)
Includes bibliographical references and index.
1. Chromatin. 2. Gene expression. I. Elgin, Sarah C. R.
II. Series.
QH599.C46 1995 574.87'322—dc20 94–42634
ISBN 0 19 963576 5 (Hbk)
ISBN 0 19 963575 7 (Pbk)

Printed in Great Britain by
The Bath Press, Bath

Preface

In the twenty-plus years since the nucleosome model of chromatin structure emerged, there has been considerable progress in elucidating how that structure contributes to the regulatory process. Much of this advance is due, of course, to the new tools of molecular biology, which have allowed us to analyse the packaging and monitor the expression of individual genes, whether *in vitro* or *in vivo*. During the last few years we have gained a much more detailed understanding of the components of the nucleosome, both the histone octamer and the DNA itself, and we are beginning to see that knowledge reflected in studies of replication and assembly. These topics are discussed in the first three chapters. Our initial efforts to understand the mechanisms of gene expression from the chromatin template have focused on active and activatable genes. Some eukaryotic genes, like the heat shock genes, appear poised for expression in an accessible chromatin structure, while others appear to require a chromatin-remodelling event in the promoter region as part of activation. Once initiated, the transcribing polymerase must continue through a nucleosome array—a difficult task. These topics are discussed in Chapters 4–6. As our understanding of the functional consequences of the nucleosome array has increased, we have been emboldened to explore structure at higher levels, trying to establish molecular mechanisms to explain genetic results that have been well known for many years. The last four chapters deal with evidence for domains, boundaries, and packaging differences in the context of known epigenetic regulation in yeast, fruit flies, and mice. A complete understanding of these higher order effects will be essential to a complete understanding of gene regulation during eukaryotic differentiation and development.

In writing our chapters, we have taken as our starting reference *Chromatin* by Kensal van Holde (1989, Springer-Verlag, New York). This book provides a thorough analysis of work prior to 1988, and should be used by the graduate student as a guide to the earlier literature. Younger students desiring a broad introduction should start with the appropriate chapters in a general text (such as *Molecular biology of the cell* by B. Alberts, D. Bray, J. Lewis, M. Raff, K. Roberts, and J. D. Watson (1994, third edition, Garland Publishing Inc., New York)) or a recent monograph (such as *Chromatin structure and function* by A. Wolffe (1992, Academic Press, London)). Many excellent books are available that provide a more detailed examination of some of the topics considered here: *Understanding DNA* by C. R. Calladine and H. Drew (1992, Academic Press, London), *Structural studies of protein–nucleic acid interactions* by T. A. Steitz (1993, Cambridge University Press, Cambridge), *DNA–protein interactions* by A. A. Travers (1993, Chapman and Hall, London), and *Replication and transcription of chromatin* by R. Tsanev, G. Russev, I. Pashev, and J. Zlatanova (1992, CRC Press, Boca Raton, Florida) may be particularly useful.

It is our hope that this book will find use in discussion courses organized for and by graduate students in the area of chromatin structure/gene expression. We have tried to point out not only recent progress in the field, but also the problems and clues that indicate where the work may lead. Thus in addition to writing a chapter in their own area of experimental work, each author read and commented on one or two other related chapters. These comments were used to formulate the 'Discussion' that appears at the end of each chapter.

Using the tools of molecular biology, genetics, and biochemistry, much progress has been made in defining chromatin structure and identifying potential mechanisms for regulation. Thus, in some areas, the problems before us are well defined. However, other areas remain enigmatic: what do we really mean by 'higher order structure'? Here the situation is most intriguing, if somewhat daunting; much remains to be done. We hope that our younger colleagues will be challenged by the problems posed here, and that some will be motivated to join us in our efforts to understand the regulation of gene expression established by nucleosomal and higher order chromatin structure.

St Louis, Missouri S.C.R.E.
March 1995

Contents

The plates referred to on pages 12–15 follow page 14.

Contributors

GINA ARENTS
Department of Biology, Johns Hopkins University, Charles & 34th Street, Baltimore, MD 21218, USA.

HORACE R. DREW
Division of Biomolecular Engineering, CSIRO, P.O. Box 184, North Ryde 2113, New South Wales, Australia.

JOEL C. EISSENBERG
Department of Biochemistry, St Louis University School of Medicine, St Louis, MO 63104, USA.

SARAH C. R. ELGIN
Department of Biology, Washington University, Box 1229, One Brookings Drive, St Louis, MO 63130, USA.

GARY FELSENFELD
Laboratory of Molecular Biology, NIDDK, National Institutes of Health, Bethesda, MD 20892, USA.

FRANK GROSVELD
Department of Cell Biology and Genetics, Erasmus University, P.O. Box 1738, 3000 DR Rotterdam, The Netherlands.

MICHAEL GRUNSTEIN
Department of Biology, University of California at Los Angeles, 405 Hilgard Avenue, Los Angeles, CA 90024, USA.

GORDON HAGER
Laboratory of Molecular Virology, National Cancer Institute, National Institutes of Health, Bethesda, MD 20892, USA.

WOLFRAM HÖRZ
Institut für Physiologische Chemie, Universität München, Schillerstrasse 44, 80336 München, Germany.

RONALD A. LASKEY
Wellcome/CRC Institute, Tennis Court Road, Cambridge CB2 1QR, UK.

JOHN LIS
Department of Biochemistry, Molecular and Cell Biology, Biotech Building, Cornell University, Ithaca, NY 14853, USA.

DONAL LUSE
Department of Molecular Biology, NC20, Cleveland Clinic Foundation, 9500 Euclid Avenue, Cleveland, OH 44195, USA.

EVANGELOS MOUDRIANAKIS
Department of Biology, Johns Hopkins University, Charles & 34th Street, Baltimore, MD 21218, USA, and Biology Department, University of Athens, Athens, Greece.

RENATO PARO
Zentrum für Molekulare Biologie, Universität Heidelberg, Im Neuenheimer Feld 282, D-69120 Heidelberg, Germany.

LORRAINE PILLUS
Department of Molecular, Cellular and Developmental Biology, University of Colorado at Boulder, Porter Biosciences Building, Boulder, CO 80309, USA.

PAUL SCHEDL
Department of Biology, Princeton University, Washington Road, Princeton, NJ 08544, USA.

CATHARINE SMITH
Laboratory of Molecular Virology, National Cancer Institute, National Institutes of Health, Bethesda, MD 20892, USA.

JOSÉ M. SOGO
Institut für Zellbiologie, Eidgenössiche Technische Hochschule, Hönggerberg, 8093 Zürich, Switzerland.

JOHN SVAREN
Institut für Physiologische Chemie, Universität München, Schillerstrasse 44, 80336 München, Germany.

SHIRLEY M. TILGHMAN
Howard Hughes Medical Institute, Princeton University, Lewis Thomas Laboratory, Washington Road, Princeton, NJ 08544, USA.

KENSAL VAN HOLDE
Department of Biochemistry and Biophysics, Oregon State University, 2011 ALS Building, Corvallis, OR 97331, USA.

HUNTINGTON F. WILLARD
Department of Genetics, Center for Human Genetics, Case Western Reserve University, Cleveland, OH 44106, USA.

ALAN P. WOLFFE
Laboratory of Molecular Embryology, National Institute of Child Health and Human Development, National Institutes of Health, Bethesda, MD 20892, USA.

CARL WU
Laboratory of Biochemistry, National Institute of Child Health and Human Development, National Institutes of Health, Bethesda, MD 20892, USA.

JORDANKA ZLATANOVA
Institute of Genetics, Bulgarian Academy of Sciences, Sofia 1113, Bulgaria.

Abbreviations

aa	amino acid
ANT-C	*Antennapedia* gene complex in *Drosophila*
ARS	autonomously replicating sequence
AS	Angelman syndrome
bHLH	basic helix–loop–helix (protein)
bp	base pairs
BWS	Beckwith–Wiedemann syndrome
BX-C	*bithorax* gene complex in *Drosophila*
CAP	catabolite-gene-activating protein
CD	circular dichroism
CHO	Chinese hamster ovary (cells)
CTD	C-terminal domain, generally of RNA polymerase II
DHFR	dihydrofolate reductase
DMS	dimethyl sulfate
DREV	developmental regulator effect variegation
EBV	Epstein–Barr virus
EDTA	ethylene diaminetetraacetic acid
EM	electron microscope
5-FOA	5-fluoroorotic acid
GH1	globular domain of histone H1
H	histone (used to designate H1, H2A, H2B, H3, H4)
HM loci	the loci containing the silent copies of *HML*α and *HMRa* (*Saccharomyces cerevisiae*)
HMG	high mobility group proteins
HOS	higher order structure
HP1	heterochromatin protein 1
HS	hypersensitive site
HSE	heat shock element
HSP 70	the 70 000 u heat shock protein
HSF	heat shock transcription factor
HSH	helix–strand–helix (protein structural motif)
hsp	heat shock protein-encoding gene
ICR	internal control region (often applied to 5S RNA gene)
kb	kilobases
LCR	locus control region
LIS	lithium diiodosalicylate
LMPCR	ligation-mediated PCR
LOH	loss of heterozygosity

LTR	long terminal DNA sequence repeat, found at the ends of a retrovirus
MAR	matrix attachment region
min	minute
MMTV	mouse mammary tumour virus
MNase	micrococcal nuclease
NHCP	non-histone chromosomal proteins
NMR	nuclear magnetic resonance
nt	nucleotides
NTP	nucleotide triphosphates
OBR	origin of bidirectional replication
ORC	origin recognition complex, or origin replication complex
Pc-G	*Polycomb* group of genes (*Drosophila*)
PCNA	proliferating-cell nuclear antigen
PCR	polymerase chain reaction
PEH	paired ends of helices (protein structural motif)
PEV	position effect variegation
PIKA	polymorphic interphase karyosomal association
PRE	*Pc*-G response elements
PS	parasegment
PWS	Prader–Willi syndrome
R	ratio of apparent size, determined by gel electrophoresis, to known size of a DNA fragment in bp (Chapter 2)
Rif1	Rap1 interacting factor
RP-A	replication protein A
SAR	scaffold attachment region
scs	special chromatin structures; apparent boundaries of enhancer function
s	second
SFM	scanning force microscope
SIR	silent information regulator
ss-DNA	single stranded DNA
SV40	Simian Virus 40
TAF	TBP-associated factor
TBP	TATA-box-binding protein
TEA	Tris-EDTA-acetate, a neutral buffer used to run DNA gels
TF	transcription factor
trx-G	*trithorax* group of genes (*Drosophila*)
uH2A	ubiquitinated histone H2A
XIC, *Xic*	X inactivation centre
YAC	yeast artificial chromosomes

1 | Elements of chromatin structure: histones, nucleosomes, and fibres

KENSAL VAN HOLDE, JORDANKA ZLATANOVA, GINA ARENTS, and EVANGELOS MOUDRIANAKIS

1. Introduction: an overview of chromatin structure

The subject of this book is chromatin, its structure, and how this structure functions in regulation of the eukaryotic genome. The biochemists' operational definition of chromatin is probably the best we can do at present: chromatin is the DNA–protein complex extracted from lysed interphase nuclei. Just which of the multitudinous substances present in a nucleus will constitute a part of the extracted material will depend in part on the techniques each researcher uses. Furthermore, the composition and properties of chromatin vary from one cell type to another, during development of a specific cell type, and at different stages in the cell cycle. Nevertheless, we can provide at least a rough description of the features common to chromatin of most eukaryotic cells (1–3).

A typical eukaryotic nucleus contains about 5 pg of DNA, corresponding to about 5×10^9 bp (base pairs). Associated non-covalently with the DNA are about 1 pg each of four basic proteins called histones H2A, H2B, H3, and H4, always present in equimolar quantities. The histones are small proteins, each with molecular mass about 10^4 u. A simple calculation shows that there are about two copies of each histone molecule for every 200 bp of DNA; this fact was instrumental in understanding the subunit structure of chromatin (4, 5). In most cell types, there is also approximately one molecule per 200 bp of a larger, very lysine-rich histone of the H1 class (6).

In the period from 1970 to 1975, research in a number of laboratories led to a major insight into how these histones were organized in chromatin. Electron micrographs of extended chromatin fibres at low ionic strength revealed a 'beaded' structure (Fig. 1a) which condenses into thicker, but still 'beaded' fibres at higher salt concentrations (Figs 1b and c). Digestion of chromatin DNA with micrococcal nuclease led to the generation of a 'ladder' of DNA fragments (on electrophoretic gels) exhibiting a periodicity of about 200 bp. Continued digestion produced

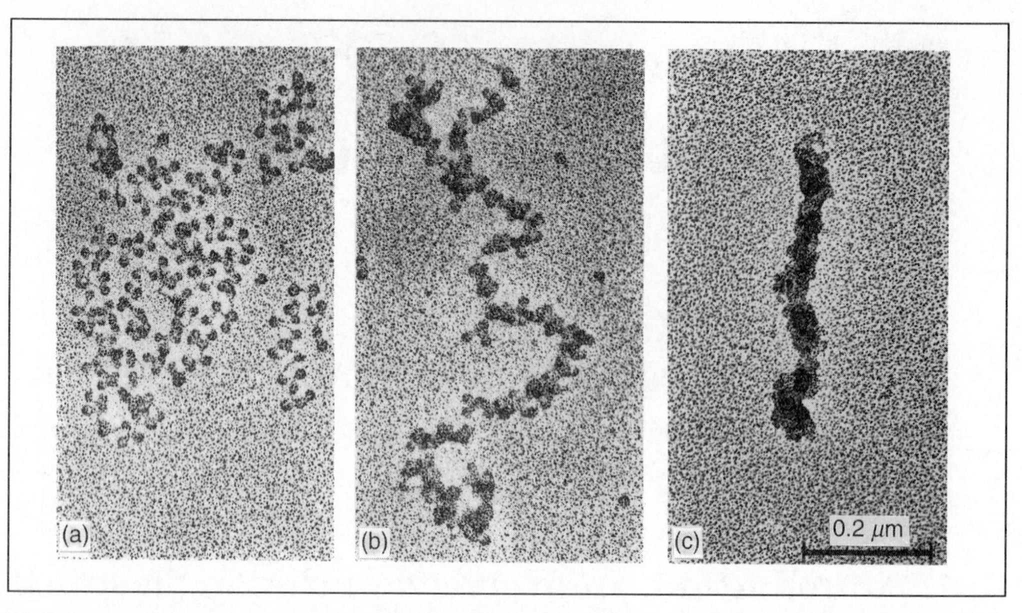

Fig. 1 Electron micrographs of rat liver chromatin fibres under three different ionic conditions: (a) 0 mM NaCl, (b) 20 mM NaCl, and (c) 75 mM NaCl. All samples were fixed in 0.1% glutaraldehyde, 5 mM TEA, 0.2 mM EDTA, pH 7.4, at the corresponding ionic strength. Adopted from (93), with permission of the authors.

chromatin particles containing about 146 bp of DNA and an octamer containing two copies of each of four histones: $(H2A–H2B)_2–[(H3)_2\ (H4)_2]$. These particles were called 'nucleosomes', or more exactly 'nucleosomal core particles,' and these histones the 'core' histones. Chromatin from different cell types will also contain varying quantities of a vast collection of other nuclear proteins referred to as 'non-histone chromosomal proteins' (NHCPs) (see Section 2.3). Neither the H1 class of histones nor the NHCP are present in the core particles. Rather, the H1 and at least some of the NHCP interact with the *linker* DNA which lies between core particles (Fig. 2). Histones of the H1 class are accordingly called 'linker' histones. In many cell types the linker DNA is about 50 bp in length; however, linker length varies from only about 20 bp (in yeast) to over 100 bp in some invertebrate sperm.

The structure of chromatin seems to be conserved with remarkable fidelity over the eukaryotic kingdom. The core histones are highly conserved proteins, and the 146 bp DNA length in the core particle is preserved from yeast to man. Even the most primitive of eukaryotes, such as trypanosomes and unicellular algae, exhibit this structure, with a remarkable exception in the dinoflagellates. But *no* prokaryotic organism has been shown to possess either true histones or the kind of chromatin structure described above. To be sure, there exist small, basic proteins specifically associated with the bacterial nucleoid (7, 8), but the relationship between these 'histone-like' proteins and eukaryotic histones remains obscure.

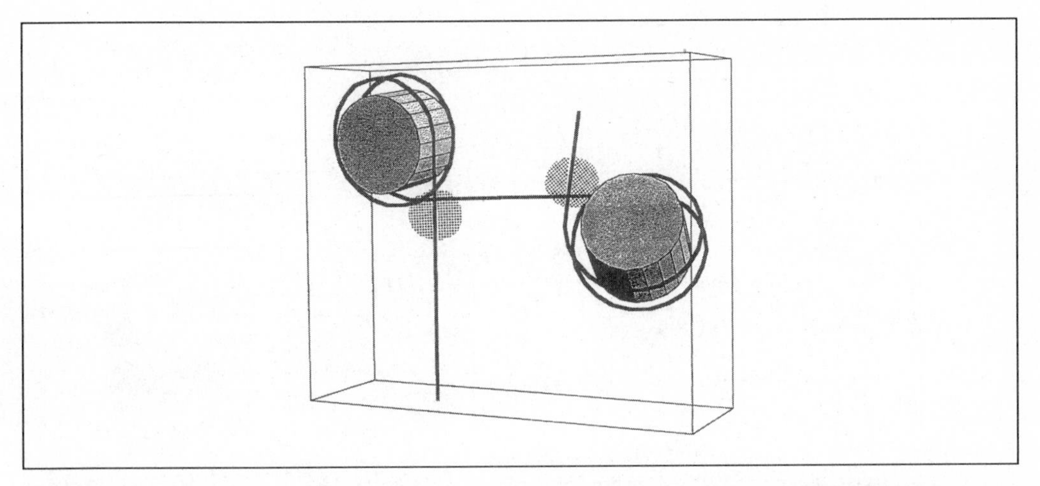

Fig. 2 A schematic view of the current model for the arrangement of histones and DNA in chromatin. Two nucleosomes in a chromatin fibre are shown. The DNA is represented as a heavy line, making 1.75 left-handed superhelical turns (146 bp) about each roughly cylindrical histone octamer. The linker DNA is 62 bp in length, corresponding to a 208 bp repeat. The dotted circles represent the globular regions of H1 molecules, binding near the DNA entry and exit for each nucleosome. The positions of the unstructured N- and C-terminals tails of H1 and of the core histones are unknown, and are not represented. Courtesy of Dr Charles Robert.

This brief overview sets the stage for the questions we wish to consider here:

- What are the properties of the histones and how do they relate to the structure and behaviour of chromatin?
- Precisely how is the histone core of the core particle organized? What kinds of interactions hold it together?
- How does the DNA interact with the histone core in a stable structure?
- Precisely where are H1-class histones located, and what is their function?
- What are the roles of NHCPs in chromatin structure?
- What accounts for the condensation of the chromatin fibre shown in Fig. 1C? What is the structure of the condensed fibres?

2. Chromosomal proteins

2.1 Histone sequences and histone variants

Over four hundred histone sequences are known (9, 10; for a detailed analysis of histone evolution, see 11). The histones are distinguished from other proteins by several features: (1) their highly basic character, (2) a domain structure in which only the central portion of the molecule adopts a structured conformation while the N- and (in some cases) C-terminal tails are unstructured (Fig. 3), (3) strong evolutionary conservation, particularly of H3 and H4, and (4) the presence of non-allelic variants of each histone (except H4) in almost every organism.

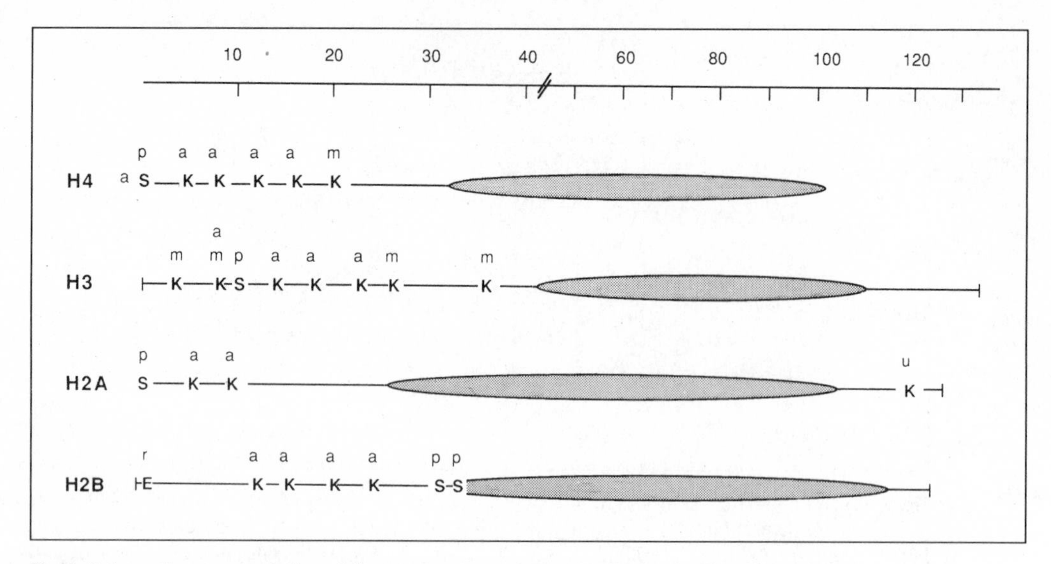

Fig. 3 Schematic representation of the structures and post-synthetic modifications of the core histones. The shaded ellipse represents the 'structured' portion of each histone; the unstructured tails are represented by lines. Code: a = acetylation, m = methylation, p = phosphorylation, u = ubiquitination, r = ADP-ribosylation. Only sites known to be modified *in vivo* are illustrated. Note that the length scale is changed between 40 and 50 residues.

It is in the linker histones that the domain structure is most dramatically exhibited. Three distinct domains are discernible: a strongly basic, unstructured fragment at the N-terminus ('nose'), a non-polar central globular domain ('head'), and again a strongly basic unfolded domain at the C-terminus ('tail') (12). The overall fold of the globular domains of histones H5 and H1 (GH5, GH1) has been determined using two-dimensional NMR (13, 14). The results agree well with the recent crystal structure of GH5 solved to 0.25 nm resolution (15). The domain consists of a three-helix bundle, with a β-hairpin at the C-terminus. Remarkably, the structure is very similar to that of the bacterial DNA-binding protein CAP, and also to the structure of the DNA-recognition motif of the *Drosophila* transcription factor HNF-3 (16). The nose–head–tail structure of the members of the H1 family is so strongly conserved that it is often used as a criterion for identification of H1 proteins. However, some proteins, identified as H1 histones by a number of other criteria, do not contain a typical globular domain, e.g., the *Tetrahymena* H1 (17), *Euplotes* H1 (18), and also one form typical of a terminally differentiated cell type (19).

Discussion of the evolutionary conservation of histones often begins with the observation that cow and pea H4s differ in only two amino acids. While dramatic, this observation gives a somewhat incorrect view of the actual situation. In fact, not all histones are highly conserved. The linker histones H1, H5, and $H1^0$ are highly variable proteins. The H1 proteins of *Tetrahymena*, for example, bear only slight resemblance to those of mammals, and the H1 histone of yeast, if it exists at all, must be so divergent that it has not as yet been positively identified. The variability of the linker histones is concentrated in the nose and tail domains, the centre being more conserved.

It is only H3 and H4 that show truly strong sequence conservation. There are only 8 amino acid differences between human and yeast H4, 19 differences between human and *Tetrahymena* H4, and 21 differences between yeast and *Tetrahymena* H4. Perhaps most remarkable is the fact that the number of residues in each of H3 and H4 has been almost *exactly* conserved in all eukaryotes (102aa (amino acids) for H4, with one exception; always 135aa for H3; see 9, 10). In contrast H2A, H2B, and the H1 proteins exhibit a considerable range of protein lengths. However, even here there are peculiar regularities: the H2A histones show their greatest length variability at the C-terminus, whereas the H2B histones diverge considerably at the N-terminus, with the C-terminus generally fixed. This pattern relates to the way in which H2A and H2B form heterotypic dimers and bind DNA in the nucleosome. Although the sequence homology among the core histones is small, they do share a common feature of tertiary structure—the 'histone fold' (see section 4).

In almost every species of eukaryote there exist non-allelic sequence variants of histones. These isohistones differ in molecular mass, amino acid sequence, and physico-chemical and immunochemical properties. The microheterogeneity is species- and tissue-specific. In general, the tissue specificity is expressed more as a difference in the relative content of various isohistones than as the presence or absence of some of them. All of the histones commonly exhibit such variants, with the exception of H4 (but even here, two variants are known in *Tetrahymena*). We shall use as an example the H1 class, which shows pronounced variance. The most well-studied tissue-specific variant is H5, which accumulates during the process of terminal differentiation of some nucleated erythrocytes and is believed to be involved in the strong chromatin compaction and inactivation that occurs in these cells. Specific subfractions of H1, such as the mammalian H1t, are often present in gametes or their precursors. Some H1 subtypes such as H1cs and H1α of *Strongylocentrotus purpuratus* are observed only during embryonic development. Qualitative and quantitative differences in H1 subtypes have also been observed between normal and malignant tissues. Especially large are the changes in the amount of $H1^0$. This histone, first described in mammalian tissues with little cell division, has been subsequently implicated in the establishment and maintenance of the terminally differentiated state.

Among the core histones, H2A exhibits the greatest diversity of variants. In mammals four iso-H2As have been recognized—two major forms (H2A.1 and H2A.2) and two minor forms (H2A.X and H2A.Z). Mammalian H2A.Z has only about 60% sequence identity with mammalian H2A.1, yet there is over 90% sequence identity between H2A.Z in mammals and in sea urchins. Thus, even though histone variants present within an organism may be quite different from one another, they may still be conserved in evolution. A remarkable fifth H2A variant, called macroH2A (mH2A) has recently been discovered in rat liver (20).

This protein is nearly three times the size of the usual H2As and appears to represent the product of a fusion between an H2A gene and a non-histone protein gene. That this unusual protein is functional is indicated by the fact that it was isolated from core particles.

Despite our ever growing knowledge of the diversity of histone variants, their function remains largely unknown. Together with histone modifications (see below), histone variation allows an enormous complexity in chromatin construction, a complexity whose meaning is still obscure to us.

2.2 Histone modification

Every histone is capable of several forms of post-translational modification. The well-recognized modifications include methylation, acetylation, phosphorylation, polyADPribosylation, and ubiquitination (see 1, 3, 21 for reviews). A remarkable aspect of this panoply of modifications is the specificity exhibited: not only are specific histones modified at specific sites, but specific patterns of modification are seen in relation to the cell cycle, in response to gene activation stimuli, etc. Fig. 3 shows the presently recognized sites for each kind of modification of core histones. Note that the sites have been strongly conserved through evolution, and that they occur almost exclusively in the unstructured N- or C-terminal domains. As with histone variation, our knowledge of the details of histone modification is much more extensive than that of the functional consequences. We consider here only a few of the better understood examples.

2.2.1 Acetylation

Higher levels of acetylation, particularly of histones H4 and H3, have long been associated with the 'opening' of chromatin for transcription. Antiserum against the epitope ε-N-acetyl lysine (the most common histone acetylation site) can selectively precipitate nucleosomes containing active gene sequences (22). Tazi and Bird (23) have observed that hyperacetylated histones tend to cluster near the 5′ ends of genes. In support of the idea that acetylation loosens chromatin structure is the observation that polyacetylation of H4 reduces the binding of the N-terminal tail to DNA (24). However, the relationship between histone acetylation and specific chromatin states may be more subtle than these results might suggest. Turner *et al.* (25), using antibodies directed against specific H4 acetylated sites, found that in polytene chromosomes acetylation at sites lysine$_5$ and lysine$_8$ occurred in 'islands' in euchromatin, whereas heterochromatic, centromeric regions were enriched in forms modified at lysine$_{12}$, and a hyperactive male X chromosome showed specific acetylation at lysine$_{16}$. In addition, defined order in successive occupancy of different acetylation sites has been observed, especially in H3 and H4 (26). These results make it clear that simply correlating *levels* of acetylation with function may be too crude a notion (see 27 for review). Genetic approaches are beginning to allow a functional analysis in yeast (see Chapter 7).

2.2.2 Ubiquitination

H2A and H2B can be modified by the covalent adduction of the small protein ubiquitin to a lysine residue near the C-terminus (Fig. 3). The amount of the modified form varies, but in many cell types ubiquitinated H2A (uH2A) comprises about 10% of all H2A; uH2B is typically present to only 1–2%. Ubiquitination of cytoplasmic proteins is widely known to serve as a signal for protein degradation (28). There is no clear evidence as to what function ubiquitination of histones serves; the modified protein is quite stable. The data as to whether uH2A/uH2B are enriched in transcriptionally active chromatin are contradictory (29–31). Because ubiquitin vanishes from chromatin when cells enter mitosis, it has been argued that ubiquitination prevents condensation of the 30 nm fibre. Note that cells entering mitosis are also shutting down transcription.

2.2.3 Phosphorylation

Although several kinds of phosphorylation sites have been reported on histones (including serine, threonine, histidine, and perhaps lysine and arginine), the most important are serine and threonine. As in other modifications, these lie exclusively on the N- or C-terminal tails (Fig. 3). Although phosphorylation of H3 is well documented, the major target of such modification is H1 (32). H1 phosphorylation might play the role of a trigger in the condensation of chromosomes during mitosis (33). Many supporting studies have utilized the slime mold, *Physarum*, a syncytium which maintains a pool of synchronized nuclei. The level of H1 phosphorylation rises from about 10 residues/molecule in early S-phase to over 20 residues/molecule at metaphase. However, exceptions to this long-held correlation between H1 phosphorylation and chromatin condensation have been reported in a number of systems in which the modification seems to correlate best with *de*condensation of chromatin (34). These exceptions call for a reconsideration of the role of H1 phosphorylation, suggesting at least that this phenomenon may be very different in different systems.

2.3 Non-histone chromosomal proteins

The non-histone chromosomal proteins can be defined as the proteins other than the histones that are associated with chromatin *in vivo*. The proteins that are found associated with isolated chromatin fall into several functional categories: chromatin-bound enzymes, high mobility group proteins, transcription factors, and scaffold proteins participating in the organization of higher order chromatin structure and metaphase chromosome structure. The enzymes associated with chromatin are those involved in DNA replication and repair, in transcription, and in post-translational modification of histones; various types of nucleases and proteases have also been detected (see Table 5–2 in ref. 1). Transcription factors constitute a vast, complex group (3, 35). The scaffold proteins will probably become an

Table 1 Mammalian HMG proteins

Subfamily	Binding to chromatin	Proposed roles
HMG1/2	Linker DNA	—DNA recombination and repair; —DNA replication and chromatin assembly; —General transcription factors; —Stabilizing loop structures
HMG14/17	Nucleosomes	—Enriched in active chromatin, localized only downstream from the starting point of transcription; —Down-regulated during differentiation; —Loosen chromatin HOS
HMGI/Y	Sequence-specific binding to A/T-rich regions; Binding to preferred regions of nucleosomes	—Chromatin condensation; —General transcription factors; —Gene amplification

important research topic, but are as yet poorly defined (36). Here we will discuss the best studied group of NHCP, the high mobility group proteins (HMGs).

The HMG proteins are operationally defined as chromatin proteins extractable with 0.35 M NaCl and soluble in 2% trichloroacetic acid or 5% perchloric acid (37). The HMG proteins possess an extremely high proportion of charged amino acids, both positive and negative (38). These are asymmetrically distributed along the polypeptide chain, giving rise, in some members of the family, to specific three-dimensional structures (39). Several protein subfamilies have been recognized within the mammalian HMGs, each characterized by strong sequence conservation among its individual members, and by similarities in structure and possibly in function (Table 1). Some organisms appear to contain HMG proteins that cannot be easily classified as belonging to any of the recognized mammalian subclasses.

HMG1/2 are a closely related pair of proteins; they comprise the majority of HMG protein in the cell. These proteins are very highly conserved in evolution (40). The molecules have tripartite structures, consisting of two internally repeated, basic, folded domains involved in DNA binding (the so-called HMG1-boxes), and a highly acidic C-terminus of very peculiar sequence, containing about 30 contiguous glutamic or aspartic acid residues. The latter region has been defined as a histone-binding domain. The unique spatial structure of the HMG1-boxes, containing four α-helices arranged in the form of a V-shaped arrowhead, has been resolved by 3-D NMR (41, 42). This is a new type of protein fold to which DNA binds. The binding can be either sequence specific (as in some transcription factors) or structure specific, as in HMG1 itself. The unifying feature of the DNA site to which this fold binds is a highly bent or otherwise distorted form. In chromatin, these proteins bind to the linker DNA. Their function is far from being understood and may be multifaceted (Table 1).

HMG14/17 are another pair of homologous proteins whose role remains elusive, as does the functional difference between the two. These proteins are less conserved in evolution than are the members of the HMG1/2 subclass, with the highest degree of conservation in the DNA-binding domain (40). The proteins exist in random coils in solution, but some ordered conformation may be imposed upon binding to chromatin. HMG14/17 bind to two sites on the core particle. Studies on the structural effect of HMG14/17 binding to chromatin have yielded conflicting results. Recent neutron-scattering studies show that HMG14 binding results in a considerable reduction in the mass per unit length of chromatin fibres (see below), which probably reflects greater spacing between neighbouring nucleosomes along the DNA strand (43). Such general loosening of the higher order structure (HOS) might be a necessary condition for transcriptional activation. A long-standing but still controversial view connects HMG14/17 with transcriptionally active chromatin (43). Some recent studies suggest an inverse correlation between HMG14/17 expression and differentiation (44, 45). Mouse fibroblasts transfected with a human HMG14 gene under the control of an inducible promoter failed to differentiate into myotubes and down-regulated the expression of myogenic differentiation markers like MyoD and myogenin when HMG14 was expressed (46). The molecular mechanisms involved, however, remain completely unknown.

HMGI/Y is another HMG subfamily of highly conserved proteins found in all vertebrates analysed so far. HMGI/Y proteins were first identified as sequence-specific DNA-binding proteins, interacting with α-satellite DNA from monkey cells. They bind selectively *in vitro* to a number of other A/T-rich sequences, all characterized by a narrow minor groove. Such permissive sequence specificity suggests the involvement of these proteins in more than one function (47) (Table 1). Binding to tandem repeats (e.g. centromeres and telomeres) may provide a reservoir from which HMGI/Y may be mobilized to other sites (48), possibly via reversible modifications. HMGI/Y also bind to preferred regions of the nucleosomal particle at the expense of both protein–protein and protein–DNA interactions (49). HMGI/Y proteins have been immunolocalized to certain A/T-rich bands in mammalian metaphase chromosomes. These proteins are phosphorylated by cell-cycle dependent cdc2-like kinases, which are also involved in the cell-cycle dependent phosphorylation of histone H1. This modification reduces the binding affinity to DNA (50). In addition to their possible role as chromatin structural components, HMGI/Y may function as general transcriptional factors (51, 52); moreover, HMGI/Y may be involved in neoplastic transformation and metastatic tumour progression (49).

3. Histone folding and histone assembly

Early work on chromatin was guided by general beliefs that under most solution conditions isolated histones existed as individual chains with ill-defined secondary structure and a high tendency to form large and non-specific aggregates (1). However, by the beginning of the 1970s a few workers began to realize that histones can form specific and stable quaternary complexes with defined stoichiometries.

For example, in 1971 Rubin and Moudrianakis (53) noted 'histone–histone interactions may involve associations of specific histone classes and may result in structures that depend on the juxtaposition of certain classes of histones' as these proteins bind to DNA. In 1972 D'Anna and Isenberg (54) observed that increasing salt concentrations caused a progressive and biphasic conformational change in purified H3 or H4, which could be attributed to an increase in the secondary structure of the proteins. The dependence of this histone folding on protein concentration suggested histone–histone interactions.

These pioneering observations were followed by spectroscopic and hydrodynamic studies (55), which established a hierarchy in the strength of the pairwise histone–histone interactions (H3–H4 $\gg$ H2A–H2B = H2B–H4 > H2A–H4), and defined the stoichiometries of the most stable complexes. By 1973, the existence of $(H3–H4)_2$ tetramers and H2A–H2B dimers as stable entities was widely recognized (1). The insight that an octamer of histones could form the core of a repeating element of chromatin, the nucleosome (4, 5), served as a powerful catalyst in the area of histone biochemistry.

Our knowledge of the non-covalent interactions between elements of the histone core is now quite complete. In 2 M NaCl at neutral pH and 4°C, the core histones form an $(H2A–H2B\text{-}H3–H4)_2$ octamer in equilibrium with its subunits (56–58). At lower ionic strength, or non-neutral pH, or higher temperature, or in the presence of urea, the histone octamer dissociates into its subunits, i.e., three thermodynamically stable entities—two H2A–H2B dimers and one $(H3–H4)_2$ tetramer (Fig. 4) (58). These steps are readily reversible. The octamer is assembled by two distinguishable sets of protein–protein interactions. The first set involves mostly hydrophobic interactions and is responsible for the internal stabilization of the H2A–H2B dimer

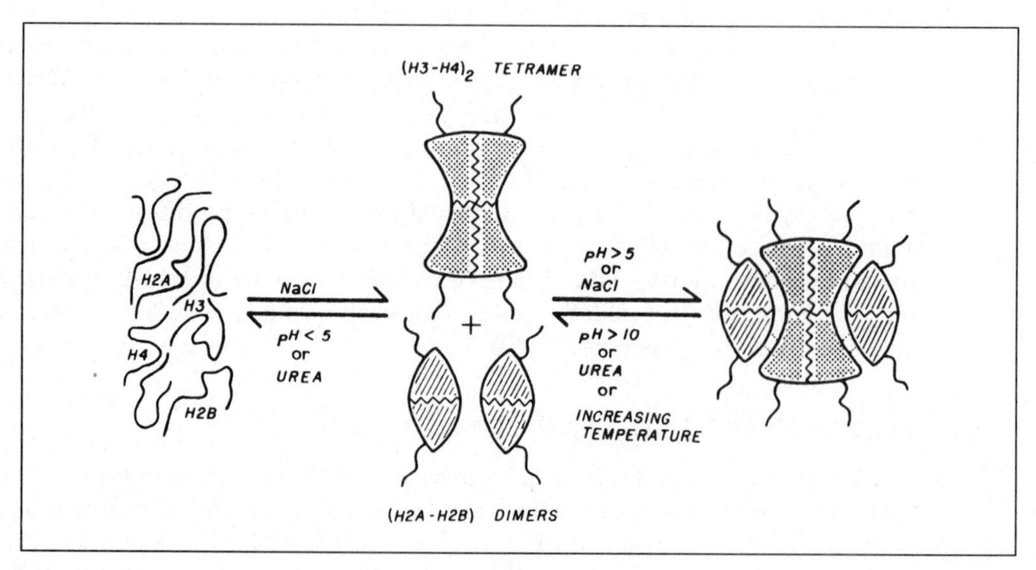

Fig. 4 The histone self-assembly scheme. The system is depicted in freely reversible equilibrium, controlled by mass action and subtle fluctuations of the chemical microenvironment. From (58).

and the $(H3–H4)_2$ tetramer. The second set involves the weak association of one $(H3–H4)_2$ tetramer with two H2A–H2B dimers to form an octamer. These weak interactions are derived primarily from hydrogen bonds between histidine and lysine, or histidine and tyrosine residues (58, 59). When the individual subunits associate, they first form an $(H2A–H2B)(H3–H4)_2$ hexamer intermediate. The binding of the first dimer to one side of the tetramer enhances by a factor of four the intrinsic binding affinity of the second dimer binding site of the tetramer (59, 60); thus, binding of dimers is co-operative.

4. Structure of the histone octamer

4.1 X-ray crystallographic studies

The first low resolution studies of the nucleosome core particle (61–63) described the histone octamer as a cylindrical wedge ~ 6.5 nm in diameter and 5.5 nm long and assigned individual histone domains in it primarily from correlations with the results of protein–DNA cross-linking studies (64) and the proposed organization of the histones within octamers in solution (58). An independent analysis of the histone octamer structure by X-ray crystallography at 0.33 nm resolution (65) revealed a particle that was also ~ 6.5 nm in diameter, but that differed considerably in length (reported to be ~ 10 nm). Despite the excellent quality of the electron density map, the complete chain tracing in this structure could not be achieved due to a peculiar technical problem arising from the location of the heavy metal atom in these crystals. Later, higher resolution data permitted a refinement of the heavy atom position and a calculation of new phases (66), which has led to our current view on octamer structure at 0.31 nm resolution (67).

The 0.31 nm resolution model accounts for about 70 per cent of the entire octamer mass and comprises the structured parts of the eight core histone chains. The 'missing' mass is attributed to the unstructured, flexible termini of the histones. Assuming that these unstructured parts are found in random configuration, one can estimate that, out of the entire mass of the octamer, 45 per cent is found in helices and 7 per cent in beta structure. These estimations are in agreement with earlier spectroscopic findings (1).

The core histone octamer is a tripartite assembly in which a central $(H3–H4)_2$ tetramer is flanked by one H2A–H2B dimer on either side (Fig. 5a). The overall shape is that of a short cylindrical wedge (Fig. 5b and c) with outer diameter of 6.5 nm and a length of 6 nm at the wide side of the wedge and ~ 1 nm at the tip. The outer shape and dimensions are in agreement with those reported earlier (61–63). Within the octamer the four histone dimers are organized into a left-handed protein supercoil with a pitch of 2.8 nm. Within each histone dimer, each histone chain associates with its heterologous partner to form an ellipsoid with one smoothly curving outer face, and this face becomes the source of the 'footprint' of that dimer on the cylindrical surface of the octamer. In forming dimers, the structured part of each histone clasps its partner in a characteristic head-to-tail

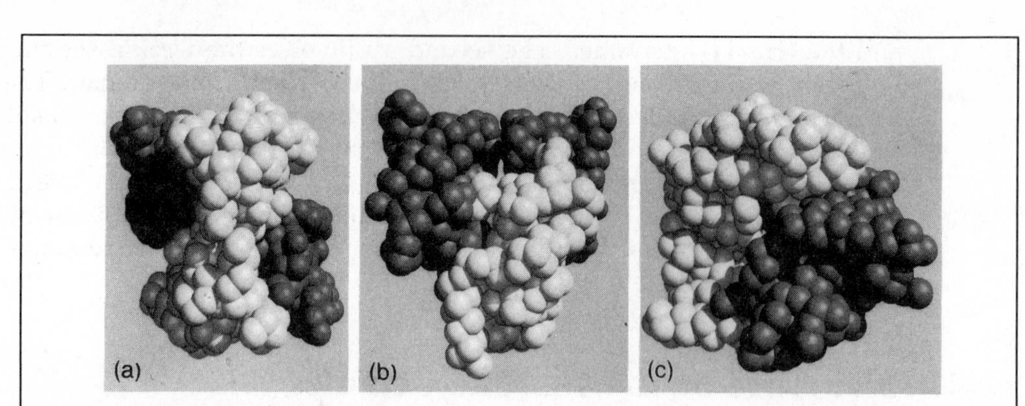

Fig. 5 Space-filling model of the histone octamer. (a) View down the molecular twofold axis, showing the tripartite nature of the histone octamer. The central $(H3–H4)_2$ tetramer (white) is flanked by the two H2A–H2B dimers (dark grey). (b) The molecular twofold axis runs from top to bottom in the plane of the paper. The H2A–H2B dimers form the wide part of the wedge. (c) View down the superhelical axis; the second dimer is the 'back side'.

handshake motif (Plate 1, opposite p. 14) (67). The extensive interface between the two partner chains of a dimer is a result of the highly extended conformation of the individual histones, i.e., the structured parts of the core histones are not globular. This is at variance with earlier assumptions and also with the suggested assignment of histone locations in the low resolution structure of the nucleosome (68).

In spite of their very low sequence homology, the four core histones contain a common tertiary structural motif, the histone fold, which comprises the major portion of their ordered structure (Plate 1, opposite p. 14). The four individual histones also utilize a comparable mode of association to form dimers, the handshake motif. Accordingly, the path of the chains in the H3–H4 dimer is nearly identical to the corresponding path in the H2A–H2B dimer. Although the four histones share both a tertiary and a dimerization motif, their structures differ from one another in two ways. First, each histone has uniquely structured, extra-fold regions. Second, each histone also contains an unstructured amino, and in the case of H2A also a carboxy, terminus.

As a result of the extended structure of the histones and the type of their pairwise association within dimers (handshake motif), each histone chain emerges at multiple places (footprints) on the surface of the octamer (Plate 2, opposite p. 14). This pattern has important implications for the histone–DNA contacts in the nucleosome. The histone fold portion of each histone chain contributes two separate patches to the potential DNA-binding surface; H2A, H2B, and H3 contribute additional potential DNA-binding sites through extra-fold structural elements (Plate 1). Accordingly, each dimer provides at least four distinct histone footprint sites to the cylindrical surface. Starting from the outermost point on the histone cylinder, and following the protein superhelical path towards the twofold axis at the 'zero' position, the structured portions of the histones make footprints on the octamer surface in the following order: $H2A^1$, $H3^2$, $H2A^1$, $H2B^1$, $H2A^1$, $H2B^1$, $H4^1$, $H3^1$, $H4^1$, and finally $H3^1$–$H3^2$ overlapping at the zero position. This sequence is in reasonable agreement with the order of histone–DNA contacts established from cross-linking experiments (64, 69).

4.2 Symmetries of the histone octamer

Closer examination of the octamer reveals a number of repeating motifs of organization of the histone chains. This repetitive pattern results from (a) the regular, spiral arrangement of the dimers within the protein supercoil, (b) the way the histone fold is utilized in the organization of each histone dimer (67), and (c) the presence of a tandem repeat of 3-D structure at about the middle of the histone fold (70).

The histone fold (Plate 1) consists of a short alpha helix (I, ~ 10 residues) followed by a variable-length loop and beta strand segment (strand A), next a long helix (II, ~ 27 residues), and again a short loop and beta strand segment (strand B), and a final short helix (III, ~ 10 residues). Thus, the histone fold has an internal tandem duplication at about the middle of the long helix II. This unit of the histone fold has been termed the helix–strand–helix (HSH) motif.

The HSH motif is the basic building block for the entire histone octamer. Due to the occurrence of the histone fold twice in each histone dimer, a pseudo-twofold axis can be placed in the centre of each dimer. These intra-dimer pseudodyads result from the near identity in structure between the two chains in each dimer and define the relationship between the four HSH motifs in a single dimer. Another set of symmetry axes relates one whole histone dimer to another; these are the inter-dimer pseudodyads. The molecular (true) twofold axis bisects the interface between one H3–H4 dimer and the other, and two inter-dimer pseudodyads relate one H3–H4 dimer to the neighbouring H2A–H2B dimer. Consequently, if H2A is rotated by 180 degrees about one of the intra-dimer pseudodyads, most of its structure can be superimposed on most of H2B and vice versa; an analogous rotation about another intra-dimer pseudodyad superimposes H3 on H4. It should be noted that all seven of the symmetry axes point directly outward from the cylindrical surface of the octamer supercoil—a necessary condition for these elements to lie on the supercoil surface with uniform spacings. The probability for the occurrence of a high degree of pseudosymmetry in the octamer, complementing the symmetry inherent in the DNA helix, has been noted earlier (71).

5. A model for the nucleosome

5.1 Octamer–DNA docking

Along the DNA double helix, a large number of pseudodyads can be placed perpendicular to the helix axis. These pseudodyad axes occur exactly half-way

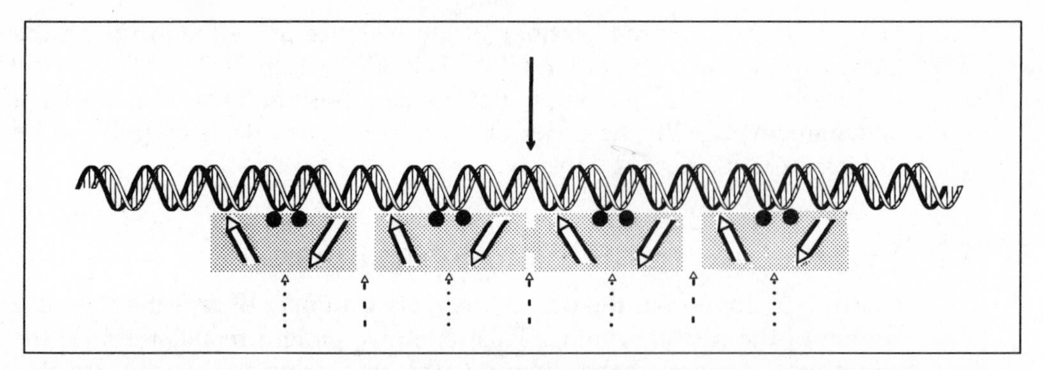

Fig. 6 Linearized view of the relations of histone and DNA symmetries in the nucleosome. Arrows show the positions of the symmetry axes of the octamer; arrows with dotted lines are intra-dimer axes; those with dashed lines are inter-dimer axes. These local pseudosymmetry axes coincide with the pseudosymmetry axes of the DNA double helix. Each histone dimer is represented by a shaded box; inside each box the parallel beta bridges are shown by two lines and the paired ends of helices are symbolized by filled circles. The locations of these repeating protein features coincide with the positions where the minor groove faces the octamer surface.

between adjacent bases and bisect the distance, at that plane, between the phosphodiester backbones, i.e., they lie in the centre of the major and minor grooves. When DNA is wrapped around the octamer, the number of pseudodyad axes operational on the nucleoprotein complex is reduced to those originating from places where the major or minor groove faces directly toward the protein. When the nucleoprotein superhelix is unrolled (Fig. 6), these symmetry axes are retained. In this linearized arrangement, it becomes easier to see that the protein and the DNA pseudodyads can be brought to exact coincidence. Furthermore, the inter-dimer pseudodyads of the octamer are found where the major groove faces inward, and the intra-dimer axes correspond to the positions where the minor groove faces inward.

As detailed elsewhere (67, 70), the histone dimers are formed by the head-to-tail association of each histone chain with its partner. In this association, a beta strand from one chain (strand A) runs parallel to a beta strand from the other chain (strand B) to form a parallel beta bridge, two per dimer. These bridges occur near the junctures of both ends of the long helix (II) with the short helices (I and III, respectively) where helix II emerges on the cylindrical face of the protein supercoil. A second type of repeating secondary structure element is also found in the pattern of footprint sites that each dimer makes on the superhelical face of the octamer, where the amino ends of helix I from both partners in a dimer point toward the surface. This has been named the paired ends of helices (PEH) motif. This motif occurs exactly half-way along the superhelical face of each dimer, and is bisected by the intra-dimer pseudodyads (70).

A closer examination of the relative spacings of the repetitive elements of the protein supercoil of the octamer with respect to the repetitive features of the DNA double helix is instructive. The average number of DNA helix turns per single turn

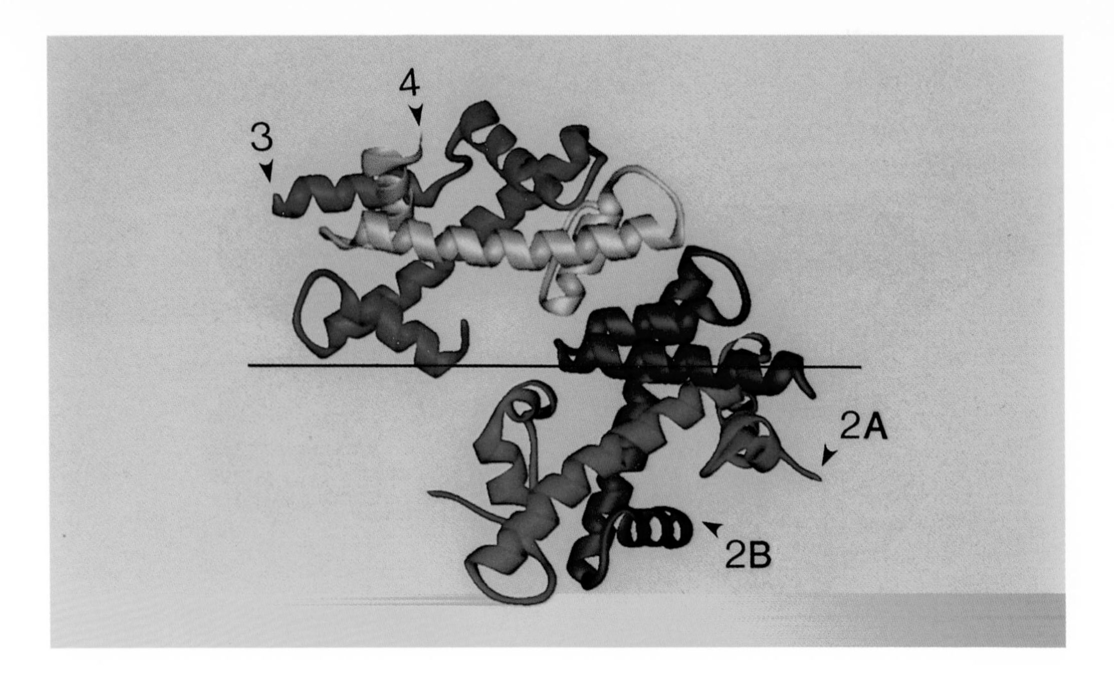

Plate 1 Ribbon model of the four histones. The view is approximately down the suprecoil axis. The H3–H4 pair is above and to the left; the H2A–H2B dimer is at the lower right. The amino-most end of each chain is marked by an arrow. Each histone contains the three helices and two strands of the histone fold, in addition to extra-fold structures. The horizontal black line marks the course of the molecular twofold axis. From (67).

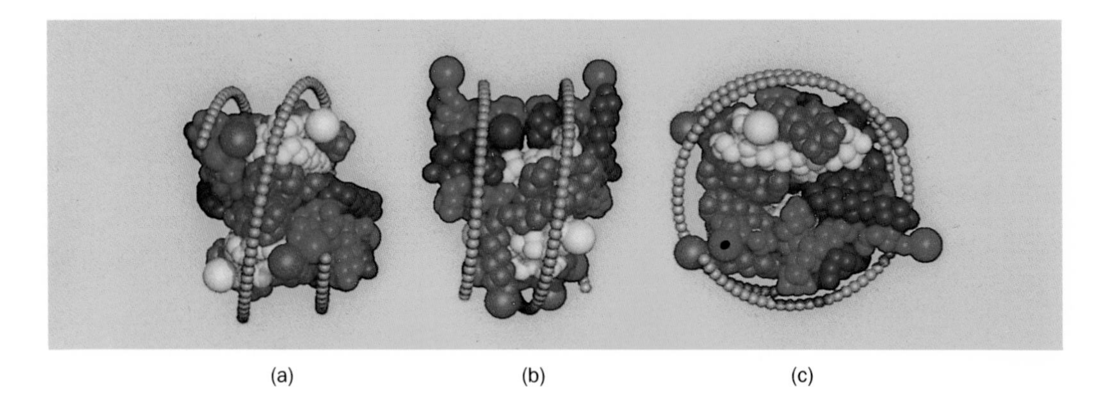

(a) (b) (c)

Plate 2 Histone 'footprints' and the axis of the DNA supercoil in the nucleosome. The footprints of each histone chain on the octamer surface are shown. H3 is shown as green, H4 as white, H2A as light blue, and H2B as dark blue. The axis (not full-width DNA) of the DNA path in the nucleosome is shown by a gray beaded line. Note the alternating pattern of the 'origins' of the amino ends (large spheres); the C-end of H2A is marked by a blue sphere with a black center dot. The views correspond to Fig. 5. From (70).

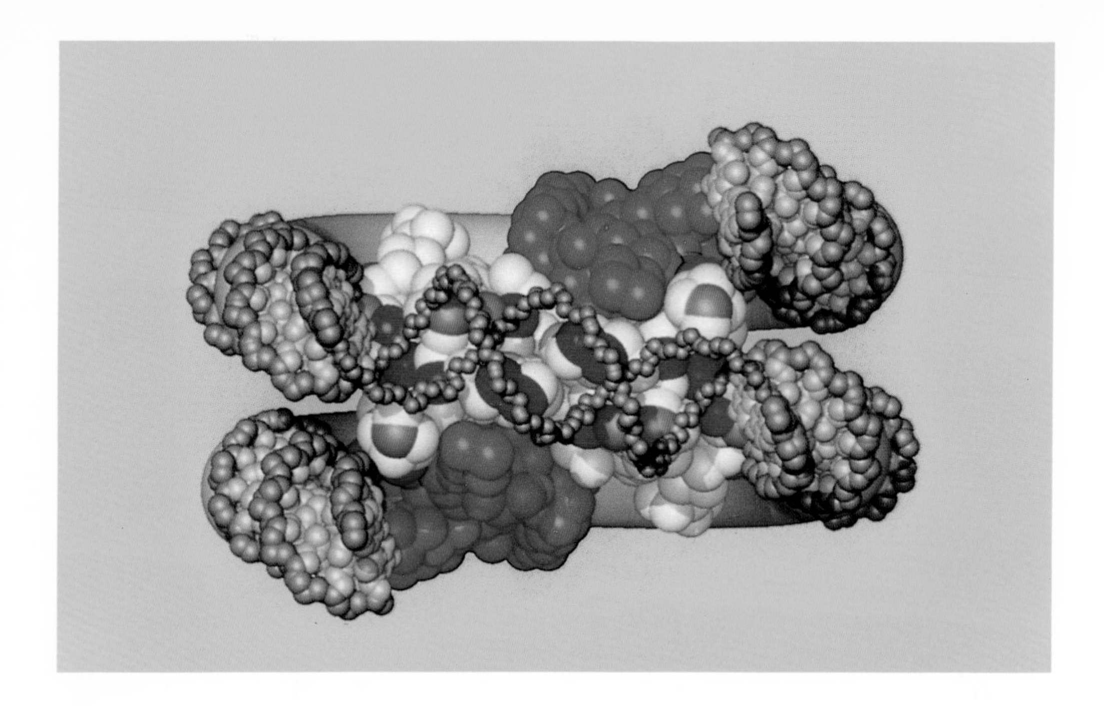

Plate 3 Nucleosome model. A 2 nm-diameter DNA (gray) is wrapped around the histone octamer (dimers are blue, tetramer is white) in the path dictated by the symmetries of the PEHs in the octamer. In the central region of the picture the bases have been stripped away and the paths of the two phosphodiester backbones are marked by 'undersized' dots to permit visualization of the protein surface underneath. The positive charges of that surface (see text) are highlighted orange and red. View is as in Plate 2a.

of the supercoil is about 7.6 (68). Accordingly, the angular spacing between consecutive complete turns of the DNA double helix must be ~ 47 degrees. In the histone octamer structure, the average angular spacing between adjacent repeating elements, beta bridges or PEH, is also ~ 47 degrees. In fact, the occurrence of the repeating elements of the histone supercoil corresponds to those places where the minor groove faces inward toward the octamer surface (Fig. 6). In other words, the positions of the repeating structural elements of the protein spatially correlate with the repeating structural elements of the DNA. Accordingly, the locations and spacing of the repeating elements, when considered together with the pattern of the coincident symmetry axes, imply that the histone octamer interacts with the phosphate backbone of DNA by means of two different but repeating 3-D motifs arrayed in a very specific pattern (70). This pattern of the octamer surface comprises the amino ends of most of its alpha helices (positive end of helix dipole) and a special subset of lysine and arginine residues, all found on the DNA-docking face along a left-handed spiral path (70). The ordered array of these positive charges presents a pattern with remarkable congruity to the footprint of the DNA backbone on the octamer surface (Plate 3, opposite).

Since the N-terminal regions of all four histones and the C-terminal region of H2A are not imaged in the X-ray structure, we can only comment on the 'origins' of the termini—the identifiable positions where these regions start to unfold beyond the octamer surface. These provide a most interesting pattern, for the five origins of the termini are found mostly on alternate sides of the path of the protein supercoil. This can be demonstrated by representing the nucleosomal DNA by a ~.5 nm diameter string wrapped around the histone octamer, centred on the path of the protein supercoil (Plate 2). The alternating pattern of occurrence of these origins to either side of the axis of the DNA path fosters the proposal that *in vivo* the true termini probably 'hug' the double helix like pairs of arms.

5.2 Evolutionary and energetic significance of nucleosome structure

The crystallographic analysis of the histone octamer established that the majority of the masses of its three thermodynamic subunits are arranged in a left-handed spiral with a pitch very similar to that of the left-handed spiral of the nucleosomal DNA. Furthermore, the surface topography of the octamer has repetitive features in a pattern highly complementary to the surface of the double helix (70). This congruence may be indicative of complementarity in the selective processes during the evolution of the double helix and the core histones for the following reasons.

The histone octamer serves as an articulated spool around which the DNA is compacted approximately sixfold in the form of the nucleosome and thus alters (mostly hinders) the access of relevant enzymes and control factors to the genetic information in the helix. During the normal cycling of the genetic material through *on* and *off* states, the histone–DNA interactions must be modulated, sometimes even completely disrupted. Since almost all the DNA in the nucleus is in the form of nucleosomes, if the energetics of these anticipated transformations were costly their impact on the overall energy balance of the cell would be overwhelming. Thus, for these two macromolecular assemblies, there would have been evolutionary advantage for the selection of structural attributes that minimize the energy cost of their disruption. The selective pressures must be more stringent for the histone octamer than for the double helix since the octamer is on the 'inside' of the nucleosome assembly while the double helix has approximately half of its surface (the side away from the octamer) facing free solvent. Thus, the histone octamer needs to be multiply constrained in order to preserve 'outer' 3-D characteristics that will favour its assembly with the double helix. With such multiple constraints simultaneously imposed on the selection of a histone core fit to be selected for a nucleosome, it is easy to appreciate why the histone chains, especially their structured parts, have been so well conserved throughout the eukaryotes, where their most common (though not their only) role is to serve as 'articulate spools' of maximum DNA compaction. At the two 'ends' of the nucleosome (the relatively flat faces pierced by the superhelical axis), the histone octamer exposes to the solvent significant areas of the H2A–H2B dimers. The segments of the histone chains found in these areas are likely to be under more relaxed selective pressures; indeed, this is where more variations have been observed in histone primary structure (9). The presence of a tandem repeat within the histone fold of all four core histones led Arents and Moudrianakis earlier to propose that all core histones might have evolved from a primordial gene encoding a protein analogous to half of the size of a present-day histone, and to suggest that histone-fold-like structures may be present in simpler forms of life, especially the Archea (70).

From the standpoint of chromatin dynamics, if the nucleosome has to be transformed into a more 'open' structure during any of its functional transitions the dimer/tetramer interface will open first, and this will in turn foster a relaxation of the overall coiling of nucleosomal organization, involving both the DNA and the protein supercoils. It has been proposed (59) that such coupled conformational transitions initiate the functional transitions of chromatin that underlie the expression of the genetic activity of chromosomes. Finally, more open chromatin conformations may be selectively achieved by incorporation in the nucleosome of specialized histone subclasses having significantly different stereochemistry, or by post-translational histone modifications.

6. Chromatin structure above the core particle level

6.1 The role of H1: the chromatosome

The core particle, obtained after extensive nuclease digestion of chromatin, contains only the four core histones together with 146 bp of DNA. Milder digestion leads to the production of 168 bp particles—chromatosomes—each containing, in addition, one copy of H1 (72). Comparison of the products of micrococcal nuclease digestion of H1-depleted chromatin and of chromatin reconstituted with H1 showed that significant protection of the 168 bp DNA is observed only in samples containing the intact H1 molecule or its globular domain (73). The location of H1 close to the nucleosomal core has been directly demonstrated in protein/DNA or protein/protein cross-linking experiments. Cross-linking occurs with all core histones, and also with HMG14.

However, using covalent protein/DNA cross-linking procedures, Mirzabekov *et al.* (69) have shown that H1 protects both ends of the chromatosomal DNA only in isolated nucleosomes and in unfolded chromatin. Within nuclei, H1 interacts with the linker DNA at just one side of the nucleosome, suggesting that its globular domain interacts with the central portion of the linker, with additional H1 contacts on the linker DNA of neighbouring nucleosomes or on adjacent turns of the fibre (solenoid). Upon decondensation of chromatin, H1 is redistributed in such a manner that its globular domain becomes bound to the linkers on both sides of the same nucleosome, and this leads to the appearance of a chromatosome. Similarly, the interaction of H5 with DNA in nuclear chromatin of chicken erythrocytes differs from that in isolated mononucleosomal particles and in unfolded chromatin (74). Initial observations indicating that the linker histones bind to the entry/exit crossing point of the DNA in the nucleosome might hold true only for chromatin in its extended conformation, observed in solution at low ionic strength, and possibly occurring in transcribing chromatin.

The interactions of the linker histones with DNA molecules which cross over each other have been recently studied with the help of stable four-way junction DNA molecules, considered to be structurally similar to DNA cross-overs. The linker histones exhibit a high preference for binding to such structures over linear DNA molecules of the same overall sequence (75); this preferential binding is determined by the globular domains of the molecules (76). A similar preference for binding crossed DNA molecules has been previously reported for HMG1, another major linker DNA-binding protein (see above) (77). The two linker DNA-binding proteins might exert their presumptive functions in silencing (H1) or activating (HMG1) transcription via competition for the DNA entering and exiting the nucleosome.

6.2 Higher order structure

The DNA in a typical eukaryotic genome would be about 1 m long if displayed as an extended double-stranded molecule. The organization of DNA into nucleosomal chains would reduce this length to about 15–20 cm. However, the chain of nucleosomes must be further compacted to fit into a nucleus of a few micrometres. These higher order structures (HOS) must allow local and temporal unfolding for replication and transcription to occur, as well as be amenable to further condensation in metaphase. What is the organization of the compacted fibre and how do transient changes take place?

These questions have been extremely difficult to approach. The organization of the fibre in the nucleus itself can only be inferred from electron microscopic imaging and from some physical techniques. The difficulty in obtaining long oriented fibres of chromatin for X-ray diffraction studies has forced researchers to turn to alternative techniques like X-ray, neutron, and light scattering, linear dichroism, hydrodynamic methods, and, most of all, a variety of electron microscopic (EM) techniques. The EM images, although an invaluable tool in ultrastructural research, are still prone to artefacts (1, 78). By necessity, most EM studies have utilized chromatin fragments isolated from lysed nuclei. However, the structure of fibres isolated in this way may be irreversibly damaged (79). Further, the studies are performed in ionic environments which may or may not adequately mimic the salt milieu of the nuclear sap. Finally, most of these studies require the use of relatively dilute solutions in which inter-fibre interactions are absent, unlike the actual situation in the nucleus. Recently, the conventional transmission EM has been supplemented by powerful analytical tools like tomographic reconstruction of images or by less sample-destructive methods like cryoelectron microscopy, in which specimens are examined in vitrified ice without staining or shadowing.

No available method can provide unambiguous information as to how the nucleosomal chain is folded to form the HOS. On the basis of the existing data two general classes of HOS have been suggested: regular helical arrangements of some kind, and discontinuous structures, in which globular clusters of nucleosomes—'superbeads'—are linearly arranged to form the thick, condensed fibre. Many models have been suggested for the helical arrangement of nucleosomes (1, 3, 80, 81). Unfortunately, none of the helical models is compatible with all of the existing data, and the data available are often quite controversial.

The actual data and the major controversies surrounding the issue of the HOS can be summarized as follows:

- The *diameter* of the fibre is around 30 nm, accounting for the name '30 nm fibre'. Measurements vary from around 25 nm to around 45 nm depending on the source of chromatin, on the preparation of the sample, and the method of investigation. A major unresolved issue is whether or not the diameter of the fibre depends on the linker length; experimental results supporting both possibilities have been reported.
- *Mass per unit length*. The density of the fibre is usually expressed as the number of nucleosomes per defined fibre length. Most of the data show a transition from about one nucleosome/10 nm at 0 salt to 6–8 nucleosomes/10 nm at 60–100 mM salt or at mM concentrations of Mg^{2+}.
- *Orientation of the nucleosomes*. Several kinds of evidence (X-ray diffraction from oriented fibres, electric dichroism) suggest that the flat faces of the nucleosomes are approximately parallel to the fibre axis, but may exhibit considerable variability in tilt.
- *Location and organization of the linker DNA*. This parameter of the fibre structure has been the matter of greatest controversy. Reliable data remain scarce. The contribution of the linker DNA to the low resolution X-ray-scattering pattern is

negligible; therefore, these patterns mainly reflect the intrinsic features of the core particles and their mutual arrangements in the fibre. Some information about the location and structure of the linker DNA has been acquired by biochemical methods. Analyses of micrococcal nuclease digestion patterns of chromatin from a variety of sources (69) suggested that the linker DNA is organized in a manner very similar to the organization of the core DNA and thus follows, together with the latter, a continuously supercoiled path.

In a recent study, chromatin fibres of different conformations were digested with micrococcal nuclease under extremely mild conditions, in the absence of added divalent ions (82). While linker DNA was readily accessible to nucleolytic attack in the extended low salt conformations, it was almost completely protected against digestion in the condensed high salt conformation; fibres of intermediate degrees of compaction were digested to intermediate degrees. This suggests that access to the linkers is limited by local steric hindrance due to high compaction rather than by linker internalization to the centre of the fibres, as predicted by some models.

- *Role and location of the linker histones*. Histone H1 is essential for 30 nm fibre formation. H1-depleted chromatin or reconstituted oligonucleosomes lacking H1 can undergo some compaction upon increasing the ionic strength, but no fibers similar to those of H1-containing chromatin have been reported. Re-addition of isolated structural domains of H1 has shown that the C-terminal tails, as well as the globular domains, of the H1 molecules are required for chromatin to condense (83).

The location of the linker histones in the HOS is an unresolved issue. Most studies have used immunochemical approaches, with controversial results (84). Some authors assert that the linker histones are internalized upon condensation of the fiber, whereas others interpret their data as indicating no change in the location of these molecules. Recently, immobilized proteases like trypsin and chymotrypsin have been used in an alternative approach to this problem (85). It was found that, while the N- and C-terminal portions of histone H1 remained accessible to trypsin upon fibre condensation, the tails of H5 became significantly inaccessible in the condensed fibre.

Application of bifunctional cross-linking agents to chromatin and nuclei results in the formation of H1 or H5 homopolymers. This has suggested that the linker histones are located in the centre of the compacted fibre. In fact, all it shows is that linker histone molecules are situated closely enough to each other to allow cross-linking to occur. Similar cross-linked products are formed at both low ionic strength, when the fibre is in an extended conformation, and at high ionic strength, when the fibre is highly compacted (86). In either case, very few cross-linked products larger than the hexamer are found. Other experiments revealed discrete H1 homopolymers, in this case integral multiples of 12 H1 molecules (87). The structure that determines this cross-linking pattern exists not only in nuclei, but also in the extended nucleosomal chain. Thus, the organization of the nucleosomal chain itself may include the structural features necessary and sufficient to determine the HOS of chromatin.

7. Final comments

7.1 Important questions

The form and function of HOS remain a key question. Some new ultrastructural data were recently obtained with the help of the scanning force microscope (SFM), under conditions much less harsh than those used in traditional EM procedures. The SFM images reveal that the fibres exist as irregular 3-D arrays of nucleosomes, of average diameter about 34 nm, even at low ionic strength (Fig. 7). Such a structure can be modelled using extended and rigid linkers with fixed exit–entrance

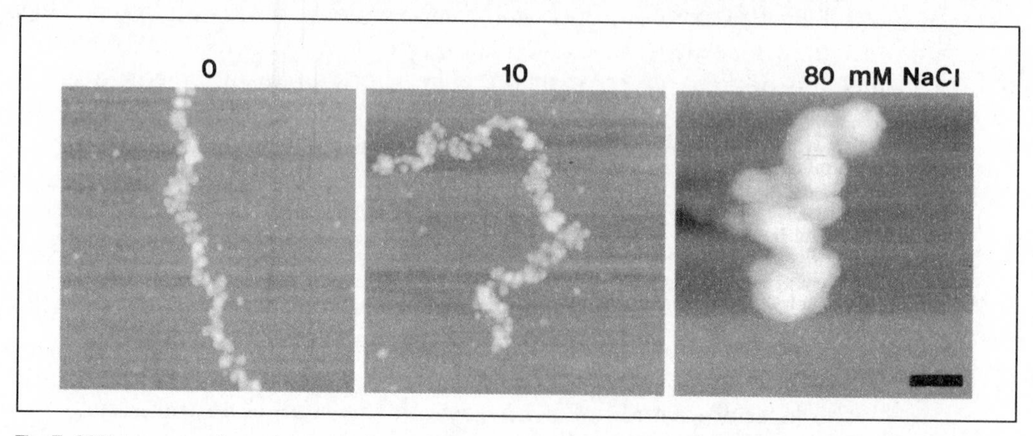

Fig. 7 SFM images of long chromatin fibres fixed with glutaraldehyde in 5 mM TEA–HCl (pH 7.0), in buffers containing (left to right) 0 mM NaCl, 10 mM NaCl, and 80 mM NaCl. The sample on the left includes purified mononucleosomal particles added as an internal control for measurements of fibre widths. From (90), with permission.

angles of the DNA in the nucleosome. If the length of the linkers varies, these structures are *not* regular helices (88, 89), and may not be expected to condense into *any* regular helical structures. As the salt concentration is increased, the fibres condense to highly compact, irregularly segmented structures (90) (Fig. 7), reminiscent of superbead fibres reported earlier (1, 3). Intrinsic irregularities of the structural organization of the condensed fibres have, in fact, been reported in the literature previously (78, 88, 91–93), but have, in our opinion, never received due attention.

7.2 Discussion

In discussion with other book authors, it was pointed out that the detailed picture of the histone octamer obtained from crystallographic studies (Plates 2 and 3) (67, 70) supports previous rationales concerning histone change over evolution. We observe that the extent of evolutionary change in the histones, as well as the number of variants, increases in the order H4, H3 $<$ H2A, H2B $<$ H1. This appears reasonable in that H3 and H4 form the core of the histone octamer, while H2A and H2B are much more exposed on the octamer 'faces'. H1 interacts with many different environments outside the nucleosome. The new structural studies also illuminate the contacts between the core histones and the DNA that stabilize the nucleosome. These contacts are extensive, and independent of the histone N-terminal tails. It has been suggested that there seems to be a basic paradox in the fact that histones are acetylated very specifically at highly conserved sites, but such acetylation seems to have little effect on the stability of isolated core particles. However, these modifications are in the N-terminal tails. It seems likely that the tails interact with other structural components *in vivo*, and thus do not affect the stability of the core particle itself, but may have profound affects on access and on higher order structures (see Chapter 2).

Despite the commonly accepted use of a regular model for 30 nm fibre, the evidence for such a regular structure is very slim. Most electron micrographs demonstrate only very limited regions that might be called regular supercoils. Diffraction from chromatin fibres gives hints of regular spacing and orientation, but the data are ambiguous. A key issue in considering past and present models is the degree of regularity in linker lengths, and the components of chromatin that set linker length. While local DNA sequence will have an affect on nucleosome positioning (see Chapter 2), other structural components may also play a role.

The fact that H1 binds to the linker DNA has led to the suggestion that the species- and tissue-specific differences in average linker lengths could be explained by differences in H1 structure. Experiments designed to resolve this issue, however, have failed to show a direct relation between changes in the H1 complement and changes in linker lengths. Replacement of H1 by H5 *in vivo* does not lead to a change in the repeat length in chromatin (94), but only increases its stability. It is possible that non-histone proteins such as HMG14 and 17 could play an important role in setting nucleosome spacing (95). Building from our knowledge of the nucleosome core to the higher order structures that exist *in vivo* is clearly one of the important challenges of the next several years.

References

1. van Holde, K. E. (1988) *Chromatin*. Springer-Verlag, New York.
2. Wolffe, A. P. (1992) *Chromatin: structure and function*. Academic Press, London.
3. Tsanev, R., Russev, G., Pashev, I., and Zlatanova, J. (1992) *Replication and transcription of chromatin*. CRC Press, Boca Raton, Florida.
4. Kornberg, R. D. (1974) Chromatin structure: a repeating unit of histones and DNA. *Science*, **184**, 868.
5. van Holde, K. E., Sahasrabuddhe, C. G., and Shaw, B. R. (1974) A model for particulate structure in chromatin. *Nucleic Acids Res.*, **1**, 1579.
6. Zlatanova, J. and van Holde, K. (1992) Histone H1 and transcription: still an enigma. *J. Cell Sci.*, **103**, 889.
7. Drlica, K. and Rouviere-Yaniv, J. (1987) Histone-like proteins in bacteria. *Microbiol. Rev.*, **51**, 301.

8. Pettijohn, D. E. (1990) Bacterial chromosome structure. In *Nucleic acids and molecular biology*, Vol. 4. Eckstein, F. and Lilley, D. M. J. (ed.). Springer-Verlag, Berlin, Heidelberg, p. 153.
9. Wells, D. and McBride, C. (1989) A comprehensive compilation and alignment of histones and histone genes. *Nucleic Acids Res.*, **17** (suppl.), r311.
10. Wells, D. and Brown, D. (1991) Histone and histone gene compilation and alignment update. *Nucleic Acids Res.*, **19** (suppl.), 2173.
11. Thatcher, T. H. and Gorovsky, M. A. (1994) Phylogenetic analysis of the core histones H2A, H2B, H3 and H4. *Nucleic Acids Res.*, **22,** 174.
12. Hartman, P. G., Chapman, G. E., Moss, T., and Bradbury, E. M. (1977) Studies on the role and mode of operation of the very lysine-rich histone H1 in eukaryotic chromatin. The three structural regions of the histone H1 molecule. *Eur. J. Biochem.*, **77,** 45.
13. Clore, G. M., Gronenborn, A. M., Nilges, M., Sukumaran, D. K., and Zarbock, J. (1987) The polypeptide fold of the globular domain of histone H5 in solution. A study using nuclear magnetic resonance, distance geometry and restrained molecular dynamics. *EMBO J.*, **6,** 1833.
14. Cerf, C., Lippens, G., Muyldermans, S., Segers, A., Ramakrishnan, V., Wodak, S. J., Hallenga, K., and Wyns, L. (1993) Homo- and heteronuclear two-dimensional NMR studies of the globular domain of histone H1: sequential assignment and secondary structure. *Biochemistry*, **32,** 11345.
15. Ramakrishnan, V., Finch, J. T., Graziano, V., Lee, P. L., and Sweet, R. M. (1993) Crystal structure of globular domain of histone H5 and its implications for nucleosome binding. *Nature*, **362,** 219.
16. Clark, K. L., Halay, E. D., Lai, E., and Burley, S. K. (1993) Co-crystal structure of the HNF-3/fork head DNA-recognition motif resembles histone H5. *Nature*, **364,** 412.
17. Wu, M., Allis, C. D., Richman, R., Cook, R. G., and Gorovsky, M. A. (1986) An intervening sequence in an unusual histone H1 gene of *Tetrahymena thermophila*. *Proc. Natl Acad. Sci. USA*, **83,** 8674.
18. Hauser, L. J., Treat, M. L., and Olins, D. E. (1993) Cloning and analysis of the macronuclear gene for histone H1 from *Euplotes eurystomus*. *Nucleic Acids Res.*, **21,** 3586.
19. Azorin, F., Olivarez, C., Jordan, A., Perez-Grau, L., Cornudella, L., and Subirana, J. A. (1983) Heterogeneity of the histone-containing chromatin in sea cucumber spermatozoa. Distribution of the basic protein f_0 and absence of non-histone proteins. *Exp. Cell Res.*, **148,** 331.
20. Pehrson, J. R. and Fried, V. A. (1992) MacroH2A, a core histone containing a large nonhistone region. *Science*, **257,** 1398.
21. Wu, R. S., Panusz, H. T., Hatch, C. L., and Bonner, W. M. (1986) Histones and their modifications. *CRC Crit. Rev. Biochem.*, **20,** 201.
22. Hebbes, T. R., Thorne, A. W., and Crane-Robinson, C. (1988) A direct link between core histone acetylation and transcriptionally active chromatin. *EMBO J.*, **7,** 1395.
23. Tazi, J. and Bird, A. (1990) Alternative chromatin structure at CpG islands. *Cell*, **60,** 909.
24. Hong, L., Schroth, G. P., Matthews, H. R., Yau, P., and Bradbury, E. M. (1993) Studies of the DNA binding properties of histone H4 amino terminus. *J. Biol. Chem.*, **268,** 305.
25. Turner, B. M., Birley, A. J., and Lavender, J. (1992) Histone H4 isoforms acetylated at specific lysine residues define individual chromosomes and chromatin domains in *Drosophila* polytene nuclei. *Cell*, **69,** 375.
26. Thorne, A. W., Kmiciek, D., Mitchelson, K., Sautiere, P., and Crane-Robinson, C. (1990) Patterns of histone acetylation. *Eur. J. Biochem.*, **193,** 701.

27. Turner, B. M. (1991) Histone acetylation and control of gene expression. *J. Cell Sci.*, **99,** 13.
28. Finley, D. and Chau, V. (1991) Ubiquitination. *Annu. Rev. Cell Biol.*, **7,** 25.
29. Bonner, W. M., Hatch, C. L., and Wu, R. S. (1988) Ubiquitinated histones and chromatin. In *Ubiquitin*. Rechsteiner, M. (ed.). Plenum Press, London, p. 157.
30. Dawson, B. A., Herman, T., Haas, A. L., and Lough, J. (1991) Affinity isolation of active murine erythroleukemia cell chromatin: uniform distribution of ubiquitinated histone H2A between active and inactive fractions. *J. Cell. Biochem.*, **46,** 166.
31. Nickel, B. E., Allis, C. D., and Davie, J. R. (1989) Ubiquitinated histone H2B is preferentially located in transcriptionally active chromatin. *Biochemistry*, **28,** 958.
32. Th'ng, J. P. H., Guo, X.-W., and Bradbury, E. M. (1992) Histone modifications associated with mitotic chromosome condensation. In *Molecular and cellular approaches to the control of proliferation and differentiation*. Stein, G. S. and Lian, J. B. (ed.). Academic Press, San Diego, etc., 381.
33. Bradbury, E. M., Inglis, R. J., and Matthews, H. R. (1974) Control of cell division by very lysine rich histone (F1) phosphorylation. *Nature*, **247,** 257.
34. Roth, S. Y. and Allis, C. D. (1992) Chromatin condensation: does histone H1 dephosphorylation play a role? *Trends Biochem. Sci.*, **17,** 93.
35. Zlatanova, J. (1992) Transcriptional regulation, eukaryotes. In *Encyclopedia of microbiology*, Vol. 4. Lederberg, J. (ed.). Academic Press, New York, p. 265.
36. Zlatanova, J. S. and van Holde, K. E. (1992) Chromatin loops and transcriptional regulation. *Crit. Rev. Eukaryotic Gene Expr.*, **2,** 211.
37. Johns, E. W. (1982) History, definitions and problems. In *The HMG chromosomal proteins*. Johns, E. W. (ed.). Academic Press, London, p. 1.
38. Johns, E. W. (ed.) (1982) *The HMG chromosomal proteins*. Academic Press, London.
39. Bustin, M., Lehn, D. A., and Landsman, D. (1990) Structural features of the HMG chromosomal proteins and their genes. *Biochim. Biophys. Acta*, **1049,** 231.
40. Ner, S. S. (1992) HMGs everywhere. *Curr. Biol.*, **2,** 208.
41. Weir, H. M., Kraulis, P. J., Hill, C. S., Raine, A. R. C., Laue, E. D., and Thomas, J. O. (1993) Structure of the HMG box motif in the B-domain of HMG1. *EMBO J.*, **12,** 1311.
42. Read, C. M., Cary, P. D., Crane-Robinson, C., Driscoll, P. C., and Norman, D. G. (1993) Solution structure of a DNA-binding domain from HMG1. *Nucleic Acids Res.*, **21,** 3427.
43. Graziano, V. and Ramakrishnan, V. (1990) Interaction of HMG14 with chromatin. *J. Mol. Biol.*, **214,** 897.
44. Crippa, M. P., Nickol, J. M., and Bustin, M. (1991) Developmental changes in the expression of high mobility group chromosomal protein. *J. Biol. Chem.*, **266,** 2712.
45. Shakoori, A. R., Owen, T. A., Shalhoub, V., Stein, J. L., Bustin, M., Stein, G., and Lian, J. B. (1993) Differential expression of the chromosomal high mobility group proteins 14 and 17 during the onset of differentiation in mammalian osteoblasts and promyelocytic leukemia cells. *J. Cell. Biochem.*, **51,** 479.
46. Pash, J. M., Alfonso, P. J., and Bustin, M. (1993) Aberrant expression of high mobility group chromosomal protein 14 affects cellular differentiation. *J. Biol. Chem.*, **268,** 13632.
47. Radic, M. Z., Saghbini, M., Elton, T. S., Reeves, R., and Hamkalo, B. A. (1992) Hoechst 33258, distamycin A, and high mobility group protein I (HMG-I) compete for binding to mouse satellite DNA. *Chromosoma*, **101,** 602.
48. Russnak, R. H., Candido, P. M., and Astell, C. R. (1988) Interaction of the mouse chromosomal protein HMG-I with the 3′ ends of genes *in vitro*. *J. Biol. Chem.*, **263,** 6392.

49. Reeves, R. and Nissen, M. S. (1993) Interaction of high mobility group-I(Y) nonhistone proteins with nucleosome core particles. *J. Biol. Chem.*, **268,** 21137.
50. Reeves, R., Langan, T. A., and Nissen, M. S. (1991) Phosphorylation of the DNA-binding domain of nonhistone high-mobility group I protein by cdc2 kinase: reduction of binding activity. *Proc. Natl Acad. Sci. USA*, **88,** 1671.
51. Thanos, D. and Maniatis, T. (1992) The high mobility group protein HMGI(Y) is required for NF-κB-dependent virus induction of the human IFN-β gene. *Cell*, **71,** 777.
52. Fashena, S. J., Reeves, R., and Ruddle, N. H. (1992) A poly(dA-dT) upstream activating sequence binds high-mobility group I protein and contributes to lymphotoxin (tumor necrosis factor-β) gene regulation. *Mol. Cell. Biol.*, **12,** 894.
53. Rubin, R. L. and Moudrianakis, E. N. (1972) Cooperative binding of histones to DNA. *J. Mol. Biol.*, **67,** 361.
54. D'Anna, J. A. and Isenberg, I. (1972) Fluorescence anisotropy and circular dichroism study of conformational changes in histone IIb2. *Biochemistry*, **11,** 4017.
55. Isenberg, I. (1979) Histones. *Annu. Rev. Biochem.*, **48,** 159.
56. Kornberg, R. D. and Thomas, J. O. (1974) Chromatin structure; oligomers of the histones. *Science*, **184,** 865.
57. Thomas, J. O. and Kornberg, R. D. (1975) An octamer of histones in chromatin and free in solution. *Proc. Natl Acad. Sci. USA*, **72,** 2626.
58. Eickbush, T. H. and Moudrianakis, E. N. (1978) The histone core complex: an octamer assembled by two sets of protein–protein interactions. *Biochemistry*, **17,** 4955.
59. Benedict, R. C., Moudrianakis, E. N., and Ackers, G. K. (1984) Interactions of the nucleosomal core histones: a calorimetric study of octamer assembly. *Biochemistry*, **23,** 1214.
60. Godfrey, J. E., Eickbush, T. H., and Moudrianakis, E. N. (1980) Reversible association of calf thymus histones to form the symmetrical octamer $(H2AH2BH3H4)_2$: a case of a mixed associating system. *Biochemistry*, **19,** 1339.
61. Finch, J. T., Lutter, L. C., Rhodes, D., Brown, R. S., Rushton, B., Levitt, M., and Klug, A. (1977) Structure of a nucleosome core particle of chromatin. *Nature*, **269,** 29.
62. Klug, A., Rhodes, D., Smith, J., Finch, J. T., and Thomas, J. O. (1980) A low resolution structure for the histone core of the nucleosome. *Nature*, **287,** 509.
63. Bentley, G. A., Lewit-Bentley, A., Finch, J. T., Podjarny, A. D., and Roth, M. (1984) Crystal structure of the nucleosome core particle at 16Å resolution. *J. Mol. Biol.*, **176,** 55.
64. Shick, V. V., Belyavsky, A. V., Bavykin, S. G., and Mirzabekov, A. D. (1980) Primary organization of the nucleosome core particles. Sequential arrangement of histones along DNA. *J. Mol. Biol.*, **139,** 491.
65. Burlingame, R. W., Love, W. E., Wang, B.-C., Hamlin, R., Zuong, N. H., and Moudrianakis, E. N. (1985) Crystallographic structure of the octameric histone core of the nucleosome at a resolution of 3.3Å. *Science*, **228,** 546.
66. Wang, B.-C., Rose, J., Arents, G., and Moudrianakis, E. N. (1994) The octameric histone core of the nucleosome: structural issues resolved. *J. Mol. Biol.*, **236,** 179.
67. Arents, G., Burlingame, R. W., Wang, B.-C., Love, W. E., and Moudrianakis, E. N. (1991) The nucleosomal core histone octamer at 3.1Å resolution: a tripartite protein assembly and a left-handed superhelix. *Proc. Natl Acad. Sci. USA*, **88,** 10148.
68. Richmond, T. J., Finch, J. T., Rushton, B., Rhodes, D., and Klug, A. (1984) The structure of the nucleosome core particle at 7Å resolution. *Nature*, **311,** 532.
69. Bavykin, S. G., Usachenko, S. I., Zalensky, A. O., and Mirzabekov, A. D. (1990) Structure of nucleosomes and organization of internucleosomal DNA in chromatin. *J. Mol. Biol.*, **212,** 495.

70. Arents, G. and Moudrianakis, E. N. (1993) Topography of the histone octamer surface: repeating structural motifs utilized in the docking of nucleosomal DNA. *Proc. Natl Acad. Sci. USA*, **90,** 10489.
71. Carter, C. W., Jr. (1978) Histone packing in the nucleosome core particle of chromatin.
Proc. Natl Acad. Sci. USA, **75,** 3649.
72. Simpson, R. T. (1978) Structure of the chromatosome, a chromatin particle containing 160 base pairs of DNA and all the histones. *Biochemistry*, **17,** 5524.
73. Allan, J., Hartman, P. G., Crane-Robinson, C., and Aviles, F. X. (1980) The structure of histone H1 and its location in chromatin. *Nature*, **288,** 675.
74. Mirzabekov, A. D., Pruss, D. V., and Ebralidse, K. K. (1989) Chromatin superstructure-dependent crosslinking with DNA of the histone H5 residues Thr1, His25 and His62. *J. Mol. Biol.*, **211,** 479.
75. Varga-Weisz, P., van Holde, K., and Zlatanova, J. (1993) Preferential binding of histone H1 to four-way helical junction DNA. *J. Biol. Chem.*, **268,** 20699.
76. Varga-Weisz, P., Zlatanova, J., Leuba, S. H., Schroth, G. P., and van Holde, K. (1994) The binding of histones H1 and H5 and their globular domains to four-way junction DNA. *Proc. Natl Acad. Sci. USA*, **91,** 3525.
77. Bianchi, M. E., Beltrame, M., and Paonessa, G. (1989) Specific recognition of cruciform DNA by nuclear protein HMG1. *Science*, **243,** 1056.
78. Tsanev, R. and Tsaneva, I. (1986) Molecular organization of chromatin as revealed by electron microscopy. In *Methods and achievements in experimental pathology*. Jasmin, G. and Simard, R. (ed.). S. Karger, Basel, p. 63.
79. Giannasca, P. J., Horowitz, R. A., and Woodcock, C. L. (1993) Transitions between *in situ* and isolated chromatin. *J. Cell Sci.*, **105,** 551.
80. Thoma, F. (1988) The role of histone H1 in nucleosomes and chromatin fibers. In *Architecture of eukaryotic genes*. Kahl, G. (ed.). VCH, Weinheim, p. 163.
81. Freeman, L. A. and Garrard, W. T. (1992) DNA supercoiling in chromatin structure and gene expression. *Crit. Rev. Eukaryotic Gene Expr.*, **2,** 165.
82. Leuba, S. H., Zlatanova, J., and van Holde, K. (1994) On the location of linker DNA in the chromatin fiber. Studies with immobilized and soluble micrococcal nuclease. *J. Mol. Biol.*, **235,** 871.
83. Allan, J., Mitchell, T., Harborne, N., Bohm, L., and Crane-Robinson, C. (1986) Roles of H1 domains in determining higher order chromatin structure and H1 location. *J. Mol. Biol.*, **187,** 591.
84. Zlatanova, J. S. (1990) Immunochemical approaches to the study of histone H1 and high mobility group chromatin proteins. *Mol. Cell. Biochem.*, **92,** 1.
85. Leuba, S. H., Zlatanova, J., and van Holde, K. (1993) On the location of histone H1 and H5 in the chromatin fiber. Studies with immobilized trypsin and chymotrypsin. *J. Mol. Biol.*, **229,** 917.
86. Thomas, J. O. and Khabaza, A. J. A. (1980) Cross-linking of histone H1 in chromatin. *Eur. J. Biochem.*, **112,** 501.
87. Itkes, A. V., Glotov, B. O., Nikolaev, L. G., Preem, S. R., and Severin, E. S. (1980) Repeating oligonucleosomal units. A new element of chromatin structure. *Nucleic Acids Res.*, **8,** 507.
88. Woodcock, C. L., Grigoryev, S. A., Horowitz, R. A., and Whitaker, N. (1993) A chromatin folding model that incorporates linker variability generates fibers resembling the native structures. *Proc. Natl Acad. Sci. USA*, **90,** 9021.

89. Leuba, S. H., Yang, G., Robert, C., Samori, B., van Holde, K., Zlatanova, J., and Bustamante, C. (1994) Three-dimensional structure of extended chromatin fibers as revealed by tapping-mode scanning force microscopy. *Proc. Natl Acad. Sci. USA*, in press.
90. Zlatanova, J., Leuba, S. H., Yang, G., Bustamante, C., and van Holde, K. (1994) Linker DNA accessibility in chromatin fibers of different conformations: a reevaluation. *Proc. Natl Acad. Sci. USA*, **91,** 5277.
91. Rattner, J. B. and Hamkalo, B. A. (1979) Nucleosome packing in interphase chromatin. *J. Cell Biol.*, **81,** 453.
92. Trifonov, E. N. (1990) DNA as chromatin organizer. In *Theoretical biochemistry and molecular biology*. Beveridge, D. L. and Lavery, R. (ed.). Adenine Press, Albany, NY, p. 377.
93. De Murcia, G. and Koller, T. (1981) The electron microscopic appearance of soluble rat liver chromatin mounted on different supports. *Biol. Cell*, **40,** 165.
94. Sun, J.-M., Ali, Z., Lurz, R., and Ruiz-Carrillo, A. (1990) Replacement of histone H1 by H5 *in vivo* does not change the nucleosome repeat length of chromatin but increases its stability. *EMBO J.*, **9,** 1651.
95. Tremethick, D. J. and Drew, H. R. (1993) High mobility group proteins 14 and 17 can space nucleosomes *in vitro*. *J. Biol. Chem.*, **268,** 11389.

2 | DNA structure: implications for chromatin structure and function

ALAN P. WOLFFE and HORACE R. DREW

1. Introduction

DNA has a simple yet intricate structure that serves as the template for the assembly of a chromosome. Each chromosome has one DNA molecule, a long unbranched double helix consisting of two antiparallel polynucleotide chains. The structure is stabilized by base-stacking interactions and hydrogen bonding between the bases: G (guanine) being paired to C (cytosine) and A (adenine) to T (thymine). Physical studies using X-ray diffraction indicate that, under conditions of physiological ionic strength, DNA is a somewhat regular helix with a diameter of 2 nm, making a complete turn every 3.0 to 3.4 nm. This particular DNA structure (B form) has approximately 10.5 bp/turn of the helix, with a variation due to base sequence of 9.9 to 11.1 bp/turn. Every base pair is rotated approximately 34° around the axis of the helix relative to the next base pair, resulting in a twisting of the two polynucleotide strands around each other. A double helix is formed that has a minor groove (very approximately 1.2 nm across) and a major groove (very approximately 2.2 nm across). The geometry of the major and minor grooves of DNA is crucial in determining the interactions of proteins with the DNA backbone sugars and phosphates. The double helix is usually right handed.

The properties of individual base pair steps differ. One of these properties can be conveniently described as propeller twist. This is where the planes of the two bases in any pair are twisted relative to each other. This minimizes their contact with water and prevents the otherwise flat base pairs from sliding easily over each other. Adjacent base pairs stacked together within the double helix can have their positions relative to each other described by a set of parameters known as twist, slide, and roll. Twist corresponds to the rotation of one base pair relative to the other about the axis that runs vertically through, or near the centres of two neighbouring base pairs. Roll describes the rolling open of two adjacent base pairs along their long axes. Slide describes the relative sliding of neighbouring base pairs along their long axes. With this basic information we can now discuss recent progress towards understanding DNA structure in detail, especially with respect to curvature. The curvature of DNA is an important variable in determining the wrapping of DNA by proteins within chromatin.

2. DNA curvature in chromatin and gels

Consider the work of God: for who can make straight, which he has made crooked?
Ecclesiastes 7: 13

Studies in the mid 1980s by Drew and Travers (1) and by Satchwell *et al.* (2) showed that different dinucleotides would prefer to occupy different locations in the DNA as it curves about the histone proteins of chromatin, or into a tight circle without protein. These dinucleotide preferences are listed in Table 1, and have been confirmed by others (3–5). Looking at Table 1, we see that the GpC (or GC/GC, reading 5′ to 3′ on both strands) step has the strongest preference of all—27 per cent—to lie with its minor groove along the outside of a curved DNA, at a location of high roll. The AA/TT step has the second strongest preference (20 per cent) to lie with its minor groove along the inside of a curved DNA, with low roll. Other dinucleotide steps show lesser preferences; some trinucleotides such as GGC/GCC or AAA/TTT show stronger preferences than others (2).

These results seem consistent with many histone–DNA reconstitution studies (6, 7) and with studies of DNA curvature about proteins such as the 434 repressor or catabolite-gene activating protein (CAP) (8–10). Nevertheless they seem inconsistent with most previous estimates of DNA curvature in gels made by measurement of electrophoretic velocity (11–13). In particular, most models to explain DNA curvature in gels (13–15) rely almost entirely on the AA/TT step or AAA/TTT trimer to produce such curvature, without any definite role for the GC step. This has long been a concern to some workers (15).

New studies by X-ray crystallography (16, 17), gel electrophoresis (18, 19), and theoretical chemistry (20, 21) go a long way to reconcile these discrepancies. Dickerson and colleagues (16, 17) have shown beyond any doubt that the GC step, at least as GGCC/GGCC or AGCT/AGCT, is highly curved in many different crystalline forms of DNA; see Fig 1. The central GGCC/GGCC sequence curves by about 20° to the right, thereby closing the major groove and opening the minor groove, as expected from earlier work (1, 2, 22). The same curvature had been seen as early as 1985 with the crystal structure, of GGGGCCCC, shown in Fig. 1b. But its significance was not then fully realized, and the authors drew two vertical lines (see Fig. 1) as helical axes for either the GGGG or the CCCC end, thereby omitting the 20° bend in the middle (23).

A similar crystal structure of the double-helical sequence GGATGGGAG (24) had its relevance cast into doubt by solution studies using NMR or CD spectroscopy; yet it also was recently confirmed using X-ray crystallography (25) as the actual structure of the DNA when bound to the zinc-finger protein GLI in a biologically meaningful complex.

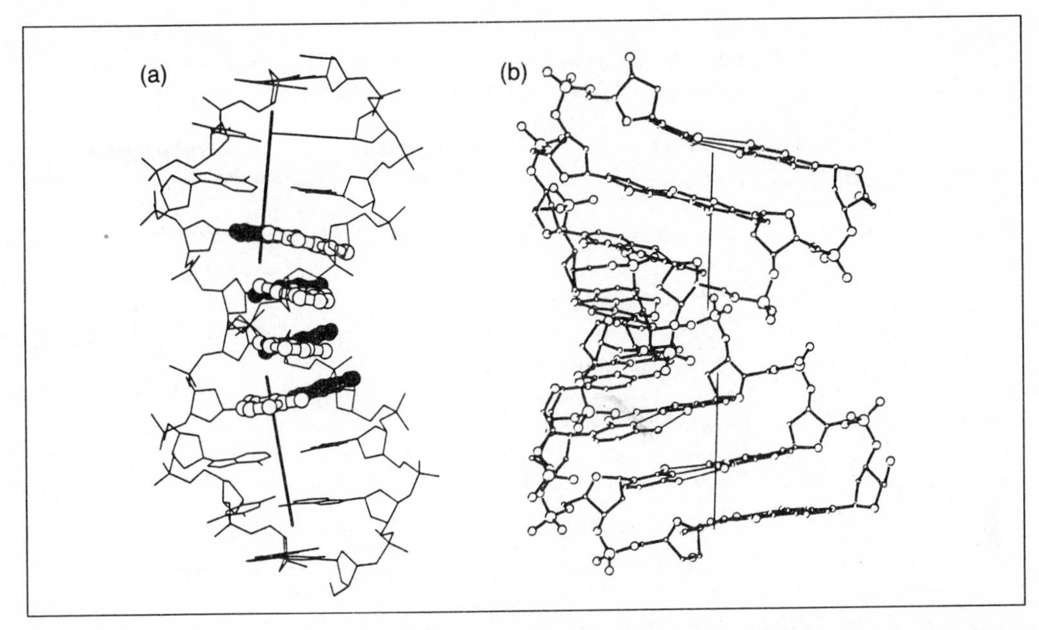

Fig. 1 (a) The crystal structure of CATGGCCATG. Note how the DNA curves to the right into the major groove, at the central GGCC bases. Courtesy of M. Kopka. (b) The crystal structure of GGGGCCCC. Again, the DNA curves into the major groove at the central GGCC bases, even though the twist of the helix is less than that shown for (a) (11.2 versus 10.2 bp per turn). The vertical lines are best-helix-axes for either the GGGG or CCCC end, omitting the central GC step. Courtesy of M. J. McCall.

If GGCC/GGCC sequences bend so as to close the major groove, how do AA/TT sequences bend? Naively, one might think that these sequences would do the opposite, and bend so as to close the minor groove, since they prefer to lie with their minor grooves along the inside of curved DNA (Table 1). However, many detailed X-ray studies of the AA/TT sequence as AAAA/TTTT or AATT/TTAA have shown that these steps usually remain perfectly unbent and straight, contrary to what one might suppose (26–28). The minor groove in runs of AA/TT is more narrow than usual, but this is caused by the high propeller twist of A/T base pairs and not by their roll, and so it does not contribute to curvature (22). Therefore, it is simply the difference in roll angle between GC/GC and AA/TT sequences that causes DNA with a periodically repeating sequence such as $(GGCCAAAAAA)_n$ to bend (15, 22).

For example, suppose that GC sequences have a mean roll per step of +6°, while AA/TT sequences have a typical roll of 0°, as compared to +3° as a broad average for mixed-sequence DNA. Then either GC/GC or AA/TT sequences alone will cause the DNA to bend slightly when spaced at intervals of 10 bp in a single molecule, yet both GC/GC and AA/TT sequences together will cause the DNA to bend a lot, when located every 10 bp in a single molecule, and also about 5 bp from one another locally.

Do the data from gel electrophoresis agree with the data from X-ray crystal-

Table 1 Dinucleotide preferences for curvature in chicken nucleosomal DNA, at a twist of 10.2 bp/turn[a]

Dinucleotide step	Percentage variation in occurrence	Preferred minor-groove location
GC	27	Out
CG	14	Out
GG/CC	12	Out
TG/CC	8	Out
GT/AC	6	Out
AG/CT	4	Out
GA/TC	1	In
AT	6	In
TA	13	In
AA/TT	20	In

[a] From (2, 5).

lography? Yes, as the result of recent work they agree precisely. Fig. 2 shows the first of two studies by Brukner *et al.* (18) which demonstrates that AAAAA and GGGCCC sequences curve, or vary in roll angle, in opposite directions by about the same amount. Part (a) shows the five related DNA segments constructed, in order to vary the distance *d* between a segment of AAAAA and a segment of GGGCCC. Part (b) shows how these five DNA molecules, when ligated end-to-end as multimers of a 42 bp repeat, run through a polyacrylamide gel. The small multimers of size 84 or 126 bp show little curvature, as measured by the apparent increase in *R*, which remains close to 1.0. But the large multimers of 168 and 210 bp

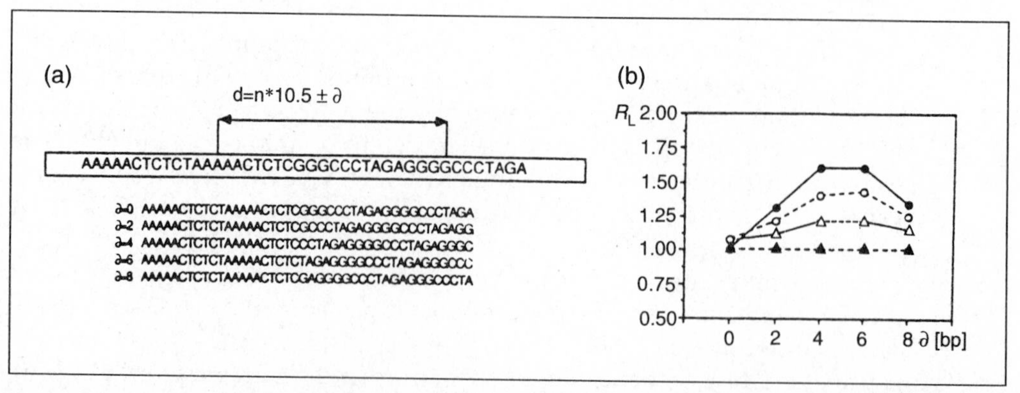

Fig. 2 The sequences AAAAA and GGGCCC contribute to helix curvature in opposite directions. (a) Design of five DNA molecules to test for the relative directions of curvature of AAAAA and GGGCCC segments. (b) Gel-running speeds of these five molecules in polyacrylamide, after ligation to form multimers, in terms of apparent relative size R_L in base pairs versus marker DNA of known size. Code: ▲ = 84 bp, △ = 126 bp, ○ = 168 bp, ● = 210 bp. Courtesy of I. Brukner.

show greatly increased apparent size in those cases where the respective centres of the AAAAA and GGGCCC sequences lie $10n + 4$ to 6 bp apart. Thus, these AAAAA and GGGCCC sequences contribute to helix curvature in opposite directions, or are 180° out of phase with one another. This is precisely as expected from nucleosomal work (Table 1) and X-ray crystal structures (summarized above).

In a second study, Brukner *et al.* (19) have shown that the curvature of GGGCCC sequences is apparently greater (when measured by gel running) in the presence of 10 mM magnesium ion than in its absence: see Fig. 3. This might explain why the curvature of GGGCCC sequences has been noted so often in crystals of DNA, which contain magnesium, and in the high ionic strength environment of DNA which is bound to the histone proteins, but not in low salt gels run in EDTA.

Finally, new theoretical calculations by Hunter (21) based on an earlier, well-documented model for the stacking of porphyrin rings (20) have independently confirmed the X-ray and gel-running results. Figs 4a and b show plots of base-stacking energy for the AA/TT and GC/GC steps respectively, versus their possible variations in terms of roll and slide. The AA/TT step shows a single energy minimum precisely at (roll, slide) = (0°, 0 nm), as seen in crystal structures (26–28). The GC/GC step shows strong contours of energy that lie parallel to the roll axis, and allow high roll at moderate-to-high slide (left-hand side of the plot)—also seen in crystals (16, 17, 23). These stacking energies are partly electrostatic in origin, as was first noted long ago by Bugg *et al.* (29). More studies are in progress to analyse the existing crystal structures of DNA and DNA–protein complexes in terms of these calculated energies, in the hope of gaining refined values (C. Calladine and M. Elhassan, personal communication).

We conclude that both GC/GC and AA/TT sequences will play a major role in

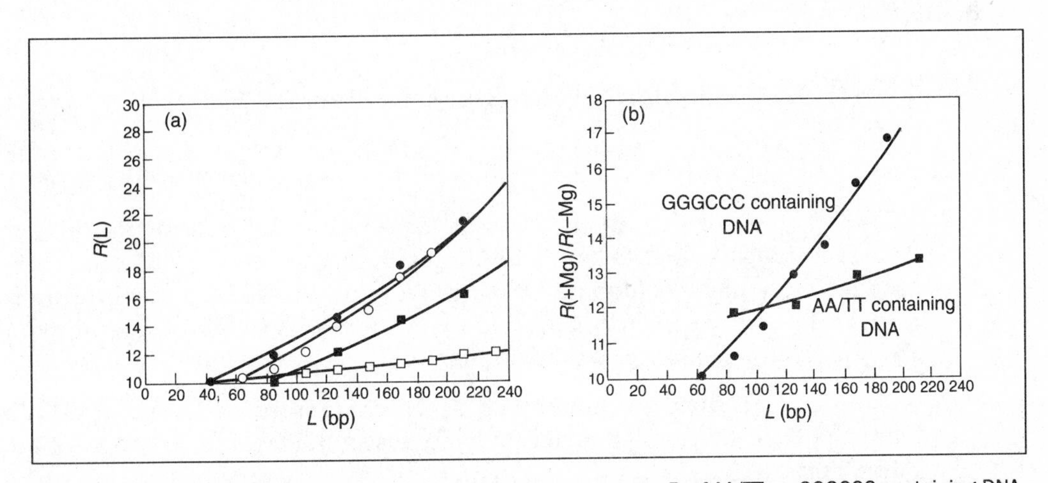

Fig. 3 Influence of magnesium ion on the apparent gel-running sizes, *R*, of AA/TT- or GGGCCC-containing DNA. Each vertical axis shows apparent relative size by gel mobility of DNA; the horizontal axis its true length in bp. (a) Primary data. Code: AA/TT-containing DNA without (■) and with (●) Mg^{2+}; GGGCCC-containing DNA without (□) and with (○) Mg^{2+}. (b) Relative mobilites in the presence or absence of Mg^{2+}. Courtesy of I. Brukner.

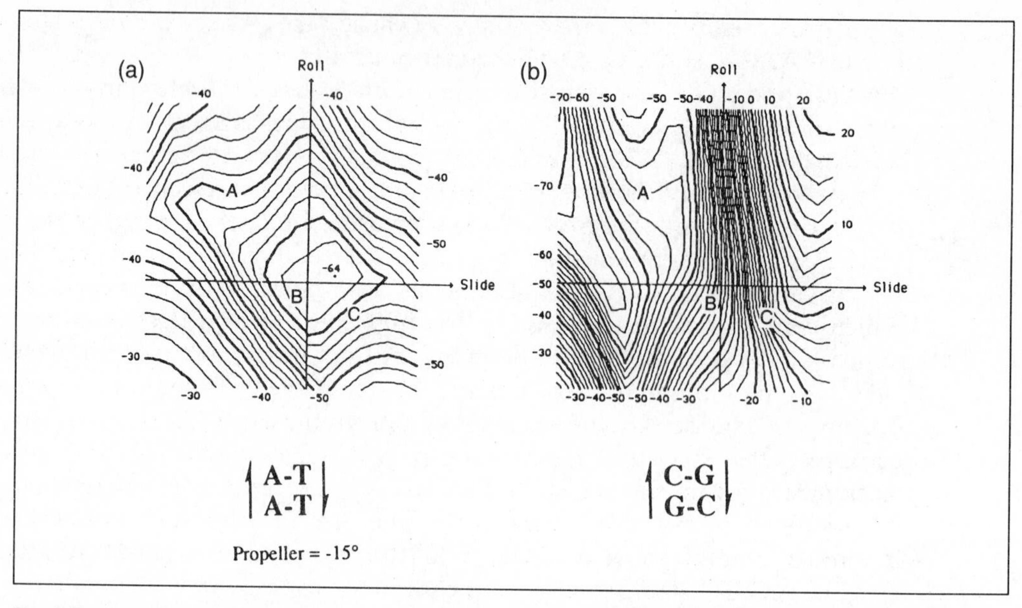

Fig. 4 Base-stacking energies plotted as a function of the coordinates of the bp step in roll and slide for: (a) the AA/TT step, or (b) the GC/GC step. Code: A, B, and C mark the coordinates of the well-known A, B, and C DNA helices from fibre diffraction studies. Courtesy of C. Hunter.

determining how the DNA bends about proteins in chromatin and elsewhere, such as in transcription complexes. There is not space here to describe all the many examples of such behaviour that are already known.

3. DNA untwisting at TATA sequences

Who would have imagined without seeing the structures involved, that a basic component of the transcription apparatus could so drastically affect the structure of its binding site? (30)

This quotation refers to the recently determined X-ray structure of a complex between transcription factor TATA-box-binding protein (TBP) and the TATA sequence to which it binds on DNA, unwinding the DNA by one-third of a helix turn. In fact, there are many well-known references to the ease of unwinding of DNA at TATA sequences, for example:

- using UV spectroscopy, Gotoh and Tagashira (31) showed that TA steps in DNA are the least thermally stable, and discussed their result in terms of biological function;
- using NMR spectroscopy, Patel *et al.* (32) showed easy base-pair melting of TATA-like sequences at room temperature;
- using nucleases as probes, Drew *et al.* (33) showed easy and co-operative melting

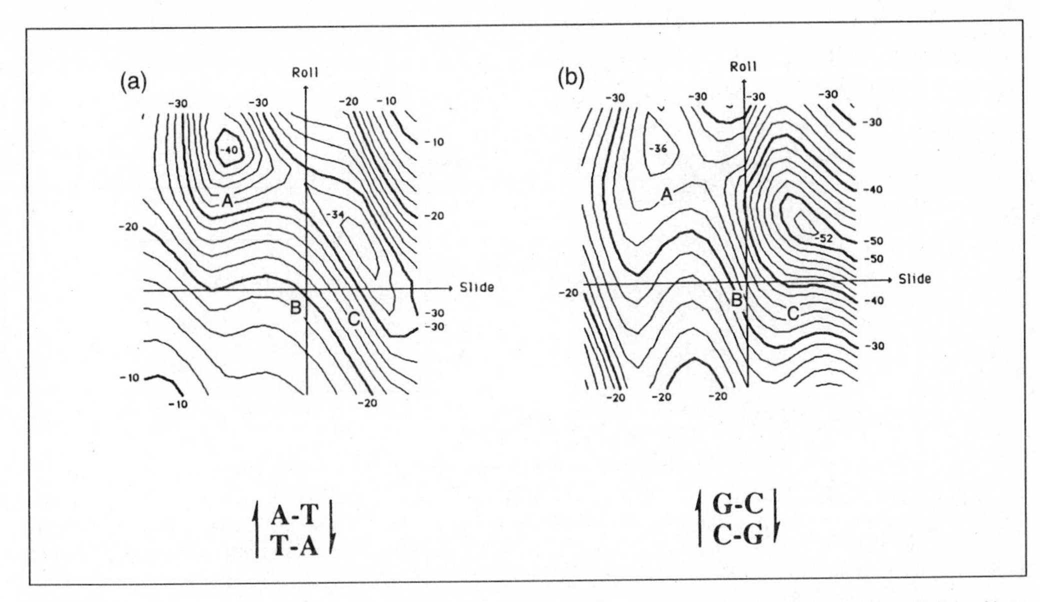

Fig. 5 Base-stacking energies for: (a) the TA/TA step, and (b) the CG/CG step in terms of roll and slide. Note the two zones of low energy for each pyrimidine–purine step, as seen also in DNA crystals. Courtesy of C. Hunter.

of TA sequences in supercoiled DNA at room temperature, and showed that a transcription-deficient mutant was impaired at helix unwinding; the general relevance of these results to transcription, recombination and replication was discussed;

- using micrococcal nuclease as a probe, Keene and Elgin (34) and Flick *et al.* (35) showed that, in higher organisms, most TATA-like sequences are located outside of coding regions, including the DNA that lies before or after any typical coding region.

Why should TATA sequences be so easy to unwind? Figs 5a and b show Hunter's calculations of base-stacking energy for the TA/TA and CG/CG steps (21). These steps are bistable, and show two low energy positions on each plot—one on the right-hand side and another on the left. This explains why DNA with many TA and/or CG steps can unwind without much cost of energy, or breaking of base pair bonds: high roll and negative A-like slide are usually coupled to low twist.

Many lines of solution evidence support the calculation of easy unwinding at TATA sequences. For example, McClellan, Lilley *et al.* (36, 37) found that the chemical osmium tetroxide reacts specifically with TATA sequences in supercoiled DNA. The osmium atom will attach to the C5–C6 double bond of thymine only when the helix is either underwound or overwound. Also Hirose *et al.* predicted directly, based on two kinds of experiment, that the TBP protein would unwind the TATA sequence. First, they showed that the protein fraction containing TBP

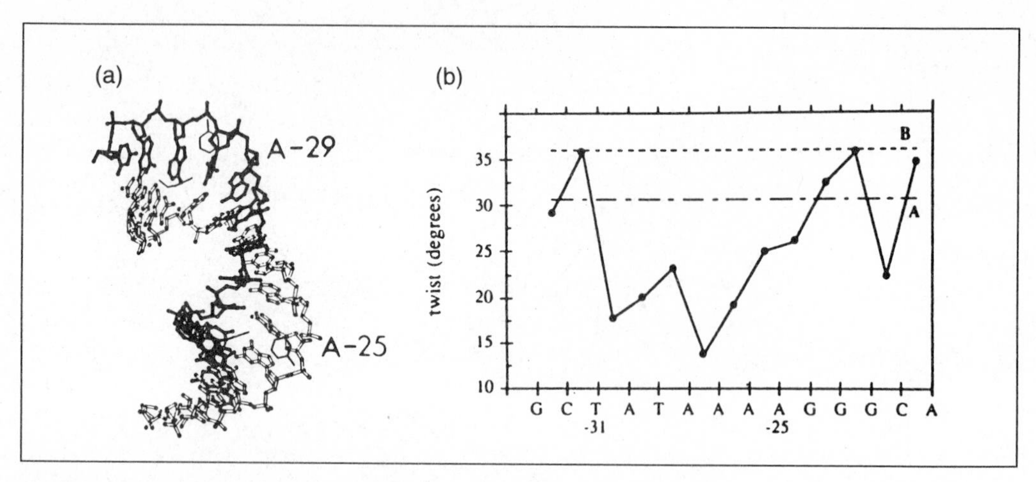

Fig. 6 Crystal structure of a GCTATAAAAGGGCA segment of DNA when bound to the TBP protein: (a) a line drawing, (b) the reduction of base-step twist at the TATAAAA sequence. Code: B and A refer to the averaged twists of fibre models for B- and A-type DNA. Reprinted with permission from (41), copyright 1993, Macmillan Magazines Ltd.

protein was solely responsible for the enhancement of promoter activity by negative supercoiling (38). Second, they showed by careful measurement of linking number *Lk* in 2-D agarose gels that the protein TBP would reduce *Lk* by about 0.5 turn (39).

Fig. 6a shows the actual, unwound structure of a TATAAAA sequence when bound to the TBP protein in a crystal (40, 41). The DNA is unwound by about 110°, or one-third of a helix turn, without breaking of A/T base pairs. Fig. 6b is a plot of twist versus location in the DNA; there one can see that the local twist between adjoining base pairs decreases from 35° in the flanking regions, to 15° or 20° in the TATAAAA segment. This unwinding is accompanied by a large positive roll at most steps. A similar unwinding was seen in the crystal structure of the GATATC recognition site bound with Eco RV (42). It is not yet known whether this extensive unwinding of the DNA at TATA-like sequences, upstream of genes, is transmitted downstream to where the base pairs separate completely, so as to bind RNA polymerase and start making RNA by Watson–Crick pairs. There is some evidence that unwinding of the TATA sequence by the TBP protein may be important in initiating replication, as well as transcription (43, 44).

Much progress has been made recently at understanding the critical role that DNA structure plays in biology. However, there is a gap (a massive gap, to be truthful) between the knowledge of DNA structure held today by crystallographers and biophysical scientists, and the application of that detailed knowledge to real biological systems. In the chromatin field, it has now been recognized that DNA structure can play a key role in locating the histone proteins on DNA in relation to the binding sites of transcription factors, so we will next discuss how the histone proteins are assembled or disassembled around active genes.

4. DNA structure in the nucleosome

The nucleosome is now thought to contribute not only the basic structural unit for packaging DNA within chromatin but also the structural framework within which the transcriptional machinery operates most effectively to regulate gene expression. Nucleosomes are positioned somewhat with respect to DNA sequence *in vivo*. This selective positioning takes two forms: (1) translational positioning describes where a particular DNA sequence starts to be associated with histones and where it finishes the association; (2) rotational positioning describes where one groove (minor or major) of a particular sequence within the double helix is facing towards the solution and the other groove (major or minor) is facing towards the histones. Data on the structure of DNA in the nucleosome suggest how particular DNA sequences might influence translational and rotational positioning. Both translational and rotational positioning of histone–DNA contacts are known to influence the association of transcription factors with DNA (45–47). Since the interaction of many transcription factors with DNA has been studied initially without histones being present, it is important to know how the structure of DNA is altered due to association with the histones.

Structural studies that examine the general features of DNA associated with the core histones make use of core particles prepared after micrococcal nuclease digestion of nuclei. These contain only the 146 bp of DNA that make the strongest histone–DNA contacts. A core particle has a disc-like shape 11 nm in diameter and 5.6 nm in height. The DNA is wrapped in 1.75 turns of a left-handed superhelix around the histone core (48; Fig. 7a). The structure has a pseudodyad axis of symmetry that passes through the centre of the nucleosomal DNA. However, the winding of DNA around the core histones is not uniform. The crystal structure suggests that the bending of DNA is most severe between one and two, and three and four helical turns to either side of the dyad axis (centre of the DNA) (Fig. 7b). The rise of the DNA superhelix is also not uniform, being gently inclined to either end and very steep in the centre of the core particle (Fig. 7a). These studies show the general path of DNA in the core particle, but they do not have sufficient resolution to determine other important details of DNA structure such as any tendency for base pairs within the structure to unstack and the number of base pairs per turn at any given region of the core particle. These local features have important consequences for where the histones will position themselves with respect to DNA sequence and also for the interaction of *trans*-acting factors with nucleosomal DNA.

Information about the solution structure of core particle DNA is available from studies with enzymes and chemical agents that cleave or modify DNA. Singlet oxygen is a sensitive probe for structural deformations in DNA which are associated with an increase in roll angle between adjacent base pairs. This reagent preferentially reacts with core particle DNA about 1.5 turns to either side of the core particle dyad (49). DNA in these regions has been proposed to be kinked such that base stacking changes, opening up base planes towards the major groove

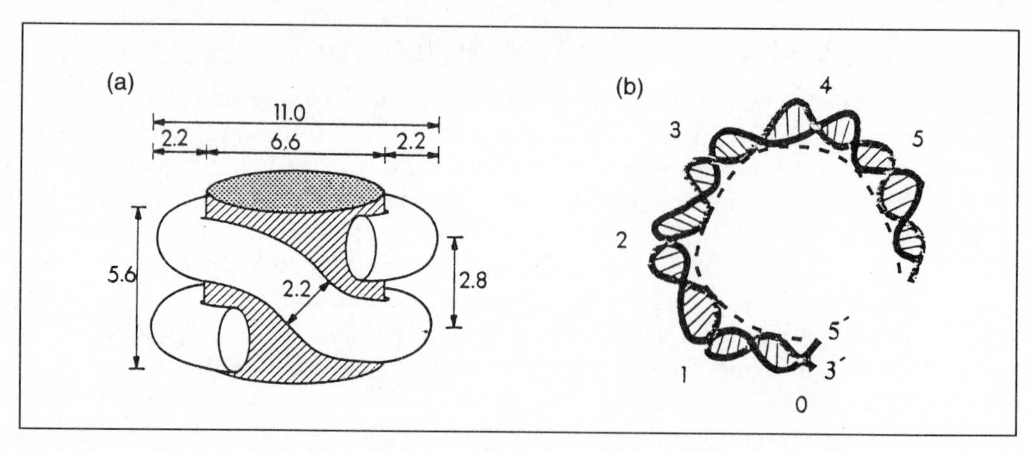

Fig. 7 The organization of DNA in the nucleosome. (a) The dimensions of the nucleosome: the histone core is represented by the hatched cylinder, DNA by the open tube. This view overlooks the centre of nucleosomal DNA; at this point the DNA superhelix rises very steeply. (b) The path of one turn of DNA. Numbers refer to turns of the DNA helix away from the dyad axis. The dotted line represents the surface of the core histones around which DNA is wrapped.

(49, 50). DNA that is more easily distorted might preferentially occur ± 1.5 helical turns from the dyad axis; indeed, unstable TA steps (see Fig. 5) are preferentially found at these sites within the nucleosome core (12). These same sites are biologically important since they are where HIV integrase prefers to integrate within the nucleosome (51). The HIV integrase also makes preferential use of the DNA that is severely bent between three and four turns from the dyad axis of the core particle. The enzyme binds to DNA through the major groove and distorts DNA as a step in its reaction mechanism—hence the preference for DNA that has a wide major groove or that is already severely distorted in the nucleosome. The HIV integrase reaction is a useful example of how the association of DNA with histones can actually facilitate a process requiring DNA as a substrate.

The number of base pairs per turn of DNA (or helical periodicity) changes when DNA is wrapped around the core histones. DNA in solution normally has an average of 10.5 bp/turn; however, in the core particle it was suggested that it may be overwound to an average of 10.2 bp/turn (6, see also 52). Importantly, direct measurements demonstrate that although this overwinding exists it is not uniform over the whole length of DNA in the core particle. Hydroxyl radical cleavage of DNA in nucleosomal core particles indicates that distinct local regions with different helical periodicities are present (53). Quantitative analysis shows that the central three turns of DNA in the core particle have a different number of base pairs per turn (10.7) to those in the remainder of the structure (10.0) (Fig. 8). Importantly, this would lead the two segments of DNA with a 10.0 bp/turn helical periodicity to be out of phase with each other by 2 bp. The junctions between these regions are ± 1.5 turns to either side of the dyad axis and correspond to the sites of DNA distortion in the nucleosome detected by singlet oxygen and HIV integrase.

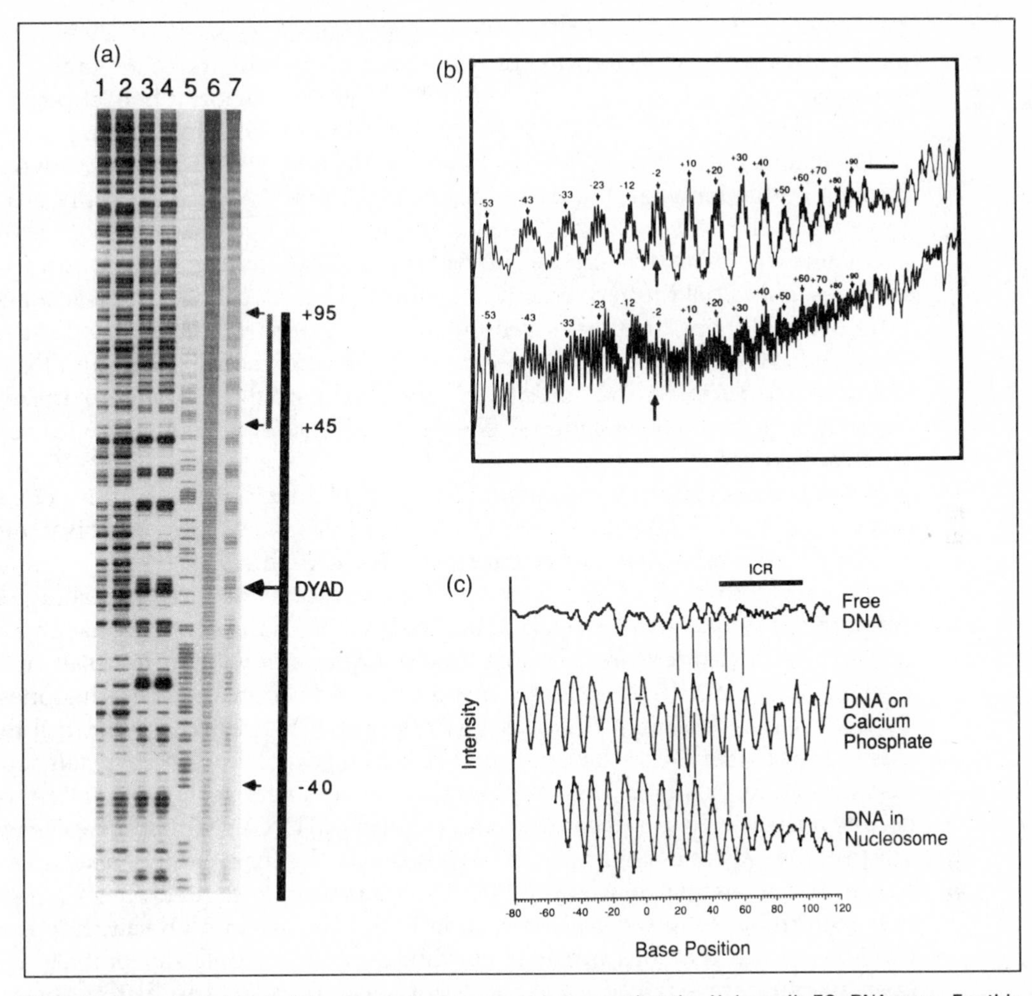

Fig. 8 Hydroxyl radical footprints of a nucleosome core incorporating the *X. borealis* 5S rRNA gene. For this experiment, a 583 bp *Eco* RI-*Hha*I fragment of pXP-10 (125) was radiolabelled on the non-coding strand of the 5S RNA gene. (a) The autoradiogram shows cleavage patterns for naked 5S DNA generated by DNase I (lanes 1 and 2) and by hydroxyl radical (lane 6) as well as for DNA in the nucleosome core generated by DNase I (lanes 3 and 4) and by hydroxyl radical (lane 7). Lane 5 contains G + A markers generated by chemical cleavage of 5S DNA. DNA fragments were separated by electrophoresis on a 6% polyacrylamide gel containing 7 M urea. The footprint of the nucleosome core that is positioned over the beginning of the 5S rRNA gene is indicated by the black bar on the right of the autoradiograph. The internal control region (ICR) of the 5S rRNA gene is indicated by the stippled bar covering positions +45 to +95. (b) Densitometer scans of the hydroxyl radical footprint of the 5S nucleosome core. Cleavage patterns for the non-coding strand of 5S DNA are shown. The upper tracing is the hydroxyl radical footprinting of the nucleosome core; the lower is the pattern of hydroxyl radical cleavage of this same DNA when free in solution. Nucleotide positions are numbered relative to the start of the 5S rRNA gene. Note that there is no zero position in this figure. (c) Plots of the hydroxyl radical cleavage frequency at each nucleotide of the non-coding strand of the *X. borealis* 5S rRNA gene free in solution, bound to calcium phosphate, and assembled in a nucleosome core. The horizontal bar above the curve indicates the approximate position of the ICR. The vertical lines are included to highlight the change in helical periodicity between DNA bound to calcium phosphate and in the nucleosome core. They also show how the periodic features in the cleavage pattern of naked DNA correlate with the other two patterns. Reprinted with permission from (88).

Crystallographic analysis of the histone octamer suggests that DNA at the dyad axis with a helical periodicity of 10.7 bp/turn would make the most effective contacts with the histones, whereas outside of this region a helical periodicity of 10.0 bp/turn would be optimal (54). DNA at the dyad axis is relatively straight and is associated predominantly with histones H3 and H4 (55, 56); however, it also forms a bridge between the two turns of DNA that wrap most tightly around the histones in the core particle. The rise of the superhelix is very steep at the dyad axis compared with the gentle inclination of DNA that is bent around the core histones with a helical period of 10.0 bp/turn. Theoretical calculations suggest that DNA with 10.0 bp/turn forms a more stable structure when distorted into a 80 bp circle than does DNA of 10.5 bp/turn (57). The necessity of having DNA with a different structure in the middle of the core particle is likely to make strong contributions to the translational positioning of the core histones with respect to DNA sequence.

Several investigators have examined the preference of the histone octamer for association with different defined DNA sequences. The assay used is competitive exchange with mixed-sequence core particles at high salt (> 0.8 M). This has the disadvantage that all of the histone–DNA contacts made at physiological ionic strength are unlikely to be made at the high salt exchange conditions, and that the preferences measured are not true free energies relevant to low salt in the cell nucleus (58). Nevertheless, DNA is still wrapped around the core histones under the high salt conditions (59); local variation in helical periodicity around the dyad axis is apparent and translational and rotational positioning occur at salt concentrations > 0.8 M (59, 60). This implies that some selective histone–DNA contacts occur even under high salt conditions. Although all DNA sequences can be wrapped around the histone octamer (61–64), use of the competitive exchange assay suggests that core particle length DNA that has flexible sequences is favoured over that with rigid sequences for interaction with histones at high salt. Importantly, a DNA sequence that has intrinsic curvature or anisotropic flexibility is favoured over a uniformly straight molecule for association with the histone octamer, as shown in Fig. 9, where the energies indicated refer to high salt solution (44, 64–66). These results are in excellent agreement with the preferred features of DNA sequence within core particles determined by a sequencing strategy (2). In spite of this progress using artificial DNA within the core particle, it is a sobering fact that naturally occurring nucleosome-positioning sequences still have more affinity for the histone octamer than do synthetic curves (64, 65, 67). The reason for this probably lies in the necessity to have local regions of DNA flexibility, curvature, and rigidity within the core particle dependent on the exact position of the DNA sequence with respect to the histones.

Based on our limited knowledge of DNA in the nucleosome we might expect: (1) deformable DNA such as TA sequences at ± 1.5 and 3.5 turns from the dyad axis, (2) AA/TT sequences lying with their minor grooves facing the histone octamer, (3) GC/GC sequences lying with their minor grooves facing away from the histones, (4) successive AA/TT (or GC/GC sequences) having a periodicity of

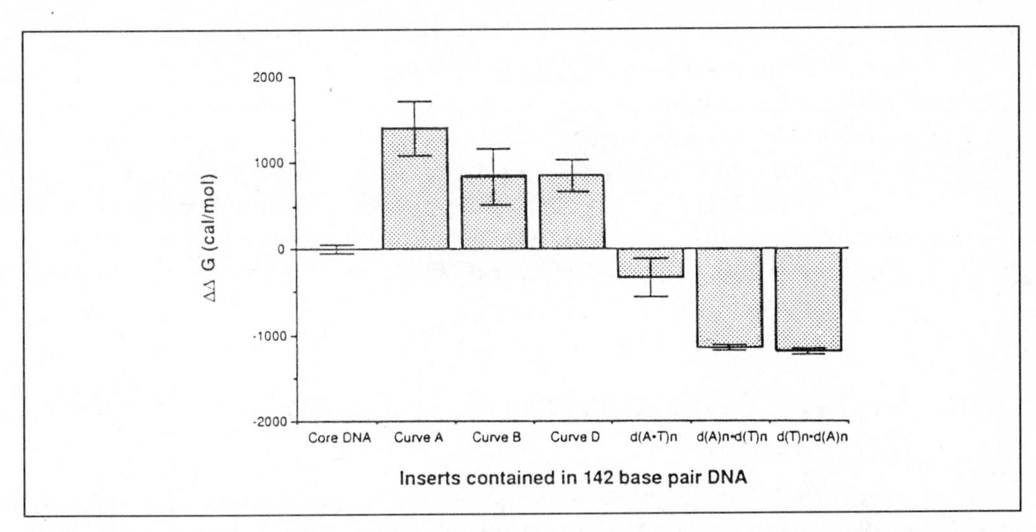

Fig. 9 Differences in Gibbs free energy change upon nucleosome formation between selected DNA fragments, at a solution reference state of high salt, about 0.6–1.0 M NaCl. Competitive reconstitution was used to determine $\Delta\Delta G$ for each of the DNA fragments used: three curved DNAs (A, B, and D), flexible DNA [(dA·T)*n*] and two rigid DNA molecules [d(A)*n*·d(T)*n* and d(T)*n*·d(A)*n*] (see (68) for details).

10.0 bp away from the dyad axis, (5) regions of curvature to either side of the dyad axis being out of phase by 2 bp, and (6) DNA at the dyad axis (32 bp) being relatively straight. So far a synthetic segment of DNA that incorporates all of these sequence features has yet to be synthesized or tested for its affinity for the histone octamer. However, simple sequences of DNA that have uniform anisotropic flexibility or curvature (65, 66) do not position histone octamers *in vivo* (68), whereas natural nucleosome positioning sequences do position octamers both *in vitro* and *in vivo* (69). Clearly we have more to learn about how the histones recognize and modulate DNA structure.

An additional complication to the determination of preferred DNA sequences for nucleosome positioning is that most studies neglect the fifth histone in the nucleosome, the linker histone, plus any other proteins present in real chromatin. The presence of a single molecule of linker histone (H1 or H5) will protect from micrococcal nuclease digestion an additional 10 bp to either side of the 146 bp within the core particle (70). At this time, it is unclear exactly how the small linker histone molecule manages to protect DNA at two widely separated sites within the nucleosome. Incorporation of the linker histone into the nucleosome does not alter the helical periodicity of DNA from that in the core particle (71, 72). Nevertheless the presence of linker histones may modify the exact position of histone octamers by unknown mechanisms (73–75). Transcription factors that have similar structures to the linker histones can also influence the positioning of core histones on DNA (76). Many other non-histone protein–DNA interactions also modulate exact nucleosome positioning especially around the regulatory regions of genes

(77, 78, see also Chapter 4). Exactly how these proteins overcome sequence-directed nucleosome positioning is not resolved.

DNA organization in the nucleosome is not simple. The histone octamer searches for favoured sequence features and chooses a set of histone–DNA interactions that cost the least amount of energy to form a stable structure. Nevertheless, all DNA in nucleosomes has many similar structural features; thus the histone octamer exerts a dominant constraint on the structure of DNA in the nucleosome (53, 64). Next we discuss how individual histones and their domains can contribute to this organization.

5. Histone contributions to the structure of DNA in the nucleosome

Many studies of the contributions of individual histones and their domains to nucleosomal structure have made use of natural nucleosome positioning sequences intrinsic to the 5S rRNA gene (79). These sequences can be reconstituted into nucleosomes by methods involving exchange of histones at high salt from donor chromatin (80) or salt–urea dialysis (81) using purified histones (82). The histone tetramer $(H3–H4)_2$ recognizes the positioning sequences of the 5S rRNA gene (53, 60). The tetramer organizes the central 120 bp of DNA identically to that found within the complete nucleosome (53, 83). Addition of the two dimers of histone H2A–H2B extend core histone–DNA interactions to 160 bp (53, 84), a result confirmed by histone–DNA cross linking (85).

Each of the core histones consists of a histone fold domain and an N-terminal tail (see Chapter 1); histone H2A also has a C-terminal tail. Each of the tail domains is positively charged and is believed to interact with the phosphodiester backbone of DNA on the outside of the nucleosome. The tail domains are the sites of many post-translational modifications, such as acetylation. Acetylation influences the stability with which the N-terminal domains of histones H3 and H4 interact with DNA in the nucleosome (86), and also the interaction of at least one transcription factor with its recognition site in the nucleosome (46). Another way of influencing the interaction of the histone tails with DNA in the nucleosome is to remove them by proteolysis with trypsin (87). This will also facilitate transcription factor access to DNA in the nucleosome (46). Removal of the histone tails does not strongly influence the recognition of the nucleosome-positioning sequence elements in the 5S rRNA gene (53, 88), the extent of bound DNA, or the helical periodicity of DNA in the nucleosome (53). Likewise, histone acetylation does not strongly influence these same parameters. Thus it would appear that removal of the histone tails alters the nature of histone–DNA contacts, rather than either their extent or the local DNA structure. It is, however, possible that modification of the histone tails might alter the overall path of DNA in the nucleosome (89). This follows from the observation that the topological change introduced into closed circular DNA by an acetylated histone octamer differs from that of an unmodified histone octamer (90).

A change in DNA path within the nucleosome on acetylation may follow from altered histone–DNA or histone–histone interactions. There is biophysical and biochemical evidence for this type of change occurring following histone acetylation (91), although the overall integrity of the nucleosome is not severely disturbed by modification or removal of the histone tails (92, 93). Subtle alterations in nucleosome conformation following histone acetylation could also contribute to an increase in transcription factor access to DNA (46). These experiments introduce the nucleosome as a structure capable of adopting several conformations, i.e., subject to allosteric effects.

6. Nucleosomal positioning and modification: influence on transcription factor–DNA interactions

Among the best-studied examples of *trans*-acting factors that interact with DNA associated with histones is the association of transcription factor (TF) IIIA with the 5S rRNA gene in chromatin. The zinc-finger protein TFIIIA continuously contacts two 10 bp segments within the 5S rRNA gene from +51 to +61 and from +81 to +91 (relative to the start of transcription at +1) (94). TFIIIA has nine zinc fingers (95) and, of these, the three N-terminal fingers make the key binding contacts between +81 and +91 (96), the three C-terminal fingers interact with the +51 to +61 region (97), and the three intervening fingers act as a linker associated with one side of the DNA helix. Such an extensive interaction with DNA would appear incompatible with the simultaneous wrapping of DNA around the histone core.

Functional studies demonstrated that the prior association of a positioned histone octamer with a 5S rRNA gene would inhibit transcription (98–100), whereas a positioned histone tetramer would allow transcription. Hydroxyl radical footprinting and DNA–protein cross-linking of the 5S rRNA gene associated with the positioned histone octamer revealed that the region of DNA that would need to make the key contacts with TFIIIA between +81 and +91 was in contact with histones H2A/H2B (54, 84, 85, Fig. 10). TFIIIA would not bind to the 5S rRNA gene in the presence of the octamer. However, TFIIIA would bind with the same affinity as its interaction with naked DNA if histones H2A/H2B were not present (101) or if the core histones were acetylated (46). Mutation of the nucleosome-positioning sequences around the start of the 5S rRNA gene led the histone tetramer to be in contact with the key binding site for TFIIIA. Under these conditions TFIIIA binding was inhibited by the tetramer alone; however, histone acetylation would again relieve any inhibition (46). These results indicate that for unmodified histones the exact extent of histone–DNA interactions, i.e., the translational positioning of the nucleosome, can be an important variable in determining transcription factor access to DNA; however, the changes in nucleosome structure caused by histone acetylation can relieve these inhibitory effects. The biological relevance of these observations lies in the fact that both chromatin structure and transcription complexes are transiently disrupted by DNA replication (see

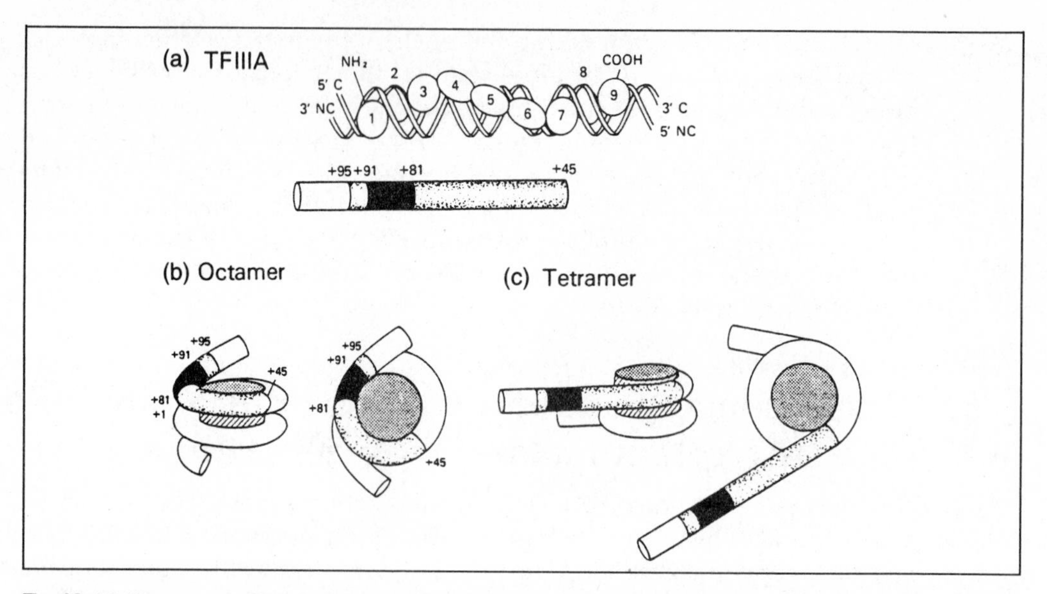

Fig. 10 (a) Histone and TFIIIA binding to a 5S RNA gene. A representation of TFIIIA showing the nine zinc-finger domains bound to a 5S RNA gene is shown. The binding site for TFIIIA is also represented as a cylinder; the speckled region is protected from DNase I cleavage. Missing contact analysis reveals the region from +81 to +91 (solid cylinder) to be an essential contact for TFIIIA. (b) When associated with a complete octamer of core histones $(H2A–H2B–H3–H4)_2$ the key contacts from +81 to +91 are in contact with the histone core, and TFIIIA cannot bind to the gene. (c) When associated with a tetramer of histones $(H3–H4)_2$ the key contacts from +81 to +91 are accessible to TFIIIA, which forms a triple complex with the gene (see (122) for details).

Chapter 3). The staging of chromatin assembly with the $(H3–H4)_2$ tetramer binding before the H2A–H2B dimers, the use of acetylated histone H4, coupled to the positioning of histone–DNA contacts on the 5S RNA gene promoter, will represent a window of opportunity for TFIIIA to bind to the gene, and thus facilitate the eventual assembly of a functional transcription complex (102). It is important to note that many different events contribute to establishing a transcriptionally active state. The exact contribution of each component will depend on the specific promoter, the DNA sequences, and the transcription factors involved.

7. Final comments

7.1 Important questions

DNA structure has a profound influence on how and why histones and transcription factors interact with the regulatory elements of genes. There is still much to be understood about these protein–DNA interactions both structurally and functionally. Future experiments are likely to uncover fundamental rules concerning the roles of DNA structure, histones, and their modification in the modulation of gene expression. Important questions will concern how DNA structure directs nucleosome positioning and the stability of histone–DNA interactions, how modification of the histones through acetylation influences the overall structural integrity of the nucleosome, and how sequestration of the linker histones influences core histone–DNA interactions both in the nucleosome and in the chromatin fibre.

7.2 Discussion

In the discussion with other book authors, it was pointed out that a protein assembly such as the histone octamer might be capable of undergoing allosteric change in response to histone modification or linker addition/removal. Histone–DNA crosslinking experiments do reveal major changes in core histone–DNA contacts at the dyad axis of the nucleosome following the association of linker histones (103). Whether modifications such as acetylation in themselves alter structure or primarily serve as signals for a more complex set of events remains to be seen.

Several methods have been used to map nucleosome structure, not always with the same results; in particular, psoralen or micrococcal nuclease reveal less extended contacts with histones than hydroxyl radical cleavage. Apparently, psoralen and micrococcal nuclease can disrupt weak histone–DNA contacts, whereas using the hydroxyl radical reagent is rather like 'spray painting' proteins bound to DNA (84). Viewed in this way, the results are reasonably congruent. All of these approaches will no doubt contribute to our attempts to analyse the transitions discussed above.

References

1. Drew, H. R. and Travers, A. A. (1985) DNA bending and its relation to nucleosome positioning. *J. Mol. Biol.*, **186**, 773.
2. Satchwell, S., Drew, H. R., and Travers, A. A. (1986) Sequence periodicities in chicken nucleosome core DNA. *J. Mol. Biol.*, **191**, 659.
3. Azizov, M. M., Ul'yanov, A. V., Kuprash, D. V., Shakhov, A. N., Gavin, I. M., and Nedospasov, S. A. (1992) Production and characterization of a library of mononucleosomal DNA from the chromatin of human cells, and its use to study the affinity of nucleosomal DNA for histones. *Doklady Biochemistry*, **322**, 415.
4. Piña, B., Barettino, D., Truss, M., and Beato, M. (1990) Structural features of a regulatory nucleosome. *J. Mol. Biol.*, **216**, 975.
5. Piña, B., Truss, M., Ohlenbusch, H., Postma, J., and Beato, M. (1990) DNA rotational positioning in a regulatory nucleosome is determined by base sequence: an algorithm to model the preferred superhelix. *Nucleic Acids Res.*, **18**, 6981.
6. Drew, H. R. and Calladine, C. R. (1987) Sequence-specific positioning of core histones on an 860 bp DNA: experiment and theory. *J. Mol. Biol.*, **195**, 143.
7. Drew, H. R. and McCall, M. J. (1987) Structural analysis of a reconstituted DNA containing three histone octamers and histone H5. *J. Mol. Biol.*, **197**, 485.
8. Koudelka, G. B., Harrison, S. C., and Ptashne, M. (1987) Effect of non-contacted bases on the affinity of 434 operator for 434 repressor and cro. *Nature*, **326**, 886.
9. Gartenberg, M. R. and Crothers, D. M. (1988) DNA sequence determinants of CAP-induced bending and protein binding affinity. *Nature*, **333**, 824.

10. Drew, H. R., McCall, M. J., and Calladine, C. R. (1990) New approaches to DNA in the crystal and in solution. In *DNA topology and its biological effects*. Cozzarelli, N. and Wang, J. C. (ed.). Cold Spring Harbor Laboratory Press, New York, p. 1.
11. Diekmann, S. and Wang, J. C. (1985) On the sequence determinants and flexibility of the kinetoplast DNA fragment with abnormal gel electrophoretic mobilities. *J. Mol. Biol.*, **186,** 1.
12. Hagerman, P. J. (1985) Sequence dependence of the curvature of DNA: a test of the phasing hypothesis. *Biochemistry*, **24,** 7033.
13. Koo, H. S., Wu, H. M., and Crothers, D. M. (1986) DNA bending at adenine–thymine tracts. *Nature*, **320,** 501.
14. Koo, H.-S. and Crothers, D. M. (1988) Calibration of DNA curvature and a unified description of sequence-directed bending. *Proc. Natl Acad. Sci. USA*, **85,** 1763.
15. Calladine, C. R., Drew, H. R., and McCall, M. J. (1988) The intrinsic curvature of DNA in solution. *J. Mol. Biol.*, **201,** 127.
16. Goodsell, D. S., Kopka, M. L., Cascio, D., and Dickerson, R. E. (1993) Crystal structure of CATGGCCATG and its implications for A-tract bending models. *Proc. Natl Acad. Sci. USA*, **90,** 2930.
17. Grzeskowiak, K., Goodsell, D. S., Kaczor-Grzeskowiak, M., Cascio, D., and Dickerson, R. E. (1993) Crystallographic analysis of CCAAGCTTGG and its implications for bending in B-DNA. *Biochemistry*, **32,** 8923.
18. Brukner, I., Dlakic, M., Savic, A., Susic, S., Pongor, S. and Suck, D. (1993) Evidence for opposite groove-directed curvature of GGGCCC and AAAAA sequence elements. *Nucleic Acids Res.*, **21,** 1025.
19. Brukner, I., Susic, S., Dlakic, M., Savic, A., and Pongor, S. (1993) Physiological concentration of magnesium ions induces a strong macroscopic curvature in GGGCCC-containing DNA. *J. Mol. Biol.*, in press.
20. Hunter, C. A. and Sanders, J. K. M. (1990) The nature of pi-pi interactions. *J. Am. Chem. Soc.*, **112,** 5525.
21. Hunter, C. A. (1993) Sequence-dependent DNA structure: the role of base stacking interactions. *J. Mol. Biol.*, **230,** 1025.
22. Calladine, C. R. and Drew, H. R. (1986) Principles of sequence-dependent flexure of DNA. *J. Mol. Biol.*, **192,** 907.
23. McCall, M. J., Brown, T., and Kennard, O. (1985) The crystal structure of d(GGGGCCCC): a model for poly(dG).poly(dC). *J. Mol. Biol.*, **183,** 385.
24. McCall, M. J., Brown, T., Hunter, W. N., and Kennard, O. (1986) The crystal structure of d(GGATGGGAG): an essential part of the binding site for transcription factor IIIA. *Nature*, **322,** 661.
25. Pavletich, N. P. and Pabo, C. O. (1993) Crystal structure of a five-finger GLI-DNA complex: new perspectives on zinc fingers. *Science*, **261,** 1701.
26. Fratini, A. V., Kopka, M. L., Drew, H. R., and Dickerson, R. E. (1982) Reversible bending and helix geometry in a B-DNA dodecamer: CGCCAATTBrCGCG. *J. Biol. Chem.*, **257,** 14686.
27. Nelson, H. C. M., Finch, J. T., Luisi, B. F., and Klug, A. (1987) The structure of an oligo(dA).oligo(dT) tract and its biological implications. *Nature*, **330,** 221.
28. DiGabriele, A. D. and Steitz, T. A. (1993) A DNA dodecamer containing an adenine tract crystallizes in a unique lattice and exhibits a new bend. *J. Mol. Biol.*, **231,** 1024.
29. Bugg, C. E., Thomas, J. M., Sundaralingam, M., and Rao, S. T. (1971) Stereochemistry

of nucleic acids and their constituents: solid-state base-stacking patterns in nucleic acid constituents and polynucleotides. *Biopolymers*, **10**, 175.
30. Short, N. (1993) The changing shape of structure. *Nature*, **366**, 203.
31. Gotoh, O. and Tagashira, Y. (1981) Stabilities of nearest-neighbour doublets in double-helical DNA determined by fitting calculated melting profiles to observed profiles. *Biopolymers*, **20**, 1033.
32. Patel, D. J., Kozlowski, S. A., Ikuta, S., Bhatt, R., and Hare, D. R. (1982) NMR studies of DNA conformation and dynamics in solution. *Cold Spring Harbor Symp. Quant. Biol.*, **47**, 197.
33. Drew, H. R., Weeks, J. R., and Travers, A. A. (1985) Negative supercoiling induces spontaneous unwinding of a bacterial promoter. *EMBO J.*, **4**, 1025.
34. Keene, M. A. and Elgin, S. C. R. (1984) Patterns of DNA structural polymorphism and their evolutionary implications. *Cell*, **36**, 121.
35. Flick, J. T., Eisenberg, J. C., and Elgin, S. C. R. (1986) Micrococcal nuclease as a DNA structural probe: its recognition sequences, their genomic distribution and correlation with DNA structure determinants. *J. Mol. Biol.*, **190**, 619.
36. McClellan, J. A. and Lilley, D. M. J. (1991) Structural alternation in alternating adenine-thymine sequences in positively supercoiled DNA. *J. Mol. Biol.*, **219**, 145.
37. Murchie, A., Bowater, R., Aboul-ela, F., and Lilley, D. M. J. (1992) Helix opening transitions in supercoiled DNA. *Biochim. Biophys. Acta*, **1131**, 1.
38. Hirose, S., Tabuchi, H., and Mizutani, M. (1992) Supercoiling of DNA facilitates TFIID-promoter interactions. In *Molecular structure and life*. Kyogoku, Y. and Nishimura, Y. (ed.). Japan Sci. Soc. Press, Tokyo, p. 229.
39. Tabuchi, H., Handa, H., and Hirose, S. (1993) Underwinding of DNA on binding of yeast TFIID to the TATA element. *Biochem. Biophys. Res. Comm.*, **192**, 1432.
40. Kim, Y., Geiger, J. H., Hahn, S., and Sigler, P. B. (1993) Crystal structure of a yeast TBP/TATA-box complex. *Nature*, **365**, 512.
41. Kim, J. L., Nikolov, D. B., and Burley, S. K. (1993) Co-crystal structure of TBP recognizing the minor groove of a TATA element. *Nature*, **365**, 520.
42. Winkler, F. K., Banner, D. W., Oefner, C., Tsernoglou, D., Brown, R. S., Heathman, S. P., Bryan, R. K., Martin, P. D., Petratos, K., and Wilson, K. S. (1993) The crystal structure of Eco RV endonuclease and of its complexes with cognate and non-cognate DNA fragments. *EMBO J.*, **12**, 1781.
43. Lue, N. F. and Kornberg, R. D. (1993) A possible role for the yeast TATA-element binding protein in DNA replication. *Proc. Natl Acad. Sci. USA*, **90**, 8018.
44. Huang, R.-Y. and Kowalski, D. (1993) A DNA unwinding element and an ARS consensus comprise a replication origin within a yeast chromosome. *EMBO J.*, **12**, 4521.
45. Wolffe, A. P. and Drew, H. R. (1989) Initiation of transcription on nucleosomal templates. *Proc. Natl Acad. Sci. USA*, **86**, 9817.
46. Lee, D. Y., Hayes, J. J., Pruss, D., and Wolffe, A. P. (1993) A positive role for histone acetylation in transcription factor binding to nucleosomal DNA. *Cell*, **72**, 73.
47. Li, Q. and Wrange, O. (1993) Translational positioning of a nucleosomal glucocorticoid response element modulates glucocorticoid receptor affinity. *Genes and Dev.*, **7**, 2471.
48. Richmond, T. J., Finch, J. T., Rushton, B., Rhodes, D., and Klug, A. (1984) Structure of the nucleosome core particle at 7Å resolution. *Nature*, **311**, 532.
49. Hogan, M. E., Rooney, T. F., and Austin, R. H. (1987) Evidence for kinks in DNA folding in the nucleosome. *Nature*, **328**, 554.

50. Crick, F. H. C. and Klug, A. (1975) Kinky helix. *Nature*, **255**, 530.
51. Pruss, D., Bushman, F. D., and Wolffe, A. P. (1994) HIV integrase directs integration to sites of severe DNA distortion within the nucleosome core. *Proc. Natl Acad. Sci. USA*, **91**, 5913.
52. Klug, A. and Lutter, L. C. (1981) The helical periodicity of DNA on the nucleosome. *Nucleic Acids Res.*, **9**, 4267.
53. Hayes J. J., Clark, D. J., and Wolffe, A. P. (1991) Histone contributions to the structure of DNA in the nucleosome. *Proc. Natl Acad. Sci. USA*, **88**, 6829.
54. Arents, G. and Moudrianakis, E. N. (1993) Topography of the histone octamer surface: repeating structural motifs utilized in the docking of nucleosomal DNA. *Proc. Natl Acad. Sci. USA*, **90**, 10489.
55. Arents, G., Burlingame, R. W., Wang, B. W., Love, W. E., and Moudrianakis, E. N. (1991) The nucleosomal core histone octamer at 3.1 Å resolution: a tripartite protein assembly and a left-handed superhelix. *Proc. Natl Acad. Sci. USA*, **88**, 10148.
56. Mirzabekov, A. D., Bavykin, S. G., Karpov, V. L., Preobrazhenskaya, O. V., Elbradise, K. K., Tuneev, V. M., Melinkova, A. F., Goguadze, E. G., Chenchick, A. A., and Beabealashvili, R. S. (1982) Structure of nucleosomes, chromatin and RNA polymerase promoter complex as revealed by DNA protein cross-linking. *Cold Spring Harbor Symp. Quant. Biol.*, **47**, 503.
57. Levitt, M. (1978) How many base pairs per turn does DNA have in solution and in chromatin? Some theoretical calculations. *Proc. Natl Acad. Sci. USA*, **75**, 640.
58. Drew, H. R. (1991) Can one measure the free energy of binding of the histone octamer to different DNA sequences by salt-dependent reconstitution? *J. Mol. Biol.*, **219**, 391.
59. Bashkin, J., Hayes, J. J., Tullius, T. D., and A. P. Wolffe. (1993) Structure of DNA in a nucleosome core at high salt concentration and at high temperature. *Biochemistry*, **32**, 1895.
60. Hansen, J. C. and van Holde, K. E. (1991) The mechanism of nucleosome assembly onto oligomers of the sea urchin 5S DNA positioning sequence. *J. Biol. Chem.*, **266**, 4276.
61. Jayasena, S. D. and Behe, M. J. (1989) Nucleosome reconstitution of core length poly(dG).poly(dC) and poly(rG-dC).poly(rG-dC). *Biochemistry*, **28**, 975.
62. Jayasena, S. D. and Behe, M. J. (1989) Competitive nucleosome reconstitution of polydeoxynucleotides containing oligoguanosine tracts. *J. Mol. Biol.*, **208**, 297.
63. Puhl, H. L., Gudibande, S. R., and Behe, M. J. (1991) Poly[d(A.T)] and other synthetic polydeoxynucleotides containing oligoadenosine tracts form nucleosomes easily. *J. Mol. Biol.*, **222**, 1149.
64. Hayes, J. J., Bashkin, J., Tullius, T. D., and Wolffe, A. P. (1991) The histone core exerts a dominant constraint on the structure of DNA in a nucleosome. *Biochemistry*, **30**, 8434.
65. Shrader, T. E. and Crothers, D. M. (1989) Artificial nucleosome positioning sequences. *Proc. Natl Acad. Sci. USA*, **86**, 7418.
66. Shrader, T. E. and Crothers, D. M. (1990) Effects of DNA sequence and histone–histone interactions on nucleosome placement. *J. Mol. Biol.*, **216**, 69.
67. Schild, C., Claret, F.-X., Wahli, W., and Wolffe, A. P. (1993) A nucleosome-dependent static loop potentiates estrogen-regulated transcription from the *Xenopus* vitellogenin B1 promoter *in vitro*. *EMBO J.*, **12**, 423.
68. Taneka, S., Zatchej, M., and Thoma, F. (1992) Artificial nucleosome positioning sequences. *Proc. Natl Acad. Sci. USA*, **86**, 7418.
69. Thoma, F. and Simpson, R. T. (1985) Local protein–DNA interactions may determine nucleosome positions on yeast plasmids. *Nature*, **315**, 350.

70. Simpson, R. T. (1978) Structure of the chromatosome, a chromatin core particle containing 160 base pairs of DNA and all the histones. *Biochemistry*, **17**, 5524.
71. Gale, J. M. and Smerdon, M. J. (1988) Photofootprint of nucleosome core DNA in intact chromatin having different structural states. *J. Mol. Biol.*, **204**, 949.
72. Hayes, J. J. and Wolffe, A. P. (1993) Preferential and asymmetric interaction of linker histones with 5S DNA in the nucleosome. *Proc. Natl Acad. Sci. USA*, **90**, 6415.
73. Meersseman, G., Pennings, S., and Bradbury, E. M. (1991). Chromatosome positioning on assembled long chromatin: linker histones affect nucleosome placement on 5S DNA. *J. Mol. Biol.*, **220**, 89.
74. Chipev, C. C. and Wolffe, A. P. (1992) Chromosomal organization of *Xenopus laevis* oocyte and somatic 5S rRNA genes *in vivo*. *Mol. Cell. Biol.*, **12**, 45.
75. Jeong, S. W., Lauderdale, J. D., and Stein, A. (1991) Chromatin assembly on plasmid DNA *in vitro*. Apparent spreading of nucleosome alignment from one region of pBR327 by histone H5. *J. Mol. Biol.*, **222**, 1131.
76. McPherson, C. E., Shim, E. Y., Friedman, D. S., and Zaret, K. S. (1993) An active tissue-specific enhancer and bound transcription factors existing in a precisely positioned nucleosomal array. *Cell*, **75**, 387.
77. Fedor, M. I., Lue, N. F., and Kornberg, R. K. (1988) Statistical positioning of nucleosomes by specific protein-binding to an upstream activating sequence in yeast. *J. Mol. Biol.*, **204**, 109.
78. Roth, S. Y., Dean, A., and Simpson, R. T. (1990) Yeast α2 repressor positions nucleosomes in TRP1/ARS1 chromatin. *Mol. Cell. Biol.*, **10**, 2247.
79. Simpson, R. T. and Stafford, D. W. (1983) Structural features of a phased nucleosome core particle. *Proc. Natl Acad. Sci. USA*, **80**, 51.
80. Tatchell, K. and van Holde, K. E. (1977) Reconstitution of chromatin core particles. *Biochem.*, **24**, 5295.
81. Camerini-Otero, R. D., Sollner-Webb, B., and Felsenfeld, G. (1976) The organization of histones and DNA in chromatin: evidence for an arginine-rich histone kernel. *Cell*, **8**, 333.
82. Simon, R. H. and Felsenfeld, G. (1979) A new procedure for purifying histone pairs H2A + H2B and H3 + H4 from chromatin using hydroxylapatite. *Nucleic Acids Res.*, **6**, 689.
83. Dong, F. and van Holde, K. E. (1991) Nucleosome positioning is determined by the $(H3\text{-}H4)_2$ tetramer. *Proc. Natl Acad. Sci. USA*, **88**, 10596.
84. Hayes, J. J., Tullius, T. D., and Wolffe, A. P. (1990) The structure of DNA in a nucleosome. *Proc. Natl Acad. Sci. USA*, **87**, 7405.
85. Pruss, D. and Wolffe, A. P. (1993) Histone–DNA contacts in a nucleosome core containing a *Xenopus* 5S rRNA gene. *Biochemistry*, **32**, 6810.
86. Cary, P. D., Crane-Robinson, C., Bradbury, E. M., and Dixon, G. H. (1982) Effect of acetylation on the binding of N-terminal peptides of histone H4 to DNA. *Eur. J. Biochem.*, **127**, 137.
87. Böhm, L. and Crane-Robinson, C. (1984) Proteases as structural probes for chromatin: the domain structure of histones. *Bioscience Reports*, **4**, 365.
88. Dong, F., Hansen, J. C., and van Holde, K. E. (1990) DNA and protein determinants of nucleosome positioning on sea urchin 5S rRNA gene sequences *in vitro*. *Proc. Natl Acad. Sci. USA*, **87**, 5724.
89. Bauer, W. R., Hayes, J. J., White, J. H., and Wolffe, A. P. (1994) Nucleosome structural changes due to acetylation. *J. Mol. Biol.*, **236**, 685.

90. Norton, V. G., Imai, B. S., Yau, P., and Bradbury, E. M. (1989) Histone acetylation reduces nucleosome core particle linking number change. *Cell*, **57,** 449.
91. Oliva, R., Bazett-Jones, D. P., Locklear, L., and Dixon, G. H. (1990) Histone hyperacetylation can induce unfolding of the nucleosome core particle. *Nucleic Acids Res.*, **18,** 2739.
92. Ausio, J. and van Holde, K. E. (1986) Histone hyperacetylation: its effects on nucleosome conformation and stability. *Biochemistry*, **25,** 1421.
93. Ausio, J., Dong, F., and van Holde, K. E. (1989) Use of selectively trypsinized nucleosome core particles to analyze the role of the histone tails in the stabilization of the nucleosome. *J. Mol. Biol.*, **206,** 451.
94. Hayes, J. J. and Tullius, T. D. (1992) Structure of the TFIIIA/5S DNA complex. *J. Mol. Biol.*, **227,** 407.
95. Miller, J., McLachlan, A., and Klug, A. (1985) Repetitive zinc-binding domains in the protein transcription factor IIIA from *Xenopus* oocytes. *EMBO J.*, **4,** 1609.
96. Hayes, J. J. and Clemens, K. R. (1992) Location of contacts between individual zinc fingers of *Xenopus laevis* transcription factor IIIA and the internal control region of a 5S RNA gene. *Biochemistry*, **31,** 11600.
97. Vrana, K. E., Churchill, M. E. A., Tullius, T. D., and Brown, D. D. (1988) Mapping functional regions of transcription factor TFIIIA. *Mol. Cell. Biol.*, **8,** 1684.
98. Tremethick, D., Zucker, D., and Worcel, A. (1990) The transcription complex of the 5S RNA gene, but not transcription factor TFIIIA alone, prevents nucleosomal repression of transcription. *J. Biol. Chem.*, **265,** 5014.
99. Almouzni, G., Méchali, M., and Wolffe, A. P. (1991) Transcription complex disruption caused by a transition in chromatin structure. *Mol. Cell. Biol.*, **11,** 655.
100. Clark, D. J. and Wolffe, A. P. (1991) Superhelical stress and nucleosome mediated repression of 5S RNA gene transcription in vitro. *EMBO J.*, **10,** 3419.
101. Hayes, J. J. and Wolffe, A. P. (1992) Histones H2A/H2B inhibit the interactions of transcription factor IIIA with the *Xenopus borealis* somatic 5S RNA gene in a nucleosome. *Proc. Natl Acad. Sci. USA*, **89,** 1229.
102. Wolffe, A. P. (1991) Developmental regulation of chromatin structure and function. *Trends Cell. Biol.*, **1,** 61.
103. Hayes, J. J., Pruss, D., and Wolffe, A. P. (1994) Contacts of the globular domain of histone H5 and core histones with DNA in a chromatosome. *Proc. Natl Acad. Sci. USA*, **91,** 7817.

3 | Chromatin replication and assembly

JOSÉ M. SOGO and RONALD A. LASKEY

1. Introduction

Chromosome replication is one of the most complex problems facing the cell. Not only must it produce exactly two accurate copies of the parental DNA, but it must copy patterns of chromosomal proteins. In this chapter we start by introducing the mechanism of eukaryotic DNA replication and some of the proteins involved. Next we consider eukaryotic origins of replication focusing on the contrast between the well-defined simple replication origins of lower eukaryotes or viruses and the paradoxically complex origins of metazoan cells. We then discuss how old nucleosomes are distributed and new nucleosomes assembled at replication forks. Finally we consider the termination of replication and the problems posed when two adjacent replication forks meet.

2. Overview of eukaryotic replication mechanism

2.1 DNA polymerases

DNA polymerases are the enzymes required for the polymerization of deoxyribonucleoside 5′ monophosphates during DNA replication. Five different types of eukaryotic DNA polymerases have been identified and classified as α, β, γ, δ, and ε (for review see 1). DNA polymerase β has been implicated in DNA excision repair mechanisms, and polymerase γ is responsible for the replication of mitochondrial DNA. The other three DNA polymerases, α, δ, and ε, are involved in chromosomal replication.

DNA polymerase α contains a DNA primase in a tight DNA-polymerase-α–primase complex. It is required for the synthesis of Okazaki fragments at the lagging strand of the replication fork and for the initiation of leading-strand synthesis, at least in simian virus 40 (SV40) (2). DNA polymerase δ appears to be the replicase involved in the synthesis of the leading strand. It possesses a 3′ → 5′ proof-reading exonuclease activity required for fidelity. Moreover, for DNA synthesis to proceed efficiently, the help of an auxiliary protein is needed; this is the proliferating-cell nuclear antigen (PCNA). Recently, a PCNA-independent DNA polymerase containing 3′ → 5′ exonuclease activity was identified. This enzyme, named polymerase ε, may possibly be involved in the gap filling reactions on the lagging strand (3). Functions for polymerase ε in DNA repair and in recombinational DNA synthesis have also been suggested (4).

The genes encoding each of the polymerases have been isolated from different organisms (yeast, mouse, calf, human), and conserved regions have been proposed as catalytic sites involved in DNA binding, dNTP (nucleotide triphosphate) binding, 3′ → 5′ exonuclease activity, etc. (for review see 5). Systematic site-directed mutagenesis and structural analysis of crystallized enzymes will reveal more data about the functional anatomy of the polymerases and their interaction with auxiliary proteins.

2.2 Helicases and auxiliary replication proteins

For initiation of replication and for movement of the replication fork, double-stranded DNA needs to be transiently melted. DNA helicases unwind the duplex DNA in a reaction using the energy of ATP hydrolysis. In organisms from *E. coli* to humans, an increasing number of helicases have been purified and identified (6), but the best characterized is the SV40 viral large T-antigen. T-antigen binds specifically to the origin sequences, oligomerizes in two hexamers (7), and by moving in opposite directions unwinds the parental DNA. With the help of replication protein A (RP-A), DNA-polymerase-α–primase enters to initiate the synthesis of both the leading and the lagging strands. The helicase moves in a 3′ → 5′ direction, indicating its association with the leading strand (8, 9) and it is presumably localized in front of the replication fork, facilitating its movement. Helicases are capable of unwinding large regions of duplex DNA working in either the 5′ → 3′ or 3′ → 5′ direction. Their involvement in repair and DNA recombination processes has been proposed.

3. Initiation of replication

3.1 Origins of replication

Activation of the origins of replication is the first crucial problem that emerges when the genome has to be replicated. The complexity of this event in higher eukaryotic chromosomes is considerable; despite numerous attempts it still remains to be resolved whether initiation sites are restricted to specific DNA sequences or to broad initiation zones. In simple genomes, e.g. in viruses (SV40, polyoma virus, Epstein–Barr virus (EBV)) or the yeast *Saccharomyces cerevisiae*, short DNA regions of a few hundred base pairs have been identified as initiation sites for replication. The organization of these simple regions is well defined, and in general they consist of an essential core sequence which requires the contribution of flanking regions for its correct function (reviewed in 10). When localized in an extrachromosomal plasmid, origin sequences allow efficient duplication in cells or *in vitro*. In yeast, replication origins were first identified as autonomously replicating sequences (ARS) and they are genetically and functionally associated with efficient chromosomal origins (11, 12). As explained below, the present picture of replication origins in higher eukaryotes is much more complex and confusing.

3.2 Initiation at simple origins

Initiation of DNA synthesis requires first, recognition of the origin by an origin-binding protein, second, unwinding of the strands by a helicase activity, and, third, stabilization of melted sequences by a single-stranded binding protein. After recruitment of DNA primase and DNA polymerase α, DNA synthesis sequentially starts on both parental strands of a bidirectional origin of replication (2). Regulatory elements (as individual or overlapping sequences) have been well characterized at the viral replication origins. Identification of specific origin-binding proteins and transcription factors that regulate origins of DNA replication have also been extensively analysed and are summarized by De Pamphilis (13–15). In *S. cerevisiae* a multiprotein complex, called the origin recognition complex (ORC), binds specifically to the ARS elements (16) throughout the cell cycle (17). Moreover, genetic evidence implicates the ORC in both DNA replication and transcriptional silencing (18–21). Data supporting the function of the origins as essential for the establishment of the repressed state, as well as the alternative model in which the ORC could directly interact with other proteins required for silencing, are reviewed in 22 and 23.

What replicates is not only DNA but chromatin. In order to ensure accessibility of the required factors to origin sequences, the nucleosomal organization at these sites is probably changed as suggested for cellular promoter sequences in transcription (24) and for enhancers on active ribosomal RNA genes (25). The observation that about 25 per cent of SV40 minichromosomes contain a nucleosome-free gap at the origin of replication led to the speculation that these minichromosomes are active in replication, although this has not yet been proved. When these sequences are reconstituted into chromatin *in vitro*, the replication efficiency drops drastically, and only recovers when the DNA is pre-incubated with T-antigen prior to chromatin assembly (26). This indicates that the origin must be free of nucleosomes in order to be functional. A similar effect is observed when transcription factors are bound at or close to the origin elements (27, 28). Whether these transcription factors have some additional stimulatory effect, besides their role as local barriers for the formation of nucleosomes, remains to be elucidated (29, 30).

Although SV40 minichromosomes have been repeatedly proposed as a model for the replication of cellular chromatin, and although many similarities do exist, caution is required in extrapolating. Perhaps the plasticity of chromatin at the viral origin could be related to the mechanisms which the virus has evolved to escape cell cycle control. However, ARS elements in *S. cerevisiae* must also be devoid of nucleosomes to facilitate replication of extrachromosomal plasmids. When the ARS element is moved into the central DNA region of a nucleosome core particle, a decrease in plasmid copy number is observed (31). On the basis of several recent publications (18–21), Newlon (22) proposed a model for the ORC in which the chromosomal origin sequences (domains A and B) are wrapped around a core protein (ORC) similar to that described in *E. coli* and bacteriophage λ. It is unlikely that such an ORC coexists with a histone octamer on the same DNA sequences. The observation that the ORC is detected throughout the whole cell cycle indicates that origin-binding proteins bind immediately after the origin DNA sequences have replicated, competing with histone octamers for association with the newly synthesized DNA. Therefore it appears that nucleosome-free origins are propagated from one cell generation to the next. Such a mechanism would imply that some origin-specific protein must remain in a functionally inactive form until the next G1 phase, in order to limit DNA replication to one round per cell cycle.

By extrapolating Simpson's data (31) and Newlon's model (22), it appears likely that the essential chromosomal origin elements remain as nucleosome-free regions. Although this assumption needs to be confirmed, a problem remains that not all ARS sequences are functional origins. One of the best-known examples is represented by the ribosomal RNA locus. Each of the 100 to 200 intergenic spacers contain an ARS element, but only one out of three to six acts as an active origin (32, 33). Several unanswered questions arise: (a) How is the chromatin state at ARS elements propagated? (b) Do all ARS elements have the same chromatin structure whether or not they will be used as initiation sites? (c) How and when are ARS elements selected to be functional? (d) Will the same ARS always be activated throughout subsequent cell divisions? *In vitro* replication systems with purified proteins will greatly facilitate the understanding of how origins are activated. Biochemical assays like footprinting analysis to determine how specific proteins are associated with specific DNA sequences should give valuable information in this respect. Direct visualization by electron microscopy of nucleoprotein complexes organized *in vivo* is limited by the methods of isolation and preparation and by the difficulty of identifying the corresponding nucleofilaments. To overcome this problem we introduced the psoralen technique (34). This technique maps precisely the positions of histone octamers in SV40 minichromosomes or in restriction fragments either by electron microscopy or by gel electrophoresis (35–38). It is possible to study origin activation by the differential accessibility of psoralen to ARS elements, depending on whether they are organized in nucleosomes or not. Replicating molecules can be fractionated by 2-D gel electrophoresis and analysed by electron microscopy or by band shifting (for details see 37). Using such techniques it should be possible to determine whether nucleosome-free initiation sites are directly transmitted from parental to daughter strands or, as has been postulated for SV40, post-replicative structural modifications are involved in these processes.

3.3 Initiation at complex origins

Little information exists about the chromatin organization of replication initiation sites in higher eukaryotes. The problems of understanding higher eukaryotic origins are clearly very complex. Not only the organization, but also the localization and identification of the DNA sequences where replication initiates are under debate. Studies have focused mostly on either embryonic cells with short doubling times, or chromosome domains consisting of amplified templates. One of the best-studied examples is the dihydrofolate reductase (DHFR) domain of Chinese hamster ovary cells (CHO), summarized in recent review articles (10, 14, 15, 39, 40). Briefly, DNA synthesis was studied by a diverse range of techniques in a 30 kilobase (kb) region downstream of the DHFR gene. The initiation site was mapped on a specific DNA sequence, the so called ori-β, 17 kb downstream of the DHFR gene. Similar results have been obtained by several methods including studying: (a) the total nascent DNA synthesized at the leading and the lagging strands ('earliest labelled fragment', 'replication origin trap,' or 'nascent DNA length'; for details see 41), or (b) only the elongated chains at the leading strands ('imbalanced DNA synthesis'), or (c) the elongated chains at the lagging strands ('Okazaki fragment distribution'). Elongation proceeded in opposite directions (bidirectionally) from a specific initiation site without any detectable impediment. Moreover, the origin of bidirectional replication (OBR) contained a series of characteristic features like bent DNA, ARS homologies, and transcription factor-binding sites, etc. (10, 40). The data and interpretations described so far seem coherent, but a problem arose when Hamlin's group used 2-D gel electrophoresis (42, 43) to analyse replicative intermediates originating in the DHFR domain (44, 45). The data revealed that DNA synthesis initiates at non-specific sequences along a broad zone 50 kb downstream from the DHFR gene. Since 2-D gel electrophoresis has been successfully applied to localizing replication origins in viruses (SV40, EBV), yeast (46, 47), and *Physarum* (48), etc., the question to be answered is: why are origins detected at defined positions in simple genomes, whereas many non-specific origins occur throughout a broad initiation zone in metazoan cells? Multiple initiation sites have also been detected in the histone gene repeats of *Drosophila* early embryos (49) and in the 'non-transcribed' spacer of human rRNA genes in tissue culture cells (50) when 2-D gel assays were used. Similarly, replicating bubbles have been mapped throughout the whole ribosomal locus of *Xenopus* embryos (51), indicating random initiation of replication. At present there is controversy over why multiple initiation sites are detected when DNA structures are analysed by 2-D gels whereas single specific origins are found when newly synthesized DNA is mapped. A clear limitation of 2-D gels is the interpretation of arcs corresponding to different classes of replicating forms (52); quantitation is difficult but required for the understanding of minor classes of replicating structures. On the other hand, assays based on the partial protection of newly synthesized DNA strands frequently require intermediate steps which could potentially be a source of error. For instance, in the imbalanced DNA synthesis method, the cells are treated with emetine, a potent inhibitor of protein synthesis. Since the function of the α-polymerase–primase is severely affected, DNA synthesis is preferentially made at the leading strands (53). Secondary effects of the drug cannot be ignored. The strand-switching assay used by Burhans *et al.* (54) is based on *in situ* DNA synthesis. Although the DNA elongation process appears to be correct, there still remains the possibility that artefacts occur during the permeabilization and manipulation of the cells. In the future a combination of different assays will conceivably help the correct interpretation of the published results. The use of model systems (52, 55) or *in vitro* replication will also help to clarify the problem.

3.4 Models of initiation of replication in eukaryotic genomes

In an attempt to resolve the puzzle set by the results of origin-mapping studies, Linskens and Huberman (56) elaborated a model which fits most published data ('the unidirectional bubble model'; 10). The postulated mechanism is divided into three steps (Fig. 1a). Initial clusters of microbubbles corresponding to stable denatured regions are randomly distributed throughout the initiation zone spanning several kb in length. In contrast to a proposal by Benbow *et al.* (57, 58; Fig. 1b) that the parental strands unwind completely over large regions to form single-stranded loops without replication forks (strand separation model), Linskens

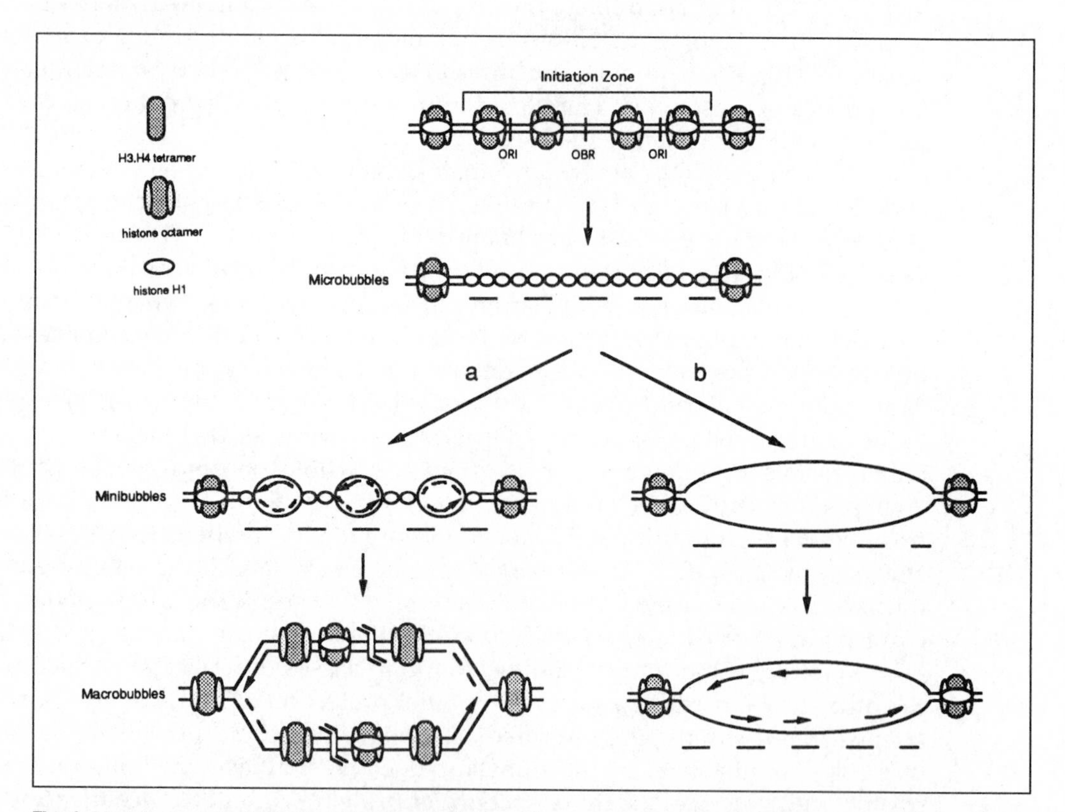

Fig. 1 Models for initiation of replication in higher eukaryotic cells. Paths suggested by: (a) Linskens and Huberman (56), and (b) Benbow *et al.* (57). Nucleosomes have been placed at positions compatible with 'stable' duplex DNA. The marked regions (dashed lines) correspond to large unusual (nuclease-hypersensitive) chromatin domains. Note that nucleosomes at replication forks are H1 depleted (for details see Fig. 3).

and Huberman (56) proposed the conversion of microbubbles to replicating minibubbles. In order to explain the data of Burhans *et al.* (53), the minibubble localized in the middle of the domain would be an OBR flanked by minibubbles replicating only unidirectionally. Replication–elongation would proceed on both sides of the central bubble in opposite directions, generating Okazaki fragments on only one strand of each minibubble. In consequence DNA synthesis is asymmetric and each site behaves like a single replication fork. The presence of minibubbles could explain non-specific initiation within a large zone, as suggested by the data of Hamlin's laboratory. Finally, fusion of the minibubbles would contribute to the formation of a macrobubble with two normal replication forks moving in opposite directions. Among the inconsistencies pointed out by the authors, such as frequency of initiation or length of the nascent leading strands (see also 10), the proposed model also implies the generation of complicated replicating restriction fragments during the second step. These non-linear structures containing more than one bubble or combinations of bubbles and forks remain to be identified on the 2-D gels. Moreover, the suggestion that microbubbles are involved in initiation arose from an observation by EM of partially denatured regions on purified DNA fragments isolated from rapidly proliferating cells (57). However, the assumption that these microbubbles represent the first step in the activation of replication origins is questionable. An alternative explanation is that these A-T-rich regions are simply denatured by the action of the formamide present in the spreading solutions.

Finally, the Linskens and Huberman model and the Benbow model are both restricted to the level of double-stranded DNA, and the organization of this DNA as a chromatin component has not been considered. What replicates is not only DNA but chromatin; therefore the question of how this model fits at the level of the chromatin fibre is more than justified. It is possible to ask whether microbubbles are restricted to the linker DNA between nucleosomes, or whether nucleosomes are first lost in order to allow the partial unwinding and stabilization of single-stranded DNA (ss-DNA). Once melting has occurred, the stabilization of separated strands can probably be maintained through the interaction with single-stranded binding proteins. The lack of nucleosomes could increase the susceptibility of these regions to degradation by nucleases. Therefore 'hypersensitive' regions should be very abundant during S-phase and easy to detect. Multiple contiguous origins (minibubbles) would necessarily create a large initiation zone free of nucleosomes. However, such large regions of disturbed chromatin appear to contradict the available EM data. Spreading of chromatin from *D. melanogaster* and *Physarum* reveals that the strands are always organized in nucleosomes irrespective of the size of the replicating bubbles (59, 60). In the models discussed above (56, 57) the conventional picture of newly replicated chromatin (see later) would be established only after the formation of the macrobubbles (> 10 kb), when the replicating domain is 'normalized' by the elongation of the two forks moving in opposite directions. We suggest therefore that it is difficult to imagine how the micro- and mini-replicating bubbles could be organized in chromatin without the establishment of large, unusual, and presumably nuclease-sensitive domains.

In contrast to the two models presented above, De Pamphilis (10) has proposed an alternative one in which initiation of replication could occur at many DNA sites, but most of them would abort. The start of DNA synthesis would consequently be restricted to a single primary bidirectional origin. It is unlikely that more than one replicating bubble is present in any individual initiation zone. Under such conditions chromatin could maintain a nucleosomal distribution (Fig. 2) similar to that described for simple genomes (e.g. in SV40) (see later). The initiation sites with disturbed nucleosomes would affect a region not longer than 200–300 bp, corresponding to 1–2 nucleosomes. Once replication forks were established, normal progression of the newly synthesized strands would occur by a transient disruption and reorganization of the nucleosomes at the elongation point.

A variant of De Pamphilis' model has been proposed recently (61). It proposes that initiation would be mediated at many sites in a broad initiation zone by polymerase α, but the product of this synthesis would be destroyed and replaced by new synthesis at one of these sites by polymerase δ and ε. Elongation is envisaged to occur only within an immobile cluster of replication forks (61), which would provide the basis of site selection. The inaccurate product of polymerase α would be erased and replaced by the higher fidelity synthesis catalysed by polymerases δ and ε held at immobile sites of clustered replication forks. Whether or not this model for events at initiation zones is correct, the clustered distribution of replication forks within the nucleus is a widespread phenomenon and requires explanation (61).

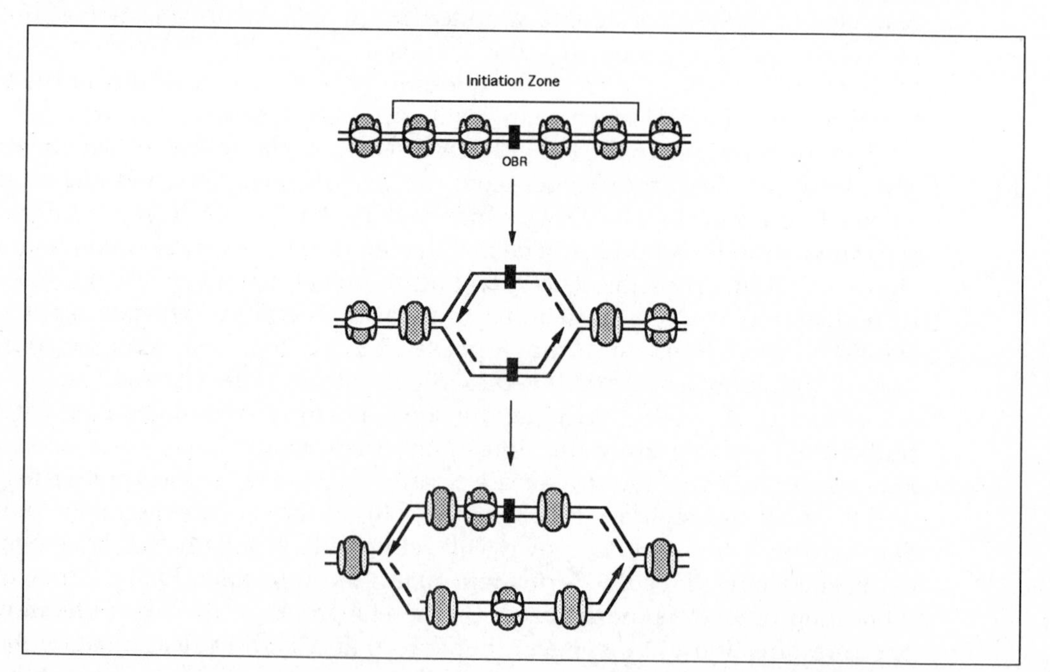

Fig. 2 Model of initiation of replication in higher eukaryotic cells according to De Pamphilis (10). Nucleosomes immediately in front and behind the elongation point are H1 depleted (see Fig. 3).

4. Progress of the replication fork through chromatin

4.1 The fate of parental histones

Replication of eukaryotic chromosomes is a co-ordinated process requiring duplication of parental DNA and its association with histones. DNA of both newly replicated strands is rapidly assembled into chromatin, and nucleosomes are visualized at the replication fork immediately behind and in front of the elongation point. Since the substrate for the multi-enzyme replication machinery is the chromatin fibre, the advance of the replication forks transiently induces structural changes. The transfer of nucleosomes from the parental DNA to the daughter strands must occur co-ordinately with the synthesis of new histones. Both 'old' and 'new' histones are involved in providing the full complement for the packing of the nascent DNA in nucleosomes. The length of the prenucleosomal DNA is found to be 170 and 300 nucleotides for the leading and lagging strands respectively (35). This asymmetry is probably related to the discontinuous DNA synthesis on the lagging strand, and it seems that only ligated Okazaki fragments are packaged into nucleosomes (62). For the transfer of parental histones to the daughter strands, different mechanisms have been proposed. The simplest model is based on the suggestion that histone octamers remain stable as independent units over successive cell generations. Consequently, these octamers are transferred from the unreplicated chromatin fibre to the sibling nascent molecules as conserved units (63). Systematic studies on stabilized bulk chromatin (aldehyde cross-linked) indicate that parental histones H3 and H4 are deposited as tetramers and do not dissociate in subsequent cell cycles. Newly synthesized histones H3 and H4 also associate with the nascent strands as new tetramers (64, 65). However, old and new H2A and H2B histones are randomly associated as dimers with the new as well as with the parental $(H3–H4)_2$ tetramers. Therefore, most of the nucleosomes behind the replication forks are hybrids formed from old and new histones (65, 66). Recently the conservation of the H2A–H2B dimers as units during the cell cycle has also been proposed (67). Re-association of old or new tetramers $(H3–H4)_2$ and dimers (H2A–H2B) as individual units in newly replicated nucleosomes obviously gives hybrid structures, but to a lesser extent than that proposed by Jackson (65) and Svaren and Chalkley (66). This point needs further clarification.

The question of whether parental histone octamers are transferred to daughter strands as conserved units or whether dissociation of histones occurs during the transfer has been re-analysed using *in vitro* replication systems. *In vitro* reconstituted or native purified minichromosomes were used as templates. The advantage lies in the direct analysis of either replicating or post-replicated minichromosomes. Since new histone synthesis does not occur *in vitro*, the fate of the parental histones could be followed without the interference of secondary effects of protein synthesis inhibitors such as cycloheximide or emetine. Bonne-Andrea *et al.* (68) took advantage of the well-established T_4 replication system. Circular DNA containing only a few nucleosomes instead of defined minichromosomes was used as the template.

Both from analysis of radioactively labelled histones segregating with replicated DNA and from electron microscopic mapping of nucleosomes by the psoralen method the authors concluded that histone octamers do not dissociate from DNA during the passage of replication forks. Similarly, others (69–73) found that the newly synthesized molecules retained the nucleosomal organization during SV40 minichromosome replication. The assembly of nucleosomes on both replicating and progeny molecules was essentially unaffected even when replication reactions contained competitor DNA. They concluded that SV40 chromosomes can be replicated without displacement of the parental histones which remain bound to the DNA during the entire replication process. Opposite findings arose from direct analysis of replicating SV40 DNA cross-linked with psoralen as chromatin. This maps single-stranded bubbles corresponding to nucleosome positions. Psoralen cross-linking revealed that the density of nucleosomes is markedly reduced on the newly replicated branches and that a substantial proportion of the replicated chromatin consists of subnucleosomal particles (26). The presence of excess protein-free competitor DNA during minichromosome replication traps the segregating histones. These data (confirmed by micrococcal nuclease digestions) suggest that the parental histones remain only loosely attached or unattached to the DNA during the passage of the replication fork. The problem is how to reconcile these contradictory results.

The bacteriophage T_4 replication assay (68) is based on a system which never normally replicates chromatin. Therefore, extrapolation of these data to eukaryotic replication is questionable. The key experiments indicating that histone octamers remain associated with replicating DNA were those in which competitor DNA was added to the replication reaction mixture. In most of the published work (70, 72, 73) limited amounts of competitor DNA were added to the reactions and no effects on post-replicative minichromosomes were detected. However, when a large excess of competitor DNA was added (26) the newly synthesized branches appeared as histone-free DNA. The results therefore indicate that parental nucleosomes are disrupted by the passage of the replication fork (35) and histones are only loosely attached to DNA during replication.

4.2 Nucleosome core assembly

When purified chromosomes replicate *in vitro* under standard conditions (in the absence of competitor DNA), the regions protected from psoralen cross-linking on the daughter strand have an average length of 80–90 nucleotides, corresponding to a subnucleosomal particle containing $(H3–H4)_2$ tetramers (26). The data support the notion that nucleosome core assembly on the newly synthesized daughter strands occurs by the formation of $(H3–H4)_2$ tetramer complexes followed by the association of H2A–H2B dimers (Fig. 3; 26, 74). Smith and Stillman (75) have identified a factor (CAF-1) that assembles nucleosome core particles on replicating DNA in mammalian cells. Although replication-independent pathways have also

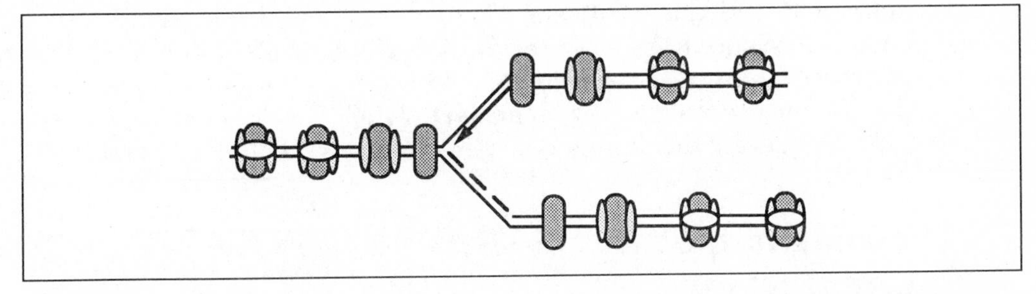

Fig. 3 Model for chromatin structure around the replication fork. In front of the fork two nucleosomes are destabilized (see text). In the nucleosome involved in replication, only the $(H3-H4)_2$ tetramer is present; the other immediately contiguous lacks histone H1. Behind the fork at first $(H3-H4)_2$ tetramers are deposited on the newly replicated DNA, then nucleosomes are completed by the addition of H2A–H2B dimers and H1.

been described (reviewed in 76), CAF-1 remains a good candidate for coupling chromatin assembly to the replication fork in all types of cells.

4.3 Segregation of nucleosomes

A loose association of parental histones at or near the replication fork facilitates their transfer to the nascent branches. Following on from the topology of the forks and the organization of the multi-enzyme replication complex, the simplest model would be the conservative segregation of the nucleosome core to the leading strands. This model was suggested by Seidman *et al.* (77) who concluded that the nucleosomes segregate exclusively to the leading strand. Cusick *et al.* (78) repeated and extended these experiments using replicating SV40 minichromosomes isolated from cells treated with cycloheximide. They clearly demonstrated that nucleosomes are equally segregated to both daughter strands corresponding to the disperse mode of segregation. A complementary approach analysed replicating SV40 chromosomes that had been psoralen cross-linked *in vivo* in the presence of cycloheximide. This approach revealed that the segregation of nucleosomes is completely random (35). These and other data from several laboratories (79–80) give compelling evidence for the random segregation of parental nucleosomes, but also for a tendency of these to cluster (35, 81). Surprisingly, the controversy over whether segregation of parental nucleosomes is conservative or random was reactivated as a consequence of the protocol developed by Handeli *et al.* (82) for mapping origins of replication in proliferating cells treated with emetine. These investigators assumed that, in the presence of this protein synthesis inhibitor, segregation of parental nucleosomes would be conservative to the leading strand. After nuclease digestion, the histone-free lagging strand would be totally degraded, and the nucleosomal DNA on the leading strand would be protected. Identification of this partially protected DNA would reveal the direction of DNA replication. Burhans *et al.* (54) demonstrated that, although the assay is valid, its proposed underlying mechanism is not correct. The pronounced asymmetry in the DNA digestion pattern induced by emetine treatment is not due to the segregation of nucleosomes, but to imbalanced DNA synthesis. The DNA synthesis on the lagging strand (synthesis of Okazaki fragments) is blocked by the action of emetine, explaining the asymmetry observed by Handeli *et al.* (82). The random segregation of $(H3\text{–}H4)_2$ tetramers is therefore the model which best fits the data available.

4.4 Coiling of the nascent DNA helix on the histone octamer core

From *in vitro* transcription data, Clark and Felsenfeld (83) proposed that the positive superhelical stress generated ahead of RNA polymerase might displace the histone octamers; these would then reassociate with the negatively supercoiled DNA behind RNA polymerase to reform a nucleosome core. A similar mechanism could be involved during DNA replication. The progressing fork could generate positive supercoils in the parental strands, displacing the histone octamers, which would reform nucleosomes behind the elongation point. Although the advancing replication machinery might generate positive supercoils ahead of the replication fork, these could be removed by the action of topoisomerase located at and in front of the replication fork (84, 85). The two nascent double helices are topologically relaxed domains and negative supercoils cannot alone account for the formation of nucleosome cores on the newly synthesized DNA (although behind the fork, the separated parental strands can be viewed as a limiting case of negative supercoiling; 86). Whether or not topoisomerase activity is needed for formation of the two DNA turns around the histone core octamer *in vivo* remains to be elucidated. However, preliminary data show that the nucleosome density on daughter strands of replicating SV40 molecules is not seriously disturbed by camptothecin (a topoisomerase I inhibitor) or VM26 (a topoisomerase II inhibitor) (84); this indicates that topoisomerases do not play a crucial role in the formation of nucleosomes on newly replicated DNA. The behaviour of extrachromosomal plasmids replicating in yeast topoisomerase mutants could help us to understand the potential role of topoisomerases in nucleosome assembly on newly replicated DNA.

4.5 Maturation of post-replicative chromatin

Chromatin assembly is not complete when $(H3\text{–}H4)_2$ tetramers and H2A–H2B dimers have been deposited on the daughter strands. Placement of histone H1 plays an important role in the formation of higher order structures of chromatin (87) and could account for the chromatin maturation process (88, 89). The lack of direct, precise assays for the presence or absence of histone H1 in short defined regions of chromatin, explains the scarce and confusing data published. A wide range of times of deposition of histone H1 has been reported, from 30 seconds to 30 minutes (90–92). We (J. M. S.) have developed a method of distinguishing indirectly between nucleosomes which contain or lack histone H1 and have used it to reinvestigate its deposition around the replication fork. Psoralen cross-linking of histone H1-depleted chromatin at pH 10 and very low ionic strength cross-links the DNA fairly continuously (93), whereas chromatin-containing H1 still cross-links mainly between nucleosomes. The results with this technique using SV40 minichromosomes suggest that the first nucleosome behind the fork frequently lacks H1 (Fig. 3). The transition of newly replicated chromatin from unstable to stable nucleosomes (S. Gasser, T. Koller, and J. M. Sogo; unpublished data) supports a previous suggestion (78, 94) that H1-dependent maturation of chromatin occurs. The association of histone H1 is an essential, though probably not a sufficient, step for chromatin maturation. Modification of histones and interaction with other proteins are additional steps to complete this process (95). In this respect, Perry *et al.* (96) provide evidence that the level of acetylation of segregated parental histones is maintained during chromatin replication. The segregation of unacetylated or acetylated histones could help to determine whether inactive or active chromatin domains are generated.

4.6 Chromatin structure in front of replicating forks

While much attention has been focused on the events following the passage of the fork, very little is known about what happens immediately ahead of the moving point (89). The difficulty of specifically labelling the DNA or histones just in front of the replication fork is a major impediment for such analysis. However, visualization of DNA cross-linked by psoralen *in vivo* and *in situ* has shown that the nucleosomes closest to the replication fork are destabilized (35). Recent studies have provided evidence (from cross-linking experiments at pH 10) that nucleosomes in contact with the replication machinery are destabilized by the lack of histone H1 and by the physical force of the advancing fork (S. Gasser, T. Koller, and J. Sogo, unpublished data). It is possible that direct contact of histones with a component of the multi-enzyme replication complex mediates the destabilization of one to two nucleosomes prior to the replication of their DNA (Fig. 3) in order to decondense the higher order structures. The unfolding of replicating chromatin could be due to phosphorylation of H1 (97). Subsequently, the passage of the replication fork itself could contribute to the destabilization of the H2A–H2B dimers (see above).

5. Termination of replication

5.1 Termination mechanisms

Segregation of newly replicated chromosomes requires resolution of the linkage between the sister chromatids to generate two separate copies for the two daughter cells. There is good evidence in yeast and mammalian cells that the activity of topoisomerase II is required during both metaphase and anaphase for successful segregation of daughter chromosomes (98–100). Both completion of DNA synthesis and decatenation of the sibling molecules appear to require DNA topoisomerase II. While both types of topoisomerase can remove the positive supercoils generated ahead of replication forks, only type II can separate the intertwined daughter chromosomes at the end of replication.

Circular SV40 minichromosomes are an excellent model for studying the termination of replication when two replication forks converge from opposite directions. Completion of DNA replication does not require a specific termination sequence (101, 102). Two pathways for the unwinding of the last DNA sequences on the parental strand have been proposed. Varshavsky and Sundin (103) suggest that unwinding of the last few helical turns and their replication might both occur without unlinking the parental strands. For every helical turn melted at the replication fork, one link between the daughter helices will be created. Completion of DNA synthesis results in the formation of catenated daughter molecules (Fig. 4a). Successively and gradually, topoisomerase II activity catalyses the decatenation process, converting each dimer into two separated supercoiled monomers. Similarly, 2 μm plasmids are accumulated as catenated dimers in yeast temperature-sensitive

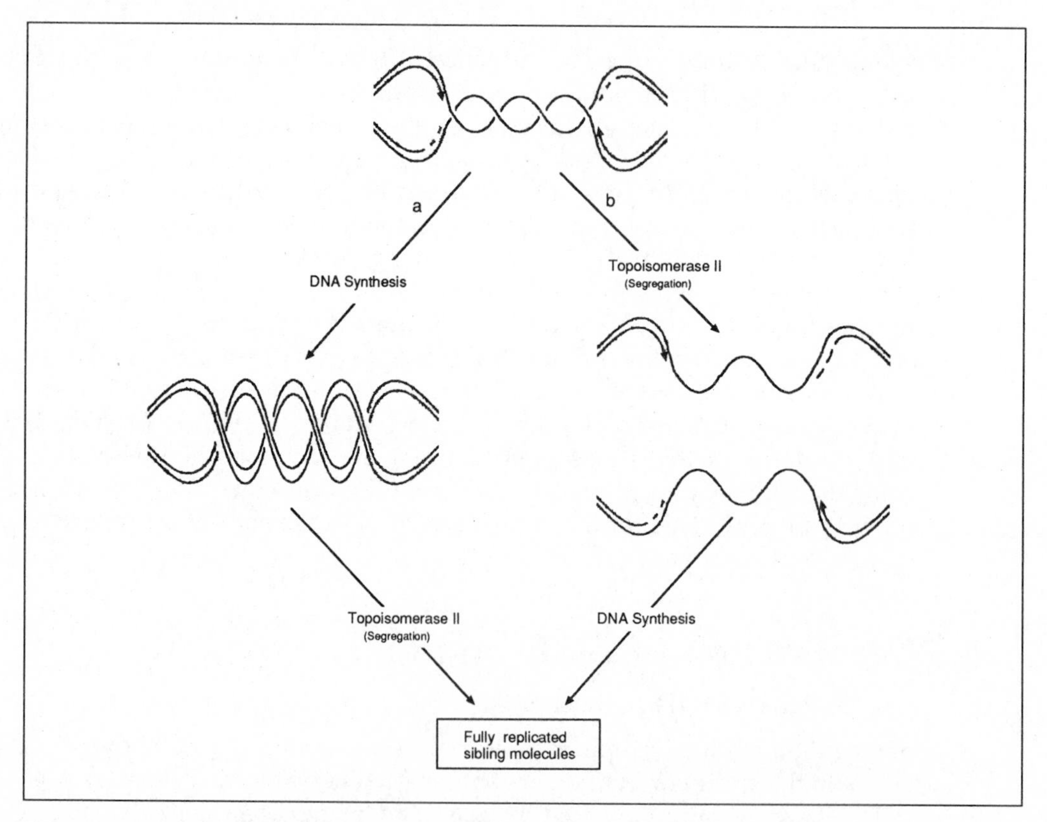

Fig. 4 Termination of replication. (a) Pathway proposed by Varshavsky and Sundin (103); full termination of replication precedes the segregation of the daughter molecules. (b) Pathway proposed by Weaver *et al.* (101); segregation of sibling molecules precedes the completion of DNA synthesis.

topoisomerase II mutants (98) or in *E. coli* cells following inactivation of topoisomerase IV (104).

An alternative pathway has been proposed by De Pamphilis *et al.* (101, 102). They proposed that DNA synthesis and topoisomerase II can act co-ordinately. During the termination process, therefore, transient catenanes between the replicating strands might be formed and resolved. Topoisomerase II is proposed to act behind the replication fork, unlinking the nascent strands as rapidly as they are generated (Fig. 4b). In summary, in the first pathway (103) full termination of replication is an obligatory step occurring before separation of the daughter strands, which in the second (101) segregation of strands is independent of the completion of DNA synthesis. A third termination pathway, in which the final replicative intermediate consists of hemicatenanes, has been described by Sogo *et al.* (35). Here, the two replicated daughter molecules are interlocked at the level of their single strands within the replication termination region.

5.2 Termination of replication on eukaryotic chromosomes

Very little is known about the fusion of replicons in the linear chromosomes of eukaryotic cells. In most cases, the sequences at which fusion occurs are probably non-specific and determined by the point of convergence of two replication forks moving in opposite directions (105). At least one clear exception exists—namely the fusion of replicons in ribosomal gene repeats. Brewer and Fangman (55) and Linskens and Huberman (56) identified a replication fork barrier near the 3′ end of yeast rRNA genes. This result has been confirmed both in human and plant cells (50, 106) and in *Xenopus* cell cultures (107).

Chromatin structure at the yeast replication fork barrier has been investigated by Lucchini and Sogo (37), who found that the nucleosomal organization on newly synthesized strands at arrested forks is similar to that of bulk unreplicated chromatin; the distance between the elongation point and the first nucleosome is shorter than the corresponding distance at normal elongating forks. DNA synthesis and ligation of Okazaki fragments are probably complete at arrested forks, with the DNA assembled into mature nucleosomes. Whether fork arrest is a consequence of specifically bound transcription termination factors or of specific replication termination complexes is still an unanswered question. It may be that such factors have a dual function, like the ORC discussed earlier. Since ribosomal loci are overrepresented in the genome, their arrested forks could be a model for further analysis of how replicons fuse in eukaryotic chromosomes.

6. Final comments

6.1 Important questions

There are many important challenges facing us in the process of achieving an understanding of chromosome replication. Although we largely understand the sequence specificity of initiation at the simple replication origins of lower eukaryotes, we do not know the mechanism of initiation at these sites, and the sites of initiation in higher eukaryotes remain paradoxical. They are clearly not random, yet they are proving difficult to define precisely. Perhaps a role of nuclear structure is important in explaining this paradox, as one of us (R. A. L.) has argued elsewhere (61).

We need to understand the respective roles of the three eukaryotic DNA polymerases that are implicated in replication: α, δ, and ε. Clues may come from the observation that one of them (α) has priming activity, while the other two have proof-reading activities to maximize fidelity.

The passage of the replication fork through chromatin also leaves many unanswered questions: How are nucleosomes displaced at the fork? How are $(H3–H4)_2$ tetramers, and perhaps H2A–H2B dimers, held together and segregated to the progeny strands? What determines the timing and location of H1 deposition? How are patterns of transcription factors transmitted to progeny strands? Finally, we need to understand more about the role of DNA topoisomerase II in termination of chromosome replication and how replication forks can pause at specific sites such as the 3′ end of rRNA genes.

The identification of simple origins in yeast and of the proteins that bind to them, the development of cell-free replication systems, and the development of nucleosome mapping techniques provide promising opportunities to answer these crucial questions.

6.2 Discussion

In discussion with other book authors it was pointed out that the mode of chromatin packaging seems to be highly dependent on the time of replication within S phase (see Chapters 7 and 8). Experiments to date suggest that the start time for a given origin is not autonomous, but reflects the domain in which the origin resides (see Chapter 7). This is a very interesting problem that deserves more attention.

Chromatin replication may impose a number of demands on patterns of protein synthesis. For example, chromatin replication requires high rates of histone synthesis. Histone synthesis during S phase is regulated at the point of transcription initiation, by mechanisms of RNA processing, and by changes in the mRNA stability. It is quite an intricate system, discussed in reference 89. In general, cell cycle regulation of synthesis of the nonhistone chromosomal proteins has not been reported. However, the differences observed in packaging genes replicated early in S phase (which generally includes active genes) and those replicated late in S phase (which generally includes inactive sequences) make it attractive to look at the synthesis of proteins specifically associated with one region or the other. Future research will no doubt include efforts to identify the cues that result in the faithful repackaging of the daughter chromatids into the chromatin form present in the parent.

Acknowledgements

We are grateful to Theo Koller for helpful comments, to Ruth Dendy for help with the manuscript, and to the Swiss National Science Foundation (31.31068.91) and the Cancer Research Campaign (SP 1961) for support.

References

1. Burgers, P., Bambara, R. A., Campbell, J. L., Chang, L., Downey, K. M., Hübscher, U., Lee, M., Linn, S. M., So, A. G., and Spadari, S. (1990) Revised nomenclature for eukaryotic DNA polymerases. *Eur. J. Biochem.*, **191,** 617.
2. Tsurimoto, T., Melendey, T., and Stillman, B. (1990) Sequence initiation of lagging and leading strand synthesis by two different polymerase complexes at SV40 DNA replication origin. *Nature,* **346,** 543.
3. Hübscher, U. and Spadari, S. (1994) DNA replication and chemotherapy. *Physiol. Rev.*, **74,** 259.
4. Jessberger, R., Produst, V. N., Hübscher, U., and Berg, P. (1993) A mammalian protein complex that repairs double-stranded breaks and deletions by recombination. *J. Biol. Chem.*, **268,** 15070.
5. Wang, T.-S. F. (1991) Eukaryotic DNA polymerases. *Ann. Rev. Biochem.*, **60,** 513.
6. Thömmes, P. and Hübscher, U. (1992) Eukaryotic DNA helicases: essential enzymes for DNA transactions. *Chromosoma,* **101,** 467.
7. Mastrangelo, J. A., Hough, P. V. C., Wall, J. S., Dodson, M., Dean, F. B., and Hurwitz, J. (1989) ATP-dependent assembly of double hexamers of SV40 T-antigen at the viral origin of DNA replication. *Nature,* **338,** 658.
8. Goetz, G. S., Dean, F. B., Hurwitz, J., and Matson, S. W. (1988) The unwinding of duplex regions in DNA by the Simian Virus 40 large tumor antigen-associated DNA helicase activity. *J. Biol. Chem.*, **263,** 383.
9. Wiekowski, M., Schwartz, M. W., and Stahl, H. (1988) Simian Virus 40 large T-antigen DNA helicase. Characterization of the ATPase-dependent DNA unwinding activity and its substrate requirements. *J. Biol. Chem.*, **263,** 436.
10. De Pamphilis, M. L. (1993) Eukaryotic DNA replication: anatomy of an origin. *Annu. Rev. Biochem.*, **62,** 29.
11. Huang, R.-Y. and Kowalski, D. (1993) A DNA unwinding element and an ARS consensus comprise a replication origin within a yeast chromosome. *EMBO J.*, **12,** 4521.
12. Deshpande, A. M. and Newlon, C. S. (1992) The ARS consensus sequence is required for chromosomal origin function in *S. cerevisiae. Mol. Cell. Biol.*, **12,** 4305.
13. De Pamphilis, M. L. (1993) How transcription factors regulate origins of DNA replication in eukaryotic cells. *Trends Cell Biol.*, **3,** 161.
14. De Pamphilis, M. L. (1993) Origins of DNA replication in metazoan chromosomes. *J. Biol. Chem.*, **268,** 1.
15. De Pamphilis, M. L. (1993) Origins of DNA replication that function in eukaryotic cells. *Curr. Opinion Cell Biol.*, **5,** 434.
16. Bell, S. P. and Stillman, B. (1992) ATP-dependent recognition of eukaryotic origins of DNA replication by a multiprotein complex. *Nature,* **357,** 128.
17. Diffley, J. F. X. and Cocker, J. H. (1992) Protein–DNA interactions at a yeast replication origin. *Nature,* **357,** 169.

18. Bell, S. P., Kobayashi, R., and Stillman, B. (1993) Yeast origin recognition complex functions in transcription silencing and DNA replication. *Science*, **262,** 1844.
19. Foss, M., McNally, F. J., Laurenson, P., and Rine, J. (1993) Origin recognition complex (ORC) in transcriptional silencing and DNA replication in *S. cerevisiae*. *Science*, **262,** 1838.
20. Micklem, G., Rowley, A., Harwood, J., Nasmyth, K., and Diffley, J. F. X. (1993) Yeast origin recognition complexes involved in DNA replication and transcriptional silencing. *Nature*, **366,** 87.
21. Li, J. J. and Herskowitz, J. (1993) Isolation of ORC6, a component of the yeast origin recognition complex by a one-hybrid system. *Science*, **262,** 870.
22. Newlon, C. S. (1993) Two jobs for the origin replication complex. *Science*, **262,** 1830.
23. Huberman, J. A. (1993) A tale of two functions. *Nature*, **367,** 20.
24. Grünstein, M. (1990) Histone function in transcription. *Annu. Rev. Cell. Biol.*, **6,** 643.
25. Lucchini, R. and Sogo, J. M. (1992) Different chromatin structures along the spacers flanking active and inactive *Xenopus* rRNA genes. *Mol. Cell. Biol.*, **12,** 4288.
26. Gruss, C., Wu, J., Koller, Th., and Sogo, J. M. (1993) Disruption of the nucleosomes at the replication fork. *EMBO J.*, **12,** 4533.
27. Cheng, L. and Kelly, T. J. (1989) Transcriptional activation nuclear factor 1 stimulates the replication of SV40 minichromosomes *in vivo* and *in vitro*. *Cell*, **59,** 541.
28. Cheng. L., Workman, J. L., Kingston, R. E., and Kelly, T. J. (1992) Regulation of DNA replication *in vitro* by the transcriptional activation domain of GALA 4-VP16. *Proc. Natl Acad. Sci. USA*, **89,** 589.
29. Go, Z.-S. and De Pamphilis, M. L. (1992) Specific transcription factors stimulate both Simian Virus 40 and Polyoma virus origins of replication. *Mol. Cell. Biol.*, **12,** 2514.
30. Huang, A. T., Wang, W., and Gralla, J. D. (1992) The replication activation potential of selected RNA polymerase II promoter elements at the Simian Virus 40 origin. *Mol. Cell. Biol.*, **12,** 3087.
31. Simpson, R. T. (1990) Nucleosome positioning can affect the function of a *cis*-acting DNA element *in vivo*. *Nature*, **343,** 387.
32. Brewer, B. J. and Fangman, W. L. (1988) A replication fork barrier at the 3′ end of yeast ribosomal RNA genes. *Cell*, **55,** 637.
33. Linskens, M. H. K. and Huberman, J. A. (1988) Organization of replication of ribosomal DNA in *S. cerevisiae*. *Mol. Cell. Biol.*, **8,** 4927.
34. Sogo, J. M., Ness, P. J., Widmer, R.-M., Parish, R. W., and Koller, Th. (1984) Psoralen-crosslinking of DNA as a probe for the structure of active nucleolar chromatin. *J. Mol. Biol.*, **178,** 897.
35. Sogo, J. M., Stahl, H., Koller, Th., and Knippers, R. (1986) Structure of replicating Simian Virus 40 minichromosomes. The replication fork, core histone segregation and terminal structures. *J. Mol. Biol.*, **189,** 189.
36. Conconi, A., Widmer, R.-M., Koller, Th., and Sogo, J. M. (1989) Two different chromatin structures coexist in ribosomal RNA genes. *Cell*, **57,** 753.
37. Lucchini, R. and Sogo, J. M. (1994) Chromatin structure and transcriptional activity around the replication forks arrested at the 3′ end of yeast rRNA genes. *Mol. Cell. Biol.*, **14,** 318.
38. Dammann, R., Lucchini, R., Koller, Th., and Sogo, J. M. (1993) Chromatin structures and transcription of rDNA in yeast *S. cerevisiae*. *Nucleic Acids Res.*, **21,** 2331.
39. Hamlin, J. L. (1992) Mammalian origins of replication. *Bioessays*, **10,** 651.

40. Held, P. G. and Heintz, N. H. (1992) Eukaryotic replication origins. *Biochim. Biophys. Acta*, **1130,** 235.
41. Vassilev, L. T. and De Pamphilis, M. L. (1992) Guide to identification of origins of DNA replication in eukaryotic cell chromosomes. *Crit. Rev. Biochem. Mol. Biol.*, **6,** 445.
42. Brewer, B. J. and Fangman, W. L. (1987) The localization of replication origins on ARS plasmids in *S. cerevisiae*. *Cell*, **51,** 463.
43. Huberman, J. A., Spotila, L. D., Nawotka, K. A., El-Assouli, S. M., and Davis, L. R. (1987) The *in vivo* replication origin of the yeast 2 μm plasmid. *Cell*, **51,** 473.
44. Vaughn, J. P., Dijkwel, P. A., and Hamlin, J-L. (1990) Replication initiates in a broad zone in the amplified CHO dihydrofolate reductase domain. *Cell*, **61,** 1075.
45. Dijkwel, P. A., Vaughn, J. P., and Hamlin, J-L. (1991) Mapping of replication initiation sites in mammalian genomes by two dimensional gel analysis. Stabilization and enrichment of replication intermediates by isolation of nuclear matrix. *Mol. Cell. Biol.*, **11,** 3850.
46. Fangman, W. L., Raghuraman, M. K., Ferguson, B. M., and Brewer, B. J. (1993) Replication initiation in yeast chromosomes. In *The eukaryotic genome—organisation and regulation*. Broda, P. M. A., Oliver, S. G., and Sims, P. F. G. (eds). Cambridge University Press, p 19.
47. Umek, R. M., Linskens, M. H. K., Kowalski, D., and Huberman, J. A. (1989) New beginnings in studies of eukaryotic DNA replication origins. *Biochim. Biophys. Acta*, **1007,** 1.
48. Diller, J. D. and Sauer, H. W. (1993) Two early replicated, developmentally controlled genes of *Physarum* display different patterns of DNA replication by two-dimensional agarose gel electrophoresis. *Chromosoma*, **102,** 563.
49. Shinomiya, T. and Ina, S. (1993) DNA replication of histone gene repeats in *Drosophila melanogaster* tissue culture cells: multiple initiation sites and replication pause sites. *Mol. Cell. Biol.*, **13,** 4098.
50. Little, R. D., Platt, T. H. K., and Schildkraut, C. L. (1993) Initiation and termination of DNA replication in human rRNA genes. *Mol. Cell. Biol.*, **13,** 6600.
51. Hyrien, O. and Méchali, M. (1993) Chromosomal replication initiates and terminates at random sequences but at regular intervals in the ribosomal DNA of *Xenopus* early embryos. *EMBO J.*, **12,** 4511.
52. Schvartzman, J. B., Martinez-Robles, M., and Hernandez, P. (1993) The migration behaviour of DNA replicative intermediates containing an internal bubble analyzed by two-dimensional agarose gel electrophoresis. *Nucleic Acids Res.*, **21,** 5474.
53. Burhans, W. C., Vassilev, L. T., Caddle, M. S., Heintz, N. H., and De Pamphilis, M. L. (1990) Identification of an origin of bidirectional DNA replication in mammalian chromosomes. *Cell*, **62,** 955.
54. Burhans, W. C., Vassilev, L. T., Wu, J., Sogo, J. M., Nallaseth, F., and De Pamphilis, M. L. (1991) Emetine allows identification of origins of mammalian DNA replication by imbalanced DNA synthesis, not through conservative nucleosome segregation. *EMBO J.*, **10,** 4351.
55. Brewer, B. J. and Fangman, W. L. (1993) Initiation at closely spaced replication origins in a yeast chromosome. *Science*, **262,** 1728.
56. Linskens, M. H. K. and Huberman, J. A. (1990) The two faces of eukaryotic DNA replication. *Cell*, **62,** 845.
57. Benbow, R. M., Gaudette, M. F., Hines, P. J., and Shioda, M. (1985) Initiation of DNA replication in eukaryotes. In *Control of animal cell proliferation.*, Vol. 1. Boynton, A. L. and Leffert, H. L. (ed.). Academic Press, New York, p. 449.

58. Benbow, R. M., Zhao, J., and Larson, D. D. (1992) On the nature of origins of DNA replication in eukaryotes. *Bioessays*, **14,** 661.
59. McKnight, S. L. and Miller, O. (1977) Electron microscopic analysis of chromatin replication in the cellular blastoderm *Drosophila melanogaster* embryo. *Cell*, **12,** 795.
60. Pierron, G. and Sauer, H. W. (1982) Physical relationship between replicons and transcription units in *Physarum polycephalum*. *Eur. J. Cell Biol.*, **29,** 104.
61. Coverley, D. and Laskey, R. A. (1994) Regulation of eukaryotic DNA replication. *Annu. Rev. Biochem.*, **63,** 745.
62. Herman, T. M., De Pamphilis, M. L., and Wassarman, P. M. (1981) Structure of chromatin at DNA replication forks: localization of the first nucleosome on newly synthesized Simian Virus 40 DNA. *Biochemistry*, **20,** 621.
63. Leffak, I. M. (1984) Conservative segregation of nucleosome core histones. *Nature*, **307,** 82.
64. Jackson, V. (1987) Deposition of newly synthesized histones: new histones H2A and H2B do not deposit in the same nucleosome with new H3 and H4. *Biochemistry*, **26,** 2315.
65. Jackson, V. (1990) *In vivo* studies on the dynamics of histone DNA interaction: evidence for nucleosome dissolution during replication and transcription on a low level of dissolution independent of both. *Biochemistry*, **29,** 719.
66. Svaren, J. and Chalkley, R. (1990) The structure and assembly of active chromatin. *Trends Genet.*, **6,** 52.
67. Yamasu, K. and Senshu, T. (1993) Conservation of the diameric unit of H2A and H2B during the replication cycle. *Exp. Cell Res.*, **207,** 226.
68. Bonne-Andrea, C., Wong, M. L., and Alberts, B. M. (1990) *In vitro* replication through nucleosomes without histone displacement. *Nature*, **343,** 719.
69. Ishimi, Y., Sugasawa, K., Hanaoka, F., and Kikuchi, A. (1991) Replication of the Simian Virus 40 chromosome with purified proteins. *J. Biol. Chem.*, **266,** 165141.
70. Krude, T. and Knippers, R. (1991) Transfer of nucleosomes from parental to replicated chromatin. *Mol. Cell. Biol.*, **11,** 6257.
71. Krude, T., De Maddalena, C., and Knippers, R. (1993) A nucleosome assembly factor is a constituent of Simian Virus 40 minichromosomes. *Mol. Cell. Biol.*, **13,** 1059.
72. Randal, S. and Kelly, T. J. (1992) The fate of parental nucleosomes during SV40 DNA replication. *J. Biol. Chem.*, **267,** 14259.
73. Sugasawa, K., Ishimi, Y., Eki, T., Hurvitz, J., Kikuchi, A., and Hanoaka, F. (1992) Nonconservative segregation of parental nucleosomes during Simian Virus 40 chromosome replication *in vitro*. *Proc. Natl Acad. Sci. USA*, **89,** 1055.
74. Fotedar, R. and Roberts, J. M. (1989) Multistep pathway for replication-dependent nucleosome assembly. *Proc. Natl Acad. Sci. USA*, **86,** 6459.
75. Smith, S. and Stillman, B. (1989) Purification and characterization of CAF-1, a human cell factor required for chromatin assembly during DNA replication *in vitro*. *Cell*, **58,** 15.
76. Laskey, R. A. and Leno, G. H. (1990) Assembly of the cell nucleus. *Trends Genet.*, **6,** 406.
77. Seidmann, M. M., Levine, A. J., and Weintraub, H. (1979) The asymmetric segregation of parental nucleosomes during chromosome replication. *Cell*, **18,** 439.
78. Cusick, M. E., De Pamphilis, M. L., and Wassarman, P. M. (1984) Dispersive nucleosome segregation during replication of SV40 chromosomes. *J. Mol. Biol.*, **178,** 249.
79. Annunziato, T. and Seale, R. (1984) Presence of nucleosomes within irregularly

cleaved fragments of newly replicated chromatin. *Nucleic Acids Res.*, **12,** 6179.

80. Jackson, V. and Chalkley, R. (1985) Histone segregation and replicating chromatin. *Biochemistry*, **24,** 6930.
81. Pospelov, V., Russev, G., Vassilev, L., and Tsanev, R. (1982) Nucleosome segregation in chromatin replicated in the presence of cycloheximide. *J. Mol. Biol.*, **156,** 79.
82. Handeli, S., Klar, A., Meuth, M., and Cedar, H. (1989) Mapping replication units in animal cells. *Cell*, **57,** 909.
83. Clark, D. J. and Felsenfeld, G. (1992) A nucleosome core is transferred out of the path of a transcribing polymerase. *Cell*, **71,** 11.
84. Avemann, K., Knippers, R., Koller, Th., and Sogo, J. M. (1988) Camptothecin, a specific inhibitor of type I DNA topoisomerase, induces DNA breakage at replication forks. *Mol. Cell. Biol.*, **8,** 3026.
85. Snapka, R. (1986) Topoisomerase inhibitors can selectively interfere with different stages of Simian Virus 40 DNA replication. *Mol. Cell. Biol.*, **6,** 4221.
86. Wang, J. C. (1991) DNA topoisomerases: why so many? *J. Biol. Chem.*, **2656,** 6659.
87. Thoma, F., Koller, Th., and Klug, A. (1979) Involvement of histone H1 in the organization of the nucleosome and of the salt-dependent superstructures of chromatin. *J. Cell. Biol.*, **83,** 403.
88. De Pamphilis, M. L. and Bradley, M. K. (1986) Replication of SV40 and Polyoma virus chromosomes. In *The papovaviridae*, Vol. 1. Salzman, N. P. (ed.). Plenum Publishing Corporation, New York, p. 99.
89. Van Holde, K. E. (1988) Chromatin. In Springer Series in *Molecular biology*. Rich, A. (ed.). Springer Verlag, New York.
90. Klempnauer, K. H., Fanning, E., Otto, B., and Knippers, R. (1980) Maturation of newly replicated chromatin of SV40 and its host cell. *J. Mol. Biol.*, **136,** 359.
91. Stefanovsky, V., Dimitrov, S., Russanova, V., and Pashev, I. (1990) Histones H1 and H4 are present near the replication fork. *Mol. Biol. Rep.*, **14,** 231.
92. Bavyykin, S., Srebreva, L., Branchev, T., Tssanev, R., Zlatanova, I., and Mirzabekov, A. (1993) Histone H1 deposition and histone–DNA interactions in replicating chromatin. *Proc. Natl Acad. Sci. USA*, **90,** 3918.
93. Conconi, A., Losa, R., Koller, Th., and Sogo, J. M. (1984) Psoralen-crosslinking of soluble and H1-depleted soluble rat liver chromatin. *J. Mol. Biol.*, **178,** 920.
94. D'Anna, J. A. and Prentice, D. A. (1983) Chromatin structural changes in synchronized cells blocked in early S phase by sequential use of isoleucine deprivation and hydroxyurea blockade. *Biochemistry*, **22,** 5631.
95. Crippa, M. P., Trieschmann, L., Alfonso, P. J., Wolffe, A. P., and Bustin, M. (1993) Deposition of chromosomal protein HMG-17 during replication affects the nucleosomal ladder and transcriptional potential of nascent chromatin. *EMBO J.*, **12,** 3855.
96. Perry, C. A., Allis, C. D., and Annunziato, T. (1993) Parental nucleosomes segregated to newly replicated chromatin are underacetylated relative to those assembled *de novo*. *Biochemistry*, **32,** 13615.
97. Roth, S. Y. and Allis, C. D. (1992) Chromatin condensation: does histone H1 dephosphorylation play a role? *Trends Biol. Sci.*, **17,** 93.
98. Di Nardo, S., Voekel, K., and Sternglanz, R. (1984) DNA topoisomerase II mutant of *S. cerevisiae*: topoisomerase II is required for segregation of daughter molecules at the termination of DNA replication. *Proc. Natl Acad. Sci. USA*, **81,** 2616.
99. Uemera, T., Ohkura, H., Adachi, Y., Morino, K., Shiozaki, K., and Yanagida, M.

(1987) DNA topoisomerase II is required for condensation and separation of mitotic chromosomes in *S. pombe*. *Cell*, **50**, 917.

100. Downes, C. S., Mullinger, A. M., and Johnson, R. T. (1991) Inhibitors of DNA topoisomerase II prevent chromatid separation in mammalian cells but do not prevent exit from mitosis. *Proc. Natl Acad. Sci. USA*, **88**, 8895.
101. Weaver, D. T., Fields-Berry, S. C., and De Pamphilis, M. L. (1985) The termination region of SV40 DNA replication directs the mode of separation for the two sibling molecules. *Cell*, **41**, 565.
102. Fields-Berry, S. C. and De Pamphilis, M. L. (1989) Sequences that promote formation of catenated intertwines during termination of DNA replication. *Nucleic Acids Res.*, **17**, 3261.
103. Varshavsky, A. and Sundin, O. (1983) Final stages of DNA replication: multiply intertwined catenated dimers as SV40 segregation intermediates. In *Mechanisms of DNA replication and recombination*. Cozzarelli, N. R. (ed.). Allan R. Liss Inc., New York, p. 463.
104. Adams, D. E., Shekhtman, E. M., Zechiedrich, E. L., Schmid, M. B., and Cozzarelli, N. R. (1992) The role of topoisomerase IV in partitioning bacterial replicons and the structure of catenated intermediates in DNA replication. *Cell*, **71**, 277.
105. Zhu, J., Newlon, C. S., and Huberman, J. A. (1992) Localization of a DNA replication origin and termination zone on chromosome III of *S. cerevisiae*. *Mol. Cell. Biol.*, **12**, 4733.
106. Hernandez, P., Martin-Parras, L., Martinez-Robles, M. L., and Schvartzman, B. (1993). Conserved features in the mode of replication of eukaryotic ribosomal RNA genes. *EMBO J.*, **12**, 1475.
107. Wiesendanger, B., Lucchini R., Koller, Th., and Sogo, J. M. (1994) Replication fork barriers in the *Xenopus* rDNA. *Nucleic Acids Res.*, in press.

4 Promoter potentiation and activation: chromatin structure and transcriptional induction of heat shock genes

JOHN LIS and CARL WU

1. Introduction

The heat shock gene regulatory circuit is an attractive model for elucidating general principles that govern inducible gene transcription and for investigating the role of chromatin and nucleosome structure in transcription activation. The degree of transcriptional stimulation is robust: transcription of the major heat shock genes can be increased over 100-fold upon heat shock (1), enhancing the detection of specific regulatory components and allowing a quantitative analysis of the role of these components. The transcriptional response is mediated by proteins that are present in induced cells (2, 3) and is extremely rapid (4), thereby facilitating kinetic and mechanistic investigations of the activation of these genes. Comprehensive studies from many laboratories provide a strong foundation of information on the primary structure of these genes, the *cis*-regulatory elements and *trans*-acting protein factors, as well as the chromatin structure of the uninduced gene and structural changes occurring after heat shock (for reviews see 5, 6, 7).

Heat shock promoters are composed of *cis*-acting elements similar to and interchangeable with those of other genes transcribed by RNA polymerase II; therefore, at least some of the principles identified in the study of heat shock genes are likely to be of general relevance. The basal promoter elements of heat shock genes (the TATA box, initiation site, and downstream sequences) are compatible with heterologous upstream enhancer elements such as those of the yolk protein YP1 gene and the salivary gland glue protein Sgs-3 gene (8, 9). These basal promoter elements are also responsive to binding sites for the yeast *GAL4 trans*-activator when *GAL4* is expressed in *Drosophila* (10). Likewise, heat shock elements (HSEs) can mediate heat-inducible transcriptional regulation from core promoter elements of a herpesvirus thymidine kinase gene that is not normally induced by heat shock (11). These examples indicate that the mechanism of regulation of the heat shock promoters is related to the regulation of other promoters that respond to metabolic or developmental cues.

In general, promoter and enhancer elements are usually situated in short regions of chromatin (< 500 bp) that lack a canonical nucleosome structure and are particularly accessible or hypersensitive to nucleolytic cleavage. The *Drosophila* heat shock promoters represent archetypes of this altered chromatin organization. The TATA box, initiation site, and downstream sequences of the major heat shock protein gene *hsp70*, the main focus of this review, are organized within hypersensitive sites in chromatin (12–15). The low molecular weight heat shock genes, including *hsp26*, also possess hypersensitive chromatin structures at their promoter sequences (16).

Molecular genetic and biochemical studies have identified several *cis*-regulatory elements and at least one transcription factor (GAGA factor) involved in the generation of the hypersensitive chromatin site (17, 18). The *cis*-regulatory elements required for chromatin alteration are also required for the assembly of a bound RNA polymerase II molecule that is paused after initiating a short transcript of 20 to 40 nucleotides (19). Accordingly, during normal, non-stress conditions, the heat shock promoter is organized in a potentiated state awaiting the binding of heat shock factor (HSF) protein. Activation of the potentiated heat shock promoter by HSF necessarily entails the release of the paused RNA polymerase II complex and the recruitment and initiation of new polymerases. Hence, the transcriptional induction of heat shock genes involves levels of regulation that are mechanistically distinct. The initial level governs the assembly of an accessible chromatin structure. This step may be linked to or be separate from the establishment of an initiation complex and its arrest shortly post-initiation. The subsequent binding of activated HSF trimers leads to the final, heat-shock-dependent escape of the paused polymerase (20). The specific removal or alteration of nucleosomes at other inducible promoters and developmental enhancers may be analogous to the pathways described for the assembly of potentiated heat shock promoters.

2. Architecture of uninduced *Drosophila* heat shock promoters: accessible chromatin sites containing bound GAGA factor, TFIID, and paused RNA polymerase II

2.1 TFIID and GAGA factor

In the absence of heat stress, the promoters of heat shock genes in multicellular eukaryotes are primed for rapid induction. The chromatin structures of uninduced heat shock genes in *Drosophila* are punctuated with nuclease-hypersensitive sites in the regulatory region of the promoter, and even in sequences immediately downstream of the transcriptional start site (12–15, 21) (Fig. 1). These hypersensitive sites are indicative of an interruption in the normal nucleosome packaging of the DNA and appear to be a consequence of the binding of specific proteins that are constitutively expressed. Two transcription factors are known to bind to heat shock promoter sequences under uninduced conditions: the GAGA factor and TBP (TATA-binding protein) (13, 14, 21, 22, 18, Giardina and J. Lis, unpublished data) (Fig. 2). GAGA factor is a constitutively expressed transcription factor that has been found to interact with GA/TC repeats present in many *Drosophila* genes, including those involved in house-keeping and developmental functions (21, 23–25). TBP is the key DNA recognition subunit of the general transcription factor TFIID (26, 27). The *in vivo* interactions of these factors are revealed by exonuclease protection and genomic footprinting assays using intact nuclei (13, 14, 21). The heat shock locus specific interaction of these two factors under normal environ-

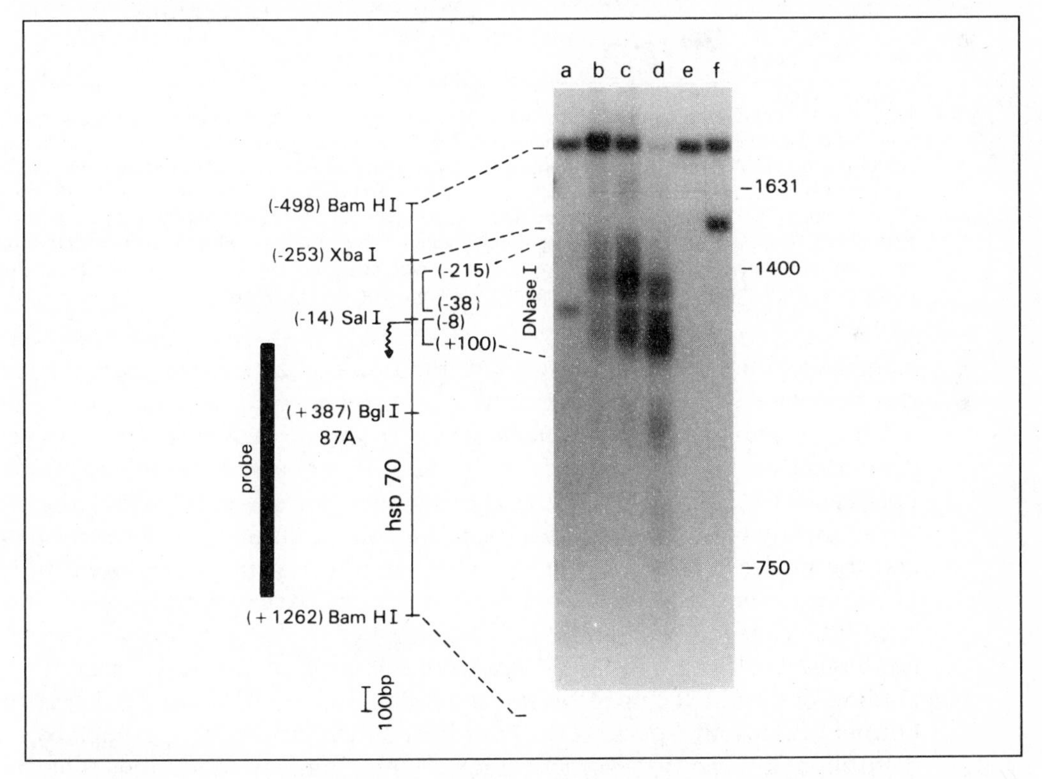

Fig. 1 DNase I hypersensitivity at the 5′ end of the non-induced *hsp70* gene. A partial restriction map of one of the two *hsp70* genes at chromosomal locus 87A is shown. The start site (+1) and direction of transcription are indicated by the arrow. Nuclei from *Drosophila* SL2 cells were isolated and digested with DNase I at 6, 8, and 4 U ml^{-1} (a separate reaction) respectively (lanes b–d). Purified DNA was restricted with *Bam*HI, electrophoresed on a 40 cm long, 1.4% agarose Tris-acetate EDTA (TAE) gel, blotted, hybridized to a ^{32}P-labelled, cloned probe fragment (solid bar), and autoradiographed. For markers, SL2 DNA was restricted partially with *Sal*I, *Bam*HI, and *Xba*I respectively, followed by complete *Bam*HI restriction and processing as above (lanes a, e, f). From (13), reprinted with permission, copyright 1984 by Macmillan Magazines Ltd.

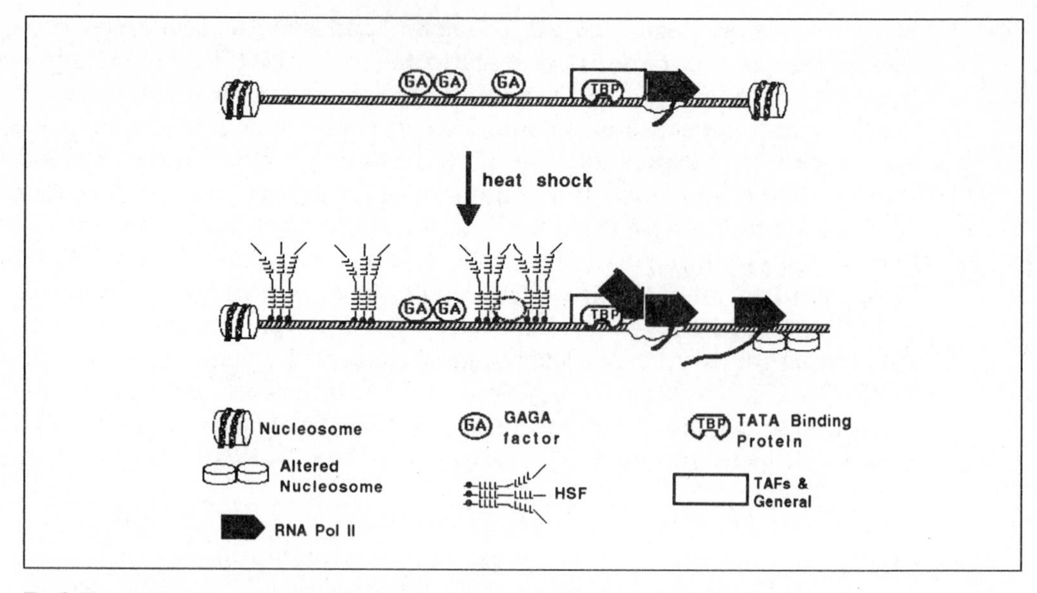

Fig. 2 The architecture of the *hsp70* promoter before and after heat shock *in vivo*. The uninduced heat shock promoter possesses a paused RNA polymerase II, a TATA box-binding protein (TBP), and an upstream factor(s) that binds to the GAGA sequence elements. TBP-associated factors (TAFs) and at least some of the general transcription factors (TFIIA, TFIIB, etc.) are also presumed to be present. Upon heat shock, HSF binds to four HSEs. A single HSF trimer is shown binding to each HSE, but the stoichiometry of binding has not been determined. Upon heat shock, DNA is apparently melted at the start site, which may indicate the entry of an additional polymerase in an open complex. The precise structure of altered nucleosomes on transcribed sequences is unknown. From (20), reproduced by permission of Cell Press.

mental conditions is also shown by immunolocalization on *Drosophila* polytene chromosomes (18, C. Wu, unpublished observations).

The accessibility of heat shock gene promoter sequences may be critically dependent on the binding of factor to GAGA sequences. Promoter mutations in upstream GA/TC repeats of the *hsp26* genes can severely reduce the nuclease hypersensitivity of the promoter region, as seen in transgenic *Drosophila* fly lines, and the nuclease hypersensitivity can be restored with sequences including GA/TC repeats from the *hsp70* promoter (28, 17). In contrast, mutations in the HSEs have little consequence on the nuclease hypersensitivity in heat shock promoters in *Drosophila* (17) or yeast (29). The consequences of the mutations suggest that binding of GAGA factor to these elements may be critical for the interruption of normal nucleosomal packing at these heat shock promoters, perhaps by effecting disruption of nucleosome organization. While there are indications that the TATA element and sequences around the initiation site may additionally contribute to nuclease hypersensitivity, precise mutational analyses of these basal elements of the heat shock promoter remain to be conducted.

Recent *in vitro* studies support a role for GAGA-factor binding to multiple sites in the generation of a nuclease-hypersensitive site (18). In a chromatin reconstitution assay using an *in vitro* nucleosome assembly system derived from *Drosophila* embryos (30), *hsp70* plasmid DNA, and recombinant GAGA transcription factor, the binding of GAGA on existing nucleosomes was found to bring about nucleosome disruption, DNase I hypersensitivity at the TATA box and heat shock elements, and a rearrangement of adjacent nucleosomes. The disruption of nucleosome organization at a specific location was measured by the loss of the 146 bp nucleosome monomer-sized DNA at the heat shock promoter upon extensive digestion with micrococcal nuclease, and by the presence of a smeared ladder of nucleosome oligomer DNA fragments produced by partial micrococcal nuclease digestion (Fig. 3). Importantly, this disruption of chromatin was found to be facilitated by ATP hydrolysis (18).

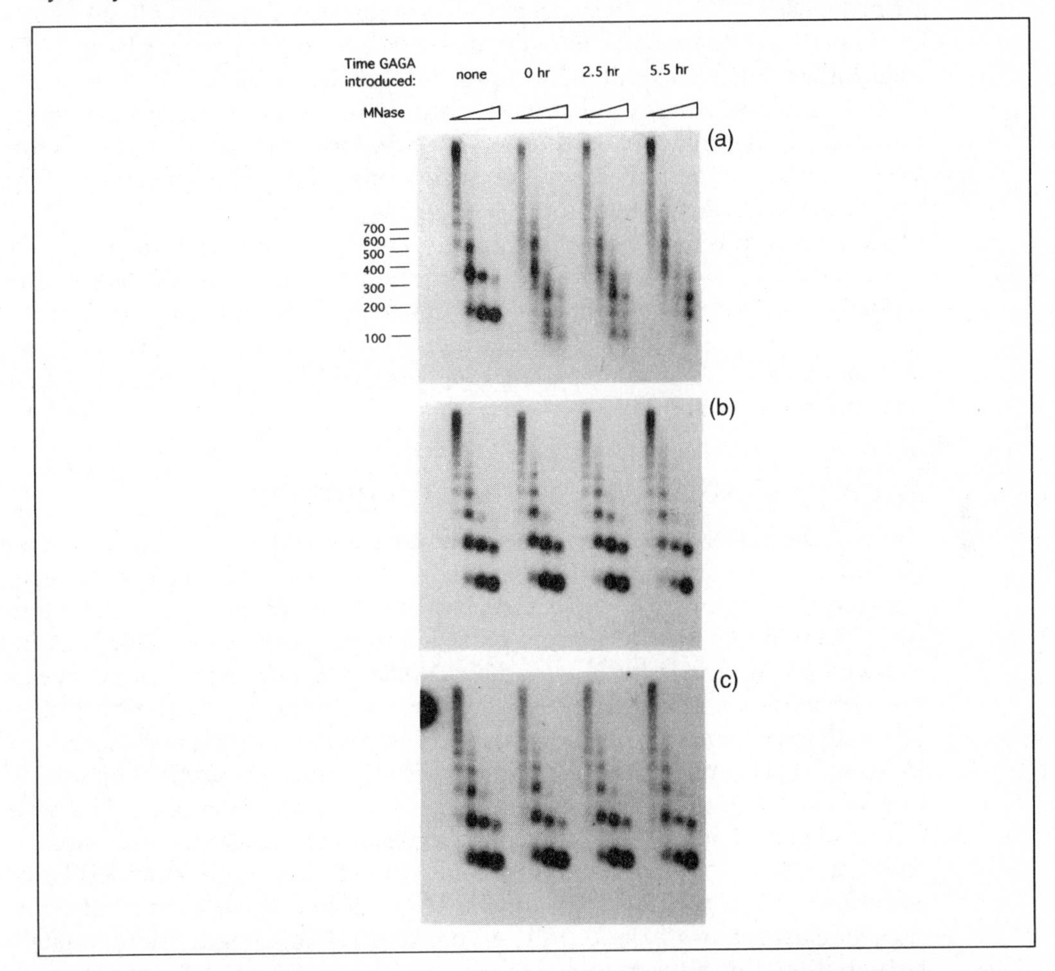

Fig. 3 GAGA-dependent chromatin disruption *in vitro*. (a–c) Micrococcal nuclease (MNase) digestion patterns of *hsp70*-plasmid chromatin reconstituted with GAGA factor as indicated. DNA blots were hybridized sequentially with oligonucleotides: (a) (−115 to −132), (b) (+1803 to +1832), and (c) (2499 to 2528 of pBluescript SK$^-$) respectively. Numbers to the left indicate size calibration (in base pairs). Sequences associated with *hsp70* promoter show a perturbation of nucleosome array. From (18), reprinted with permission, copyright 1994 by Macmillan Magazines Ltd.

The requirement for ATP hydrolysis to facilitate GAGA-mediated nucleosome disruption suggests a number of energy-dependent mechanisms to alter histone–DNA contacts. GAGA itself could bind ATP and disrupt nucleosome structure by a conformational change; however, inspection of its protein sequence fails to reveal a canonical ATP-binding motif, and no ATP-binding activity can be detected for the purified protein (T. Tsukiyama and C. Wu, unpublished observations). It is plausible that GAGA acts in concert with other components in the crude embryo extract that require ATP hydrolysis for nucleosome disruption. The evolutionary conserved proteins of the SWI2/SNF2 family are non-DNA-binding mediators that facilitate transcriptional activation by antagonizing the inhibitory effects of chromatin proteins (31, 32). As these proteins share conserved motifs with pox virus DNA-dependent ATPases (33), homologous *Drosophila* proteins such as *brahma* (34) might assist GAGA by fulfilling the ATP-dependent function. Alternatively, other ATP-dependent factors or enzymes affecting nucleosome assembly or spacing (for a review see 35) may act constitutively or be locally concentrated or activated by GAGA to modify or destabilize the nucleosome core particle such that binding of the transcription factor at multiple sites *per se* would suffice to complete the process of disruption. It is interesting that a requirement for energy in the form of glucose was observed for the nucleosome disruption observed in the absence of DNA replication at the *PHO5* promoter in *S. cerevisiae* (Schmid *et al.*, 1992; see Chapter 5, G. Hager and W. Horz). Whether there is a special property or domain unique to GAGA factor that confers the ability to effect ATP-dependent nucleosome disruption remains to be established.

2.2 A paused RNA polymerase II complex

Unstressed heat shock promoters contain not only GAGA factor and TATA-binding proteins but also an RNA polymerase II complex. This polymerase is near the start of the gene and is paused after initiating synthesis of a short transcript. An *in vivo* cross-linking technique provided initial evidence for this RNA polymerase on the uninduced *hsp70* gene (36). Nuclear run-on assays demonstrated that this polymerase is engaged in transcription, but paused near the start of the gene (37) with approximately one polymerase II molecule on each *Drosophila hsp70* gene (38) (Fig. 4). This paused RNA polymerase is distributed over a 20 bp interval (+17 to +37) with two preferred positions separated by one turn of the DNA helix—as can be seen from high resolution characterization of nuclear run-on transcripts (39) and mapping of transcription bubbles *in vivo* with the single-stranded DNA probe $KMnO_4$ (22) (Fig. 5). The *hsp26* and *hsp27* genes also show a distribution of pause sites similar to that of *hsp70*, but the pausing regions begin approximately 10 bp farther from the transcription start sites. The short RNAs associated with the paused polymerase of all three genes are progressively capped at their 5′ end as polymerase moves through the pause region (39). While the mechanism for restraining polymerase after the initial pulse of transcription is unknown, the bimodal distribution of the paused polymerase suggests that polymerase may be

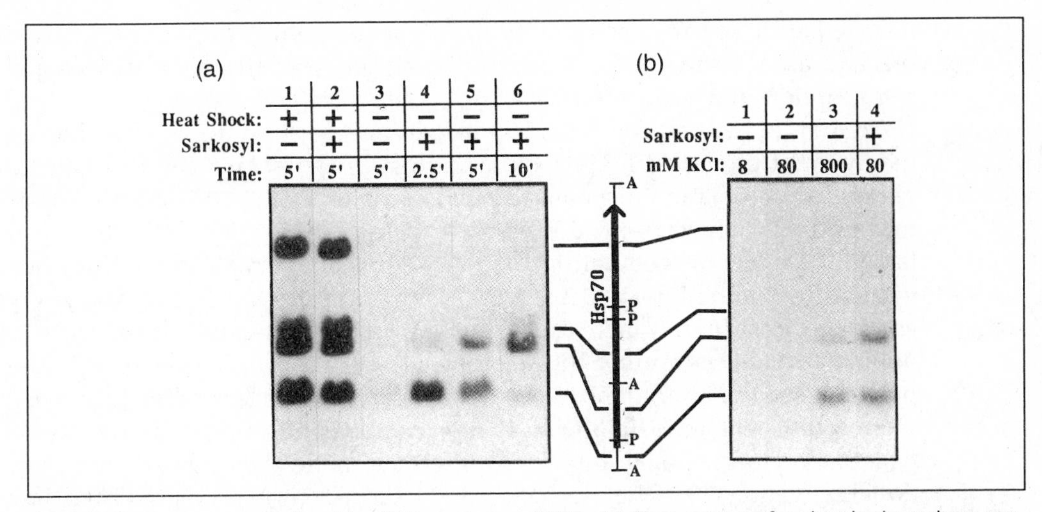

Fig. 4 Nuclear run-on analysis of *hsp70* homologous RNAs. (a) Time course of sarkosyl-released run-ons. Aliquots of ~10^8 nuclei were allowed to transcribe *in vitro* in the presence of [^{32}P]ATP under the conditions indicated, and the RNA was purified and hybridized to filters containing immobilized p70X2.6 digested with *Ava*I and *Pst*I, and visualized by autoradiography. Time indicates the duration of the run-on reaction begun by nucleotide addition. (b) Effect of salt concentration on run-on reactions; experiment is as in (a), but run-on reactions were all performed with nuclei from uninduced cells under conditions indicated in the figure. Reactions were incubated for 5 min at 22°C. A restriction map of the insert of p70X2.6 appears between the two panels. The solid arrow represents the *hsp70* transcription unit. Code: A = *Ava*I, P = *Pst*I. From (37), reproduced by permission of Cell Press.

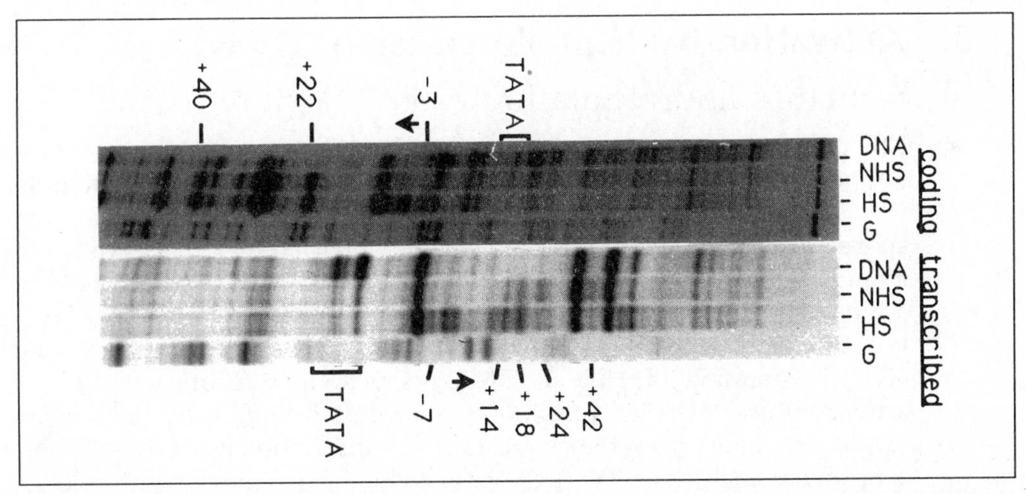

Fig. 5 $KMnO_4$-hypersensitive site mapping on the coding and transcribed strands of the *hsp70* gene in *Drosophila* cells (Kc). Non-heat-shocked cells (NHS), heat-shocked cells (HS), and purified genomic DNA (DNA) were treated with 25 mM $KMnO_4$ for 30 s at 0°C. All DNAs were then purified, cut with restriction enzyme, and treated with piperidine to cleave at the $KMnO_4$ modification sites. The DNA was amplified and radiolabelled using the ligation-mediated PCR (LMPCR) technique to display either the coding or transcribed strand of the *hsp70* 87A genes (as indicated). G ladders were prepared using dimethyl sulfate followed by piperidine cleavage and LMPCR. The *Xba*I cleavage site on the coding strand is found at the *top* of the gel; the *Cla*I cleavage site on the transcribed strand is not shown. The arrows indicate the position of the transcriptional start site. From (22), reproduced by permission of CSH Press.

tethered to a specific factor positioned on the same side of the DNA helix, or blocked as a result of the DNA's association with the first nucleosome of the transcription unit.

Analysis of hybrid promoters in transgenic *Drosophila* lines reveals that sequences upstream of the *hsp70* TATA box including GAGA sequences and HSEs can programme the formation of a paused polymerase on a non-heat-shock-gene promoter that normally displays no detectable pausing (19). While mutations in the HSE have little effect on generating the pause (but do dramatically reduce heat shock inducibility), mutations in the GAGA element reduce the level of paused polymerase by several-fold. When the extent of *hsp70* upstream sequence is less than optimal, sequences around the start site also play a role in generating paused RNA polymerase. The alterations to the hybrid *hsp70* promoter that reduce pausing also reduce the heat inducibility of the promoter, indicating that the formation of paused polymerase is an important intermediate in the pathway to full transcriptional activation (19). The mechanism by which factor binding to GAGA sequences facilitates the recruitment and initiation of the paused polymerase is unknown. GAGA-binding factor could act directly, by facilitating binding of TFIID and formation of a complete initiation complex, and/or indirectly, by clearing or restricting the position of a nucleosome on the promoter, thereby allowing access for TFIID and RNA polymerase.

3. Activation by heat shock factor (HSF)

3.1 Structure and regulation of HSF activity

HSF protein is synthesized constitutively in cells and stored in a latent form under normal conditions (2, 3). In all eukaryotes studied, with the exception of budding yeasts, where HSF binds DNA constitutively, the latent HSF is activated in response to heat shock by the acquisition of high affinity sequence-specific DNA-binding activity. This activity is accomplished by a conversion of HSF protein from a monomer to a homotrimer (40–42). When induced to trimerize, HSF becomes redistributed from non-specific sites on chromatin to discrete chromosomal targets (41). The binding of HSF to DNA is apparently insufficient to cause *trans*-activation; under certain conditions, the induced binding of the metazoan HSFs can be uncoupled from the acquisition of transcriptional activity (43–45).

Sequence analysis of cDNA clones encoding HSFs from a wide variety of species reveals two conserved regions in the N-terminal part of the HSF protein that represent the DNA-binding domain and the trimerization domain. The overall sequence of the DNA-binding domain does not show extensive similarity to any known category of DNA-binding motif, and it was therefore anticipated that a determination of the structure of HSF should reveal a new motif for specific DNA recognition. However, recent X-ray and NMR studies reveal that, despite the lack of overall sequence similarity, there is a structural relatedness of the DNA-binding domain of HSF to DNA-binding motifs exemplified by the helix–turn–helix protein family (46–48).

The trimerization domain of HSF is situated next to the DNA-binding domain and is separated by a short linker. The conserved residues in this region essentially define several arrays of hydrophobic heptad repeats or coiled-coil motifs (49, 50). Physico-chemical studies of HSF trimerization domains synthesized as ~90-residue polypeptides indicate that they form stable homotrimers in solution of predominantly α-helical character, suggesting a triple-stranded α-helical coiled-coil structure (51). How the three DNA-binding domains of the HSF trimer interact with the HSE sequence is unclear at present. HSEs are composed of a contiguous array of the alternately oriented 5 bp modular unit, AGAAn (40, 52–54). HSF interacts with a variety of HSEs of different lengths with striking co-operativity (55). High resolution structural analyses of these individual domains, of the entire protein, and eventually of protein–DNA complexes should provide key insights into the mechanics of how HSF with its three-fold symmetry interacts with 5 bp units arranged about multiple two-fold axes of symmetry, as well as the basis of the strong co-operative binding.

The regulation of the HSF monomer–trimer equilibrium requires a short region near the C-terminus that includes a conserved leucine zipper motif. Mutation of this motif leads to constitutive trimer formation and high affinity DNA-binding activity, indicating that this region participates in the suppression of trimer assembly, possibly by means of a coiled-coil interaction between the C-terminal leucine zipper and several other zipper motifs located at the N-terminal end of the protein (56). *In vitro* studies with crude extracts indicate that HSF can be activated by heat, low pH, and even polyclonal antibodies raised to the active HSF trimer, although it is unclear whether this activation in the crude extracts is exerted through the natural induction mechanism (43, 57, 58). As the HSF protein forms trimers constitutively when expressed at low growth temperatures in bacteria, intracellular components of eukaryotic cells are thought to mediate the monomer–trimer transitions (50, 56, 59). This notion is reinforced by the decrease in the induction temperature of human HSF1 to the induction temperature of *Drosophila* HSF when the human protein is expressed in fly cells (50). Possible candidates for this function are molecular chaperones, including the heat shock proteins themselves. There is some evidence for an association between HSF and the major heat shock protein HSP70, although its functional significance is unclear (60, 61; S. K. Rabindran, J. Wisniewski, L. Li, G. C. Li, and C. Wu, in preparation). Post-translational modifications, including phosphorylation, might also affect the inter-conversion between monomer and trimer. Finally, little is known about the *Drosophila* HSF *trans*-activation domain. One would like to know whether and how this domain is regulated in response to heat stress and how it interacts with components of the transcriptional machinery. The constitutively trimeric yeast HSF protein undergoes a heat-shock-dependent unmasking of its *trans*-activation domain (62, and references therein), and genetic experiments indicate that this *trans*-activation function is negatively autoregulated by the level of HSP70 (63).

3.2 Promoter architecture after heat shock

Many features of the promoter architecture of the uninduced heat shock genes persist when cells are stimulated by heat shock. Heat shock promoters remain hypersensitive to nuclease digestion, and the GAGA elements and the TATA box remain occupied (12, 21, 22). A major change is the rapid binding of HSF to the HSEs, except in budding yeasts where HSF is bound constitutively (3, 13, 14, 21). This rapid binding of HSF may require the nucleosome-free promoter in uninduced cells, since HSF (at least human HSF) fails to bind *in vitro* to HSEs packaged in nucleosomes (64), and *Drosophila* HSF appears to bind poorly to the heat shock puff sites *in vivo* when GAGA elements are deleted or substituted with point mutations (Shopland and Lis, unpublished observations). The concentration of paused polymerase does not appear to change upon heat shock even though the rate with which RNA polymerases escape from the pause into an active elongation mode must increase by more than 100-fold to account for the increase in transcription (22, 38). This indicates that escape from the pause is still the rate-limiting step during heat shock. While the transcription bubble observed at the pause site does not diminish after heat shock, it appears to extend downstream a few tens of base pairs. This is consistent with RNA polymerase creeping through a pausing region before entering into a rapid elongation mode (22) (Fig. 4). Additional melting of DNA occurs at the promoter at a region centred on the start site and may be caused by the entry of the next RNA polymerase. The persistent occupancy of the promoter by paused polymerase may also aid the entry of this next polymerase—contributing, along with other factors like GAGA and TBP, to the accessibility of the promoter region.

4. Speculations on the mechanism of *trans*-activation by HSF

Transcription of a eukaryotic gene could potentially be regulated at any of the many discrete steps. These include the modulation of accessibility of RNA polymerase to the promoter, pre-initiation complex formation, initiation, escape to a productive elongation complex, and elongation through downstream pause and termination sites. The presence of a paused RNA polymerase at the 5′ end of uninduced heat shock genes indicates that both RNA polymerase binding and initiation are rapid in comparison to the escape of polymerase from this early pause to a fully elongating complex. Therefore, HSF must act directly or indirectly to accelerate the rate of this escape from the pause.

HSF could interact directly with the paused RNA polymerase to alter its elongation properties; it could also be interacting directly with entering RNA polymerase molecules and have a role in accelerating their recruitment and initiation. During heat shock, additional melting occurs *in vivo* at the promoter and may represent the entry of the next RNA polymerase (22). This entry could be obligatorily coupled to the escape of the paused polymerase, with the recruitment of the new polymerase displacing the paused polymerase.

Alternatively, HSF may indirectly facilitate the escape of RNA polymerase from the pause, perhaps by severing interactions that tether polymerase to upstream factors. For example, the interaction between TBP and the heptapeptide repeats of the C-terminal domain (CTD) of the largest subunit of RNA polymerase II could serve as a component of such a tether (65). The extended structure of the CTD could provide the flexibility for the movement of RNA polymerase a short distance into the gene while still maintaining contact with the promoter. Since RNA polymerase II with an unphosphorylated CTD binds TBP and is the form of polymerase that enters into an initiation complex *in vitro*, HSF could switch the enzyme to the fully competent elongating mode by triggering phosphorylation of the CTD (27, 66, 67). Consistent with this hypothesis are the results from *in vivo* cross-linking experiments and immunolocalization on polytene chromosomes, which have shown that the CTD of the paused polymerase is mainly unphosphorylated, while the CTDs of the transcribing RNA polymerases are phosphorylated (66, 68).

HSF could also act directly or indirectly to remove a steric block to RNA polymerase elongation that may be caused by histone H1 and nucleosome cores. There is a precedent for nucleosomes impeding RNA polymerase II during early phases of polymerase elongation *in vitro* (69, see also Chapter 6); however, it remains to be determined whether histones are responsible for polymerase pausing on heat shock promoters *in vivo*. Finally, HSF may stimulate the recruitment of elongation factors that may be necessary for overcoming the block in elongation (70).

5. Generality of the heat shock model

Several features of heat shock promoters are common to many genes. The paused RNA polymerase present on all the *Drosophila* heat shock genes is also found on a variety of non-heat-shock-gene promoters from *Drosophila* and mammals. These include the promoters of the α- and β-1-tubulin, polyubiqitin, and glyceraldehyde-3-phosphate dehydrogenase 1 and 2 genes in *Drosophila*. These paused RNA polymerases have been detected by nuclear run-on assays (71) and, in the case of β-1-tubulin, by high resolution mapping of the transcription bubble with $KMnO_4$ (22). Krumm *et al.* (72) have also observed a paused RNA polymerase on the 5′ end of the human *c-myc* gene around position +30 using both *in vivo* $KMnO_4$ mapping and nuclear run-on assays. Likewise, Mirkovitch and Darnell (73) have used $KMnO_4$ to detect a transcription bubble at the start of the mouse transthyretin gene. Remarkably, these paused RNA polymerases in mammalian genes reside at positions that are very similar to that of *Drosophila hsp70* gene. There is, therefore, an emerging class of eukaryotic genes transcribed by RNA polymerase II whose regulation is not confined to the level of pre-initiation or initiation, but is exercised at a post-initiation step. The heat shock system should serve as a simple model to explore the post-initiation control of this class of promoters by upstream factors.

GAGA factor and HSF appear to fulfil two distinct requirements in gene regulation that may be general to all eukaryotic promoters. GAGA factor appears to function in nucleosome or histone displacement—a role consistent with its interaction with a wide variety of genes in *Drosophila*, and with its anti-repressive character when assayed by *in vitro* transcription assays (74). In contrast, *Drosophila* HSF may act only on a nucleosome-free promoter potentiated by bound GAGA factor, TFIID, and a paused polymerase. It is interesting that the constitutively trimeric yeast HSF appears capable of gaining access to the HSE and activating transcription of *hsp70* genes *in vivo* without the assistance of a GAGA-like factor (75). Perhaps *S. cerevisiae* HSF has a novel ability to disrupt chromatin structure, or the yeast chromatin template may be accessible to HSF and other transcription factors because of a difference in composition (such as the apparent lack of histone H1, or acetylation of core histones) (76). Alternatively, the yeast HSF protein may be an effective competitor of nucleosome assembly during DNA replication, when the template is transiently free of nucleosomes.

6. Final comments

6.1 Important questions

The heat shock genes are a highly inducible family of genes that are activated by heat shock and other forms of cellular stress. Their transcriptional activation is mediated by the binding of HSF to specific DNA sequence elements, HSEs. HSF activity has been shown to be controlled both at the level of DNA binding and of acquisition of the ability to stimulate transcription. Heat and a variety of agents activate HSF binding in multicellular eukaryotes; however, the mechanism of this activation remains an intriguing and open question. Possible mechanisms for regulating the monomer–trimer transitions and the acquisition of transcriptional competence include an autoregulatory circuit based on the cellular concentration of heat shock proteins, and post-translational modification of HSF by cellular enzymes. The regulation of DNA binding is not the only mode of HSF control, since the ability of constitutively bound yeast HSF to stimulate transcription is dramatically enhanced in response to heat shock without a large change in its DNA-binding activity. This acquisition of transcriptional activity in yeast and higher eukaryotes *in vivo* is accompanied by a dramatic increase in phosphorylation of HSF. That phosphorylation controls HSF activity both in yeast and at the second tier of post-DNA-binding control in higher eukaryotes is a hypothesis that should be a target for future investigations. The post-initiation mechanisms by which the DNA-bound, activated HSF increases transcription of heat shock genes by causing the release of the paused RNA polymerase, and the recruitment of new polymerases, are particularly challenging questions whose answers are likely to have relevance to a number of other transcription factors and gene promoters.

The establishment of an accessible, potentiated chromatin structure similar to that observed at heat shock promoters under non-heat-shock conditions is also likely to be a key step in the path towards the transcriptional activation of genes whose activity is subject to physiological or developmental signals. In addition to

GAGA factor and TBP, it will be important to determine which of the other subunits of the TFIID complex (the TAF proteins), and which basal transcription factors are pre-assembled at the heat shock promoter. How the GAGA factor is able to disrupt or displace nucleosome structure at the heat shock promoter and how GAGA factor and TBP collude with each other and with the basal factors to assemble a paused RNA polymerase (and perhaps further alter chromatin structure) are important questions that can feasibly be investigated using biochemical and genetic techniques. The principles to be learned regarding nucleosome displacement and assembly of the paused polymerase should provide valuable models for other investigations in the interactions between nucleosomes, sequence-specific factors, and the general transcriptional machinery.

6.2 Discussion

These questions were explored further in discussion with other authors. It is not clear where the paused polymerase is anchored. It might be possible to alter the distances between enhancers and TATA box, as well as between TATA box and the transcription start site, in order to see which element controls the position of the polymerase. A similar tethering mechanism might apply in cases where a polymerase is constitutively present on a TATA box; there is some evidence of this in the PEPCK gene (77). Such tests could provide clues to the identity of the factors to which RNA polymerase may be anchored. The general tethering mechanism being considered for generating and maintaining the pause could also play a role at other steps in transcription. Polymerase interactions with factors on the promoter could conceivably create rate-limiting steps at the progression of RNA polymerase from open complex formation, initiation, or the start of elongation. In these cases one could envision upstream factors severing these interactions or tethers by mechanisms akin to those which have been proposed above for the escape of the paused polymerase.

How the nucleosome disruption observed *in vitro* relates to the situation inside the cell is unclear at present. It is conceivable that the presence of a constitutively active GAGA factor during DNA replication might obviate the necessity for additional activities for nucleosome destabilization. However, the need for co-factors to facilitate competition against the assembly of nucleosomes should not necessarily be excluded, even for this transient period of chromatin instability. It will be interesting to determine the stage of the cell cycle at which GAGA factor is able to gain access to heat shock promoters, to investigate whether the passage of a replication fork affects the access of GAGA factor to nucleosomal DNA, and whether the binding of GAGA factor is affected by the presence of the ATP-dependent nucleosome disruption activity. No doubt a combination of biochemical and genetic studies will be needed to develop a reliable model of the mechanisms by which GAGA factor dictates chromatin structure. GAGA factor could be the first identified member of a family of NHC proteins that serve a critical role as 'antirepressors' active in chromatin assembly.

Acknowledgements

This study was supported in part by a Public Health Service grant GM25232 from the National Institutes of Health and by the Intramural Research Program of the National Cancer Institute.

References

1. Gilmour, D. S. and Lis, J. T. (1985) *In vivo* interactions of RNA polymerase II with genes of *Drosophila melanogaster*. *Mol. Cell. Biol.*, **5**, 2009.
2. Kingston, R. E., Schuetz, T. J., and Larin, Z. (1987) Heat inducible human factor that binds to a human *hsp70* promoter. *Mol. Cell. Biol.*, **7**, 1530.
3. Zimarino, V. and Wu, C. (1987) Induction of sequence-specific binding of *Drosophila* heat shock activator. *Nature*, **327**, 727.
4. O'Brien, T. and Lis, J. T. (1993) Rapid changes in *Drosophila* transcription after an instantaneous heat shock. *Mol. Cell. Biol.*, **13**, 3456.
5. Eissenberg, J. C., Cartwright, I. L., Thomas, G. H., and Elgin, S. C. R. (1985) Selected topics in chromatin structure. *Ann. Rev. Genet.*, **19**, 485.
6. Bienz, M. and Pelham, H. R. B. (1987) Mechanisms of heat-shock gene activation in higher eukaryotes. *Adv. Genet.*, **24**, 31.
7. Nover, L. (1987) Expression of heat shock genes in homologous and heterologous systems. *Enz. Microb. Tech.*, **9**, 130.
8. Garabedian, M. J., Shepherd, B. M., and Wensink, P. C. (1986) A tissue-specific transcription enhancer from the *Drosophila* yolk protein 1 gene. *Cell*, **45**, 859.
9. Martin, M., Giangrande, A., Ruiz, C., and Richards, G. (1989) Induction and repression of the *Drosophila Sgs-3* glue gene are mediated by distinct sequences in the proximal promoter. *EMBO J.*, **8**, 561.
10. Fischer, J. A., Giniger, E., Maniatis, T., and Ptashne, M. (1988) GAL4 activates transcription in *Drosophila*. *Nature*, **332**, 853.
11. Pelham, H. R. B. (1982) A regulatory upstream promoter element in the *Drosophila hsp70* heat-shock gene. *Cell*, **30**, 517.
12. Wu, C. (1980) The 5′ ends of *Drosophila* heat shock genes in chromatin are hypersensitive to DNase I. *Nature*, **286**, 854.
13. Wu, C. (1984) Two protein-binding sites in chromatin implicated in the activation of heat shock genes. *Nature*, **309**, 229.
14. Wu, C. (1984) Activating protein factor binds *in vitro* to upstream control sequences in heat shock gene chromatin. *Nature*, **311**, 81.
15. Costlow, N. and Lis, J. T. (1984) High-resolution mapping of DNaseI-hypersensitive sites of *Drosophila* heat shock genes in *Drosophila melanogaster* and *Saccharomyces cerevisiae*. *Mol. Cell. Biol.*, **4**, 1853.
16. Cartwright, I. L. and Elgin, S. C. R. (1986) Nucleosomal instability and induction of new upstream protein–DNA associations accompany activation of four small heat shock protein genes in *Drosophila melanogaster*. *Mol. Cell. Biol.*, **6**, 779.
17. Lu, Q., Wallrath, L. L., Granok, H., and Elgin, S. C. R. (1993) $(CT)_n \cdot (GA)_n$ repeats and heat shock elements have distinct roles in chromatin structure and transcriptional activation of the *Drosophila hsp26* gene. *Mol. Cell. Biol.*, **13**, 2802.

18. Tsukiyama, T., Becker, P. B., and Wu, C. (1994) ATP dependent nucleosome disruption at a heat shock promoter mediated by binding of GAGA transcription factor. *Nature*, **367,** 525.
19. Lee, H.-S., Kraus, K. W., Wolfner, M. F., and Lis, J. T. (1992) DNA sequence requirements for generating paused polymerase at the start of *hsp70*. *Genes and Dev.*, **6,** 284.
20. Lis, J. T. and Wu, C. (1993) Protein traffic on the heat shock promoter: parking, stalling, and trucking along. *Cell*, **74,** 1.
21. Thomas, G. H. and Elgin, S. C. R. (1988) Protein/DNA architecture of the DNase I hypersensitive region of the *Drosophila hsp26* promoter. *EMBO J.*, **7,** 2191.
22. Giardina, C., Perez-Riba, M., and Lis, J. T. (1992) Promoter melting and TFIID complexes on *Drosophila* genes *in vivo*. *Genes and Dev.*, **6,** 2190.
23. Soeller, W. C., Poole, S. J., and Kornberg, T. (1988) *In vitro* transcription of the *Drosophila engrailed* gene. *Genes and Dev.*, **2,** 68.
24. Biggin, M. D. and Tjian, R. (1988) Transcription factors that activate the *Ultrabithorax* promoter in developmentally staged extracts. *Cell*, **53,** 699.
25. Gilmour, D. S., Thomas, G. H., and Elgin, S. C. (1989) *Drosophila* nuclear proteins bind to regions of alternating C and T residues in gene promoters. *Science*, **245,** 1487.
26. Hoey, T., Dynlacht, B. D., Peterson, M. G., Pugh, B. F., and Tjian, R. (1990) Isolation and characterization of the *Drosophila* gene encoding the TATA box binding protein, TFIID. *Cell*, **61,** 1179.
27. Zawel, L. and Reinberg, D. (1993) Initiation of transcription by RNA polymerase II: a multi-step process. *Prog. Nuleic Acids Res.*, **44,** 67.
28. Lu, Q., Wallrath, L. L., Allan, B. D., Glaser, R. L., Lis, J. T., and Elgin, S. C. R. (1992) A promoter sequence containing $(CT)_n \cdot (GA)_n$ repeats is critical for the formation of the DNase I hypersensitive sites in the *Drosophila hsp26* gene. *J. Mol. Biol.*, **226,** 985.
29. Lee, M.-S. and Garrard, W. (1992) Uncoupling gene activity from chromatin structure: promoter mutations can inactivate transcription of the yeast *HSP82* gene without eliminating nucleosome-free regions. *Proc. Natl Acad. Sci. USA*, **89,** 9166.
30. Becker, P. B. and Wu, C. (1992) Cell free system for assembly of transcriptionally repressed chromatin from *Drosophila* embryos. *Mol. Cell. Biol.*, **12,** 2241.
31. Peterson, C. L. and Herskowitz, I. (1992) Characterization of the yeast *SWI1, SWI2,* and *SWI3* genes, which encode a global activator of transcription. *Cell*, **68,** 573.
32. Winston, F. and Carlson, M. (1992) Yeast SNF/SWI transcriptional activators and the SPT/SIN chromatin connection. *Trends Genet.*, **8,** 387.
33. Henikoff, S. (1993) Transcriptional activator components and poxvirus DNA-dependent ATPases comprise a single family. *Trends Biol. Sci.*, **18,** 291.
34. Tamkun, J. W., Deuring, R., Scott, M. P., Kissinger, M., Pattatucci, A. M., Kaufman, T. C., and Kennison, J. A. (1992) *brahma*—a regulator of *Drosophila* homeotic genes structurally related to the yeast transcriptional activator SNF2/SWI2. *Cell*, **68,** 561.
35. Almouzni, G. and Wolffe, A. (1993) Nuclear assembly, structure and function: the use of *Xenopus in vitro* systems. *Exp. Cell Res.*, **205,** 1.
36. Gilmour, D. S. and Lis, J. T. (1986) RNA polymerase II interacts with the promoter region of the noninduced *hsp70* gene in *Drosophila melanogaster* cells. *Mol. Cell. Biol.*, **6,** 3984.
37. Rougvie, A. E. and Lis, J. T. (1988) The RNA polymerase II molecule at the 5′ end of the uninduced *hsp70* gene of *Drosophila melanogaster* is transcriptionally engaged. *Cell*, **54,** 795.

38. O'Brien, T. and Lis, J. T. (1991) RNA polymerase II pauses at the 5′ end of the transcriptionally induced *Drosophila hsp70* gene. *Mol. Cell. Biol.*, **11,** 5285.
39. Rasmussen, E. and Lis, J. T. (1993) *In vivo* transcriptional pausing and cap formation on three *Drosophila* heat shock genes. *Proc. Natl Acad. Sci. USA*, **90,** 7923.
40. Perisic, O., Xiao, H., and Lis, J. T. (1989) Stable binding of *Drosophila* heat shock factor to head-to-head and tail-to-tail repeats of a conserved 5 bp recognition unit. *Cell*, **59,** 797.
41. Westwood, J. T., Clos, J., and Wu, C. (1991) Stress-induced oligomerization and chromosomal relocalization of heat-shock factor. *Nature*, **353,** 822.
42. Westwood, J. T. and Wu, C. (1993) Activation of *Drosophila* heat shock factor: conformational change associated with monomer to trimer transition. *Mol. Cell. Biol.*, **13,** 3481.
43. Larson, J. S., Schuetz, T. J., and Kingston, R. E. (1988) Activation *in vitro* of sequence-specific DNA binding by a human regulatory factor. *Nature*, **335,** 372.
44. Hensold, J. O., Hunt, C. R., Calderwood, S. K., Housman, D. E., and Kingston, R. E. (1990) DNA binding of heat shock factor to the heat shock element is insufficient for transcriptional activation in murine erythroleukemia cells. *Mol. Cell. Biol.*, **10,** 1600.
45. Jurivich, D., Sistonen, L., Kroes, R., and Morimoto, R. (1992) Effect of sodium salicylate on the human heat shock response. *Science*, **255,** 1243.
46. Harrison, C. J., Bohm, A. A., and Nelson, H. C. M. (1994) Crystal structure of the DNA binding domain of the heat shock transcription factor. *Science*, **263,** 224.
47. Vuister, G. W., Kim, S.-J., Wu, C., and Bax, A. (1994) NMR evidence for similarities between the DNA-binding regions of *Drosophila melanogaster* heat shock factor and the helix-turn-helix and HNF/forkhead families of transcription factors. *Biochem.*, **33,** 10.
48. Vuister, G. W., Kim, S.-J., Orosz, A., Marquardt, J., Wu, C., and Bax, A. (1994) Solution structure of the DNA-binding domain of *Drosophila* heat shock transcription factor. *Nature Struc. Biol.*, **1,** 605.
49. Sorger, P. K. and Nelson, H. C. M. (1989) Trimerization of a yeast transcriptional activator via a coiled-coil motif. *Cell*, **59,** 807.
50. Clos, J., Westwood, J. T., Becker, P. B., Wilson, S., Lambert, K., and Wu, C. (1990) Molecular cloning and expression of a hexameric heat shock factor subject to negative regulation. *Cell*, **63,** 1085.
51. Peteranderl, R. and Nelson, H. C. M. (1993) Trimerization of the heat shock transcription factor by a triple-stranded a-helical coiled-coil. *Biochem.*, **31,** 12272.
52. Fernandes, M., Xiao, H., and Lis, J. T. (1994) Fine structure analyses of the *Drosophila* and *Saccharomyces* heat shock factor–heat shock element interactions. *Nucleic Acids Res.*, **22,** 167.
53. Amin, J., Ananthan, J., and Voellmy, R. (1988) Key features of heat shock regulatory elements. *Mol. Cell. Biol.*, **8,** 3761.
54. Xiao, H. and Lis, J. T. (1988) Germline transformation used to define key features of heat-shock response elements. *Science*, **239,** 1139.
55. Xiao, H., Perisic, O., and Lis, J. T. (1991) Cooperative binding of *Drosophila* heat shock factor to arrays of a conserved 5 bp unit. *Cell*, **64,** 585.
56. Sarge, K., Murphy, S. P., and Morimoto, R. I. (1993) Activation of heat shock transcription by HSFI involves oligomerization, acquisition of DNA binding activity, and nuclear localization and can occur in the absence of stress. *Mol. Cell. Biol.*, **13,** 1392.
57. Rabindran, S. K., Haroun, R. I., Clos, J., Wisniewski, J., and Wu, C. (1993) Regulation of heat shock factor trimerization: role of a conserved leucine zipper. *Science*, **259,** 230.

58. Mosser, D. D., Kotzbauer, P. T., Sarge, K. D., and Morimoto, R. (1990) *In vitro* activation of heat shock transcription factor DNA-binding by calcium and biochemical conditions that affect protein conformation. *Proc. Natl Acad. Sci. USA*, **87,** 3748.
59. Zimarino, V., Wilson, S., and Wu, C. (1990) Antibody-mediated activation of *Drosophila* heat shock factor *in vitro*. *Science*, **249,** 546.
60. Abravaya, K., Myers, M. P., Murphy, S. P., and Morimoto, R. I. (1992) The human heat shock protein HSP70 interacts with HSF, the transcription factor that regulates heat shock gene expression. *Genes and Dev.*, **6,** 1153.
61. Baler, R., Welch, W. J., and Voellmy, R. (1992) Heat shock gene regulation by nascent polypeptides and denatured proteins: HSP70 as a potential regulatory factor. *J. Cell Biol.*, **117,** 1151.
62. Chen, Y., Barlev, N. A., Westergaard, O., and Jakobsen, B. K. (1993) Identification of the C-terminal activator domain in yeast heat shock factor: Independent control of transient and sustained transcriptional activity. *EMBO J.*, **12,** 5007.
63. Boorstein, W. R. and Craig, E. A. (1990) Transcriptional regulation of *SSA3*, an *hsp70* gene from *Saccharomyces cerevisiae*. *Mol. Cell. Biol.*, **10,** 3262.
64. Taylor, I. C. A., Workman, J. L., Schuetz, T. J., and Kingston, R. E. (1991) Facilitated binding of GAL4 and heat shock factor to nucleosomal templates: differential function of DNA-binding domains. *Genes and Dev.*, **5,** 1285.
65. Usheva, A., Maldonado, E., Goldring, A., Lu, H., Houbavi, C., Reinberg, D., and Aloni, Y. (1992) Specific interaction between the nonphosphorylated form of RNA polymerase II and the TATA-binding protein. *Cell*, **69,** 871.
66. O'Brien, T., Hardin, S., Greenleaf, A., and Lis, J. T. (1994) Phosphorylation of RNA Polymerase II C-terminal domain coincides with elongation from the 5′ pause of *Drosophila* genes. *Nature*, **370,** 75.
67. Laybourn, P. J. and Dahmus, M. E. (1990) Phosphorylation of RNA polymerase IIA occurs subsequent to interaction with the promoter and before the initiation of transcription. *J. Biol. Chem.*, **265,** 13165.
68. Weeks, J. R., Hardin, S. E., Shen, J., Lee, J. M., and Greenleaf, A. L. (1993) Locus-specific variation in phosphorylation state of RNA polymerase II *in vivo*: correlations with gene activity and transcript processing. *Genes and Dev.*, **7,** 2329.
69. Izban, M. and Luse, D. (1991) Transcription on nucleosomal templates by RNA polymerase II *in vitro*: inhibition of elongation with enhancement of sequence-specific pausing. *Genes and Dev.*, **5,** 683.
70. Marshall, N. F. and Price, D. H. (1992) Control of formation of two distinct classes of RNA polymerase II elongation complexes. *Mol. Cell. Biol.*, **12,** 2078.
71. Rougvie, A. E. and Lis, J. T. (1990) Post-initiation transcriptional control in *Drosophila melanogaster*. *Mol. Cell. Biol.*, **10,** 6041.
72. Krumm, A., Meulia, T., Brunvand, M., and Groudine, M. (1992) The block to transcriptional elongation within the human c-myc gene is determined in the promoter-proximal region. *Genes and Dev.*, **6,** 2201.
73. Mirkovitch, J. and Darnell, J. E. (1992) Mapping of RNA polymerase II on mammalian genes in cells and nuclei. *Mol. Biol. Cell.*, **3,** 1085.
74. Kerrigan, L. A., Croston, G. E., Lira, L., and Kadonaga, J. T. (1992) Sequence-specific transcriptional antirepression of the *Drosophila Kruppel* gene by the GAGA factor. *J. Biol. Chem.*, **266,** 574.
75. Gross, D. S., Adams, C. C., Lee, S., and Stentz, B. (1993) A critical role for heat shock

transcription factor in establishing a nucleosome-free region over the TATA-initiation site of the yeast HSP82 heat shock gene. *EMBO J.*, **12,** 3931.

76. Nelson, D. A. (1982) Histone acetylation in baker's yeast: maintenance of the hyper-acetylated configuration in log phase protoplasts. *J. Biol. Chem.*, **257,** 1565.
77. Faber, S., O'Brien, R. M., Imai, E., Granner, D. K., and Chalkley, R. (1993) Dynamic aspects of DNA/protein interactions in the transcriptional initiation complex and hormone-responsive domains of the phosphoenolpyruvate carboxykinase promoter *in vivo*. *J. Biol. Chem.*, **268,** 24976.

5 | Initiation of expression: remodelling genes

GORDON HAGER, CATHARINE SMITH, JOHN SVAREN, and WOLFRAM HÖRZ

1. Introduction

The periodic nucleosomal organization of DNA in eukaryotic cells has been appreciated for more than twenty years (1). However, despite the ubiquity of nucleosomes, the exact role which nucleosome structure plays in transcriptional regulation has been, and remains, a subject of vigorous debate. The omnipresent nature of nucleosome structure itself, which was initially an advantage in chromatin studies, has made it difficult to target studies *in vivo* to this problem.

It has been shown in numerous genes that promoter and enhancer elements are often contained within regions of DNA that are hypersensitive to nucleases within nuclei (2, 3). Such regions appear not to be organized into nucleosomes, at least as assayed by conventional methods. This correlation has suggested that nucleosomal structure over critical promoter elements would prevent factor binding and transcriptional activation (4, 5). However, it has proved difficult to determine whether the absence of regular nucleosome structure over active promoters is a prerequisite for efficient initiation of transcription, or simply reflects the results of factor-binding events. Nucleosomes are like the Clark Kent character in the Superman comic books: they never seem to be where the action is, but we are not quite sure if it is significant.

To address such questions, we and others have utilized inducible genes which undergo a chromatin transition in the promoter region upon activation (6–10). An inducible system allows one to study both repressed and activated states in the same strain or cell line, thereby making it easier to determine the function and mechanism of chromatin modifications. The yeast *PHO5* promoter (5, 11, 12), and the mouse mammary tumor virus (MMTV) (13–16) promoter represent two systems that are uniquely useful in this regard. Both genes acquire a specific nucleoprotein organization upon replication of the sequence *in vivo*, and both undergo chromatin remodelling events in the absence of DNA synthesis (17). The latter feature is particularly important; remodelling for these promoters is an active process that can be addressed without the added complications of nucleoprotein disruption and reassembly that occur during replication.

2. Chromatin structure of the yeast *PHO5* and murine MMTV promoters

Structures of the *PHO5* and MMTV promoters are presented in Fig. 1. The repressed *PHO5* promoter is organized into an array of well-defined nucleosomes which is interrupted by a short hypersensitive region of approximately 70 bp (11). This region contains one of the upstream activating sequences, UASp1, which is a binding site for Pho4 (18, 19), a *trans*-activator with a basic helix–loop–helix (bHLH) DNA-binding domain (20–23). The two nucleosomes that are proximal to the site of transcription initiation, nucleosomes −2 and −1, incorporate the vital *cis*-acting elements, UASp2 and the TATA box, respectively (18). UASp2 contains another binding site for Pho4 (19).

Regulatory sequences for the MMTV promoter extend out to −1.3 kb, beginning with the immediate proximal elements NF1/CTF, Oct-1, and sites for the basal transcription initiation complex (−30/+15). These elements are located on the leading edge of nucleosome A, and in the linker region between A and B (Fig. 1). Well-characterized steroid-receptor-binding sites are located on nucleosome B (24); these sites are bound by the glucocorticoid, progesterone, and androgen receptors, all of which activate the MMTV promoter. A secondary set of glucocorticoid-receptor (GR) binding sites are found on the 3′ side of nucleosome C (25); although not as intensively studied, these sites may also contribute to the hormone response *in vivo*.

In addition, the MMTV long terminal repeat (LTR) contains far upstream elements that contribute to the tissue-specific expression observed for this promoter. A negative transcriptional element has been characterized at position −400; this region would be associated with nucleosome C. A tissue-specific enhancer is located at the left end of the LTR (26–28). The factors binding to this enhancer are not yet well characterized, but at least four protein-binding sites have been identified in the region −1070/−975, which comprises a minimal enhancing element (S. John, unpublished data). These binding sites are located immediately upstream of nucleosome F. It is noteworthy that the position of this tissue-specific enhancer, at the left edge of nucleosome F, would place the activating complex six nucleosomes, or one turn of a solenoidal superhelix, away from the target basal promoter complex, potentially bringing the two elements into close proximity. The significance of long-range factor positioning on nucleosomal arrays is only beginning to be appreciated (2, 29), and will undoubtedly attract greater attention as techniques to address these long-range effects appear.

Upon activation of the *PHO5* promoter by phosphate starvation, the two pairs of nucleosomes that flank UASp1 are disrupted (12). The MMTV nucleosomal array also undergoes a specific structural transition upon steroid receptor activation of the promoter (Fig. 1). The regions associated with, and surrounding, *PHO5* nucleosomes −1 to −4 and MMTV nucleosome B become markedly accessible to a variety of nucleolytic reagents (DNaseI, restriction endonucleases, etc.) (12, 13, 30).

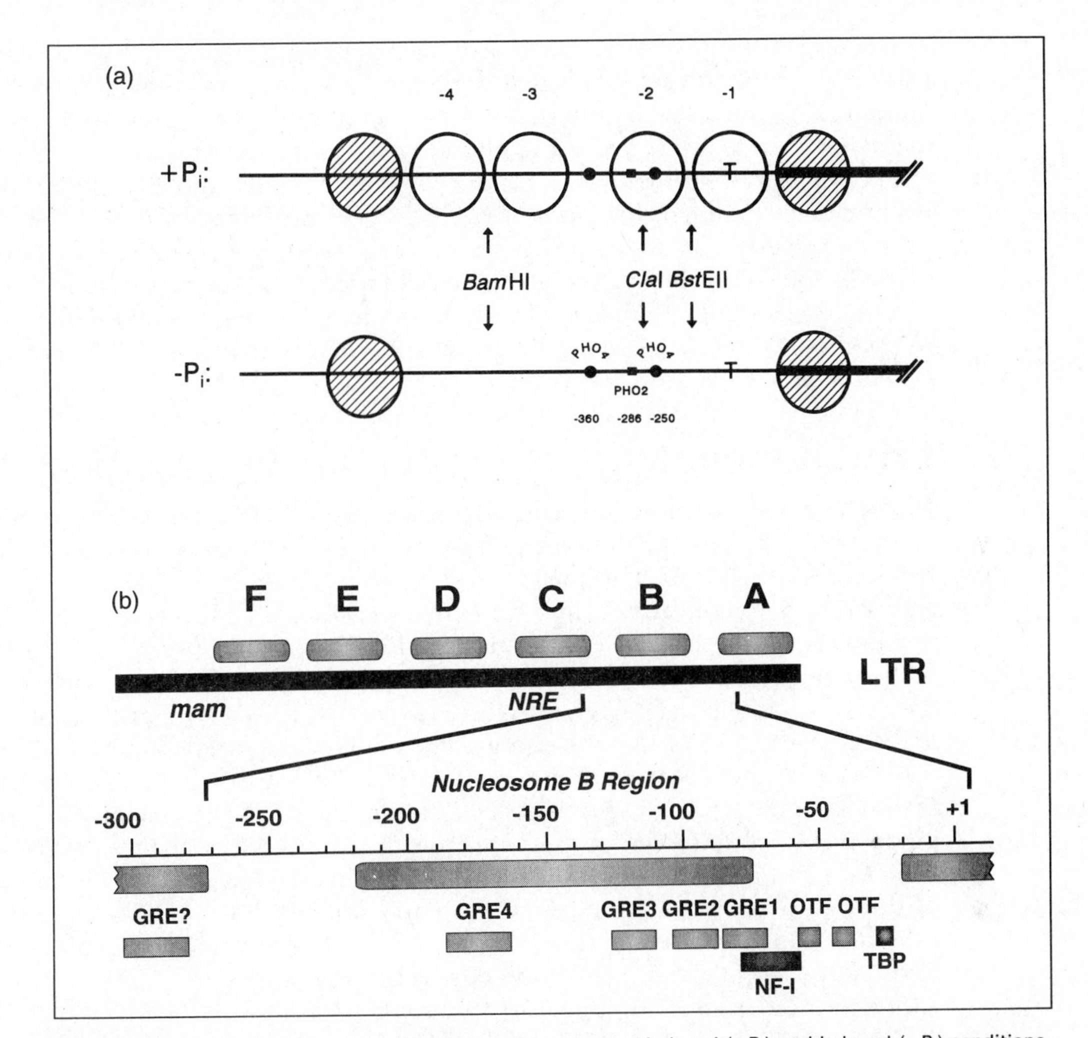

Fig. 1 (a) Chromatin structure at the *PHO5* promoter under non-induced (+P_i) and induced (−P_i) conditions. Nucleosomes −1, −2, −3, and −4 are removed upon activation (12). Code: solid circles mark the two Pho4-binding sites and the solid square a Pho2-binding site found by footprinting experiments *in vitro* (19); T denotes the TATA box (18). The locations of a *Bst*EII site at −174, a *Cla*I site at −275, and a *Bam*HI site at −542 relative to the coding sequence (solid black) are shown. (b) Chromatin structure of the MMTV promoter. The retroviral LTR accommodates an array of six nucleosomes (A–F). Histone cores are non-randomly displayed across the LTR, positioned as shown in the diagram. Three groups of *cis*-acting regulatory sequences are indicated; *mam*, a positive enhancer, contains binding sites for factors involved in tissue-specific expression of MMTV (26–28); *NRE* is a negative regulatory element that may also function in cell-specific expression; the proximal region, indicated at the bottom, contains binding sites for steroid receptors (GRE1–4)(24), constitutive factors OTF and NF1/CTF, and the basal transcription factors, TBP, etc. An additional GRE, indicated in the C region, may also function synergistically in the hormone response. The region demarcated as the *Nucleosome B Region* becomes hypersensitive to nucleolytic reagents (MPE, DNaseI, restriction endonucleases) during hormone activation of the promoter (13, 51).

The MMTV transition event has been characterized in many cell lines, with the LTR in a variety of vector and reporter gene configurations. Similarly, the ability of the *PHO5* promoter to undergo the chromatin transition is unaltered when it is maintained on an episomal yeast plasmid (31) or fused to the *lacZ* gene (32). In both cases, it is clear that DNA–histone contacts are substantially remodelled in a way that allows much easier access of nucleolytic reagents to DNA.

Both UAS elements in the *PHO5* promoter are required for chromatin disruption (31, 33), indicating the possibility that co-operative interactions are involved in chromatin disruption. Similarly, there are multiple binding sites for glucocorticoid receptor on MMTV nucleosome B, but whether more than one GR site is required for chromatin disruption is not yet known.

3. Critical importance of the chromatin transition

Under a variety of inducing conditions for the MMTV and *PHO5* promoters, substantial activation of transcription does not proceed without nucleosome disruption. The implication is that nucleosome structure inhibits binding of *trans*-acting factors to critical *cis*-acting elements. Consistent with this idea, turning off histone H4 synthesis in yeast activates the *PHO5* promoter (34, 35). Using *in vivo* footprinting experiments, we have observed binding of Pho4 to either UAS in the *PHO5* promoter only when nucleosomes are disrupted (33). Furthermore, there is direct evidence that nucleosome −2 inhibits binding of Pho4 to UASp2. Overexpression of the DNA-binding domain on Pho4 alone results in no alteration of the nucleosome structure (36). Under these conditions, the truncated Pho4 protein binds to UASp1 in the nucleosome-free region but not to the nucleosomal UASp2, even though Pho4 binds with higher affinity to UASp2 *in vitro* (19). The activation of the *PHO5* promoter by nucleosome depletion (34, 35) requires the proximal promoter containing the TATA box, but not the UAS elements. Therefore, it is also likely that nucleosome −1 inhibits TFIID binding to the *PHO5* TATA box, although there is so far no direct evidence.

Other results indicate that nucleosomes can exert a repressive effect that is not dependent upon specific positioning over critical activator elements. Replacement of nucleosome −2 in the *PHO5* promoter with a satellite DNA fragment that forms a stable nucleosome *in vitro* results in a promoter that cannot be activated by phosphate starvation (Fig. 2) (32). Pho4 is unable to activate the promoter, despite the fact that the only remaining Pho4-binding site lies in a nucleosome-free region. This binding site is sufficient for a substantial degree of activation when the DNA sequence comprising nucleosome −2 is derived from a segment of pBR322 (32), which does not form a stable nucleosome. Therefore, the presence of nucleosomes within the promoter, regardless of their positioning, may hamper interactions between UAS elements and the transcription start site that are necessary for efficient activation of transcription.

There is also direct evidence that the MMTV chromatin transition event is an important component of the complete hormone response. The chromatin structure

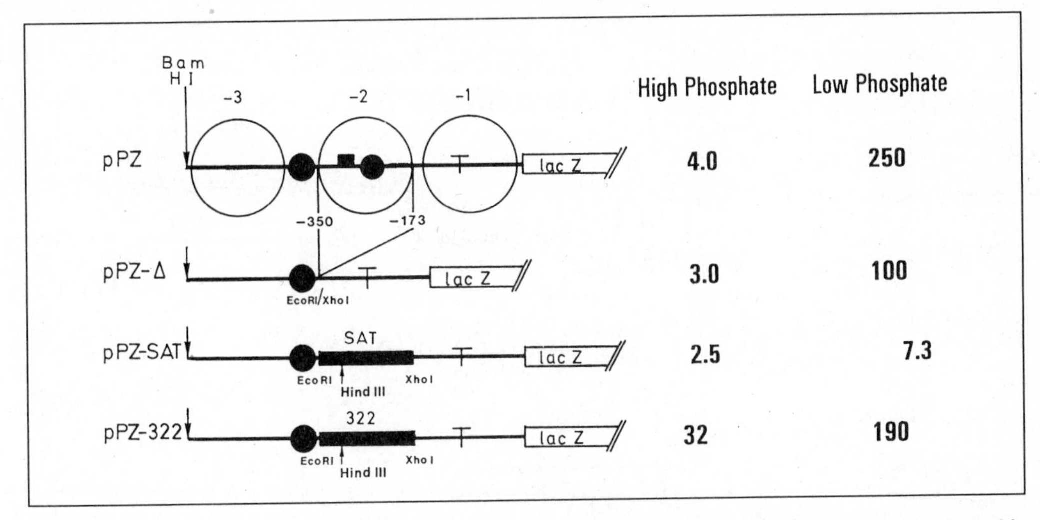

Fig. 2 *lacZ* activity of different *PHO5* promoter constructs. A series of *PHO5* promoter/*lacZ* fusion plasmids (pPZ) were constructed as described in (31). The relevant part of the *PHO5* promoter is shown schematically (see text for details). *lacZ* activities were determined after growing the cells in either high phosphate or no phosphate media. The values are listed in relative units, with the level obtained for plasmid pPZ-Δ after induction in no phosphate medium set as 100.

of the repressed promoter excludes several known components of the immediate upstream proximal elements, NF1 and Oct-1, and the basal transcription factors (TFIID, etc.). When DNA is introduced transiently into cells, factors that are normally excluded from the promoter (NF1, Oct-1) are found to be constitutively present on the template (30). The transient template exhibits neither the phased nucleosome array nor the transition in enzyme accessibility characteristic of the replicated template. It appears that some feature of the organized nucleosome array is necessary for factor exclusion from the promoter.

It has been proposed (30) that NF1 and Oct-1 load on the template as a direct result of receptor-dependent chromatin opening. Steroid receptors also interact directly with the basal machinery by protein–protein contacts; presumably, this aspect of the activation mechanism is observed in transient assays. Under this model, stable replicated templates support both loss of chromatin repression and direct receptor–basal-factor interactions, leading to the complete receptor-dependent response.

It was recently found (37, 38) that *trans*-activation at the MMTV promoter is dynamically complex, involving an initial transitory 'spike' of very high run-on transcription rate, followed by a steady state characterized by lower rates of transcription (Fig. 3). This transcription profile is mirrored by a similar appearance and loss of restriction endonuclease sensitivity in the nucleosome B region. As initial transcription rates fall to steady-state levels, the nucleosome B region loses some of its open structure, demonstrating a direct correlation between high rates of transcription and the open chromatin structure. Thus, for both the *PHO5* and

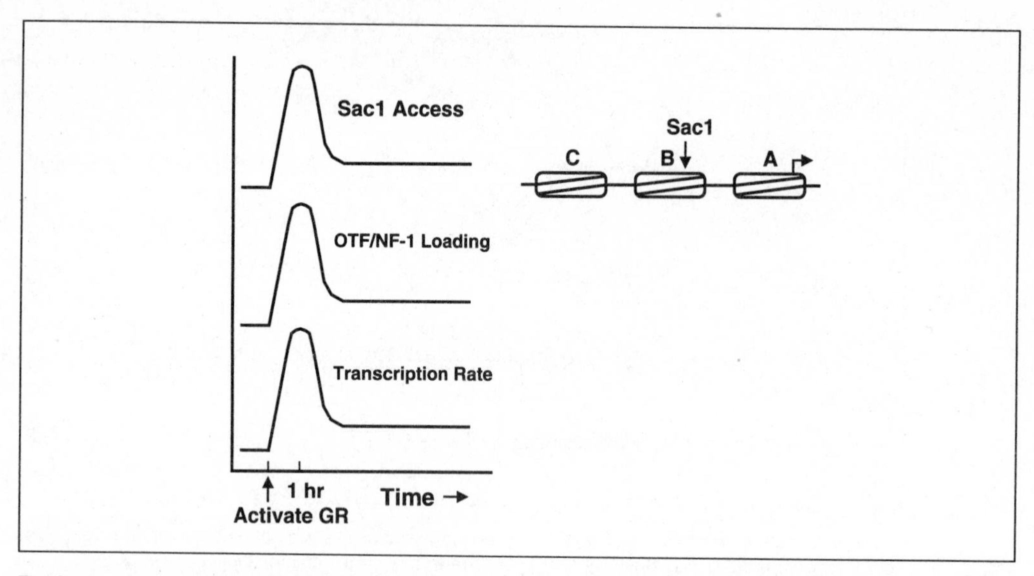

Fig. 3 The hormone-induced transition in MMTV promoter chromatin is dynamically complex, and correlates with transcription activation. The transcription rate (measured by run-on elongation in nuclei isolated from treated cells) for MMTV manifests a sharp peak in rate approximately one hour after dex stimulation (37, 38). This profile is mirrored by the access of restriction endonucleases to the nucleosome B region (shown here as *Sac*1) (37, 38). In addition, loading of some of the proximal region factors, NF1 and OTF, apparently also follows the complex binding and unloading profile (38).

MMTV promoters, there is evidence that a compacted nucleosome structure is incompatible with factor loading and promoter activation.

4. Requirements for chromatin remodelling

Since nucleosomes are generally quite stable structures, it has been suggested that a *trans*-activator could not directly modify nucleosome structure, but rather would have to bind during DNA replication, when nucleosome structure is broken down and then reassembled after passage of the replication fork (39). However, the kinetics of induction of the MMTV promoter (detected within 90 s) are incompatible with a requirement for DNA synthesis. Direct analysis has supported this interpretation (T. K. Archer and G. L. Hager, unpublished data). In the case of *PHO5*, experiments utilizing a temperature-sensitive component in the induction pathway have revealed that passage through S-phase is not required for chromatin remodelling (17).

Another formal possibility is that *trans*-acting factors do not directly modify chromatin structure, but rather some event incidental to the process of transcription is responsible for disrupting nucleosome structure. For example, binding of the transcription initiation complex or changes in supercoiling density caused by an elongating RNA polymerase could in principle bring about chromatin disruption. Indeed, there are significant changes in the chromatin structure in the coding regions of genes that are transcription dependent (40). However, such a mechanism cannot be responsible for the chromatin transition in the *PHO5* promoter. A deletion of the *PHO5* TATA box reduces transcription to background levels (18), but the chromatin transition proceeds exactly as in the wild type promoter when cells are starved of phosphate (31).

The recent observations of Smith *et al.* (37, 41) suggest that the criteria for receptor-induced remodelling of MMTV may be quite complex. Transiently expressed progesterone receptor, which is very active on transient MMTV templates, was found to be incapable of remodelling nucleosome B on stable, replicated chromatin. Remodelling activity was acquired when the progesterone receptor was introduced stably into the cell and expressed constitutively during cell division. These findings again indicate that there are additional functional requirements for chromatin remodelling over and above those necessary for simple gene activation on disorganized templates.

5. *Trans*-activators and cofactors in chromatin modification

In both systems, modification of chromatin structure appears to be initiated by binding of a *trans*-activator. For MMTV, ligand activation of the glucocorticoid receptor is necessary and sufficient to induce the chromatin transition; cofactors necessary for the structural transition have not been clearly identified. Although both Pho4 and a homeodomain protein, Pho2 (42, 43), are normally required for the transition in the *PHO5* promoter, overproduction of Pho4 in the absence of Pho2 results in fully open chromatin. Overproduction of Pho2, however, cannot compensate for loss of Pho4 (44). Pho2 might facilitate Pho4 binding (45), similar to the manner in which it facilitates binding of Swi5 to the yeast *HO* promoter (46, 47). Therefore, the glucocorticoid receptor and Pho4 are the key *trans*-activators which trigger chromatin disruption in the MMTV and *PHO5* promoters, respectively.

To modify chromatin, Pho4 and glucocorticoid receptor must first gain access to the promoter. As mentioned above, the binding affinity of Pho4 for UASp2 seems to be effectively reduced by nucleosome −2. Therefore, it seems reasonable to assume that UASp1, which lies in a constitutively nucleosome-free region, is the site through which Pho4 gains a foothold on the *PHO5* promoter. The affinity of Pho4 for the *PHO5* promoter is most likely regulated by phosphorylation (48). In contrast, the glucocorticoid receptor has been shown to bind *in vitro* to specific sites on the surface of a nucleosome with an only slightly reduced affinity compared with a non-nucleosomal binding site (49–51).

Simple binding of glucocorticoid receptor to nucleosomes reconstituted *in vitro* does not result in significant modification of nucleosome structure, certainly not to an extent that would reflect the major transition observed *in vivo* (49, 51). This would suggest that additional proteins (or enzymes) are involved. Gal4 can disrupt a nucleosome which contains Gal4-binding sites both *in vivo* and *in vitro* (52–54). The nucleosome assembly protein, nucleoplasmin, can facilitate the reaction *in vitro* (55). The recent report of an ATP-dependent active remodelling system from *Drosophila* embryo extracts (56) is an important step in developing the complex *in vitro* systems that will be required to address the mechanisms involved in chromatin transitions induced during transcription activation.

Genetic methods have identified several proteins which might be involved in modulating chromatin structure (5, 57). For example, the Swil, 2, and 3 proteins, identified in yeast as global coactivators of transcription, have been hypothesized to act through a chromatin-mediated mechanism (57, 58). The *SWI2* gene (also known as *SNF2*) has been reported to be required for derepression of the *PHO5* and *SUC2* genes (59), and its effects on *SUC2* are suppressed by a deletion of one of the two sets of genes which code for histones H2A and H2B (7).

At least some of the proteins required for modulating chromatin structure should interact with Pho4 and/or the glucocorticoid receptor. The Swi3 protein has been implicated in direct binding to glucocorticoid receptor and *SWI*1, 2, and 3 activities are all required for transactivation by glucocorticoid receptor in yeast (60). However, the mechanism by which the *SWI* complex affects transcription is unclear at this time. Activation effects have been reported with *in vitro* transcription systems on naked DNA templates (60). Human homologues of the *SWI* genes have also been shown to augment activation by glucocorticoid receptor in transient transfection assays, where the importance of chromatin is in question (61).

6. Nature of the disrupted chromatin state

In both *PHO5* and MMTV, information concerning the chromatin transition and the nature of the disrupted state derives almost exclusively from studies *in vivo* with nucleolytic reagents (DNase I, endonucleases, etc.). One study, however, has begun to address the nature of the transition using *in vivo* DNA–protein cross-linking. Bresnick *et al.* (62), using uv light cross-linking coupled with antibody selection for specific proteins, found that histone H1 was selectively depleted from the immediate proximal promoter of MMTV, including the A-B linker and the first two transcribed linkers (Fig. 4). These findings raise the interesting possibility that disruption of higher order structure(s) may be involved in the hormone-dependent transition. It is usually assumed that the template seen *in vivo* by a *trans*-activator such as a steroid receptor is the 10 nm fibre, or a simple polynucleosome array. It remains possible, however, that activating factors can interact with more complex superhelically wrapped nucleosome arrays, in which case decondensation of a compacted structure ('closed chromatin') could lead to the transition state ('open chromatin').

In vivo cross-linking, coupled with antisera to specific factors as well as H1 and the core histones, offers a new and potentially useful approach to establishing the nature of the transition state(s). UV cross-linking suffers the inherent limitation of

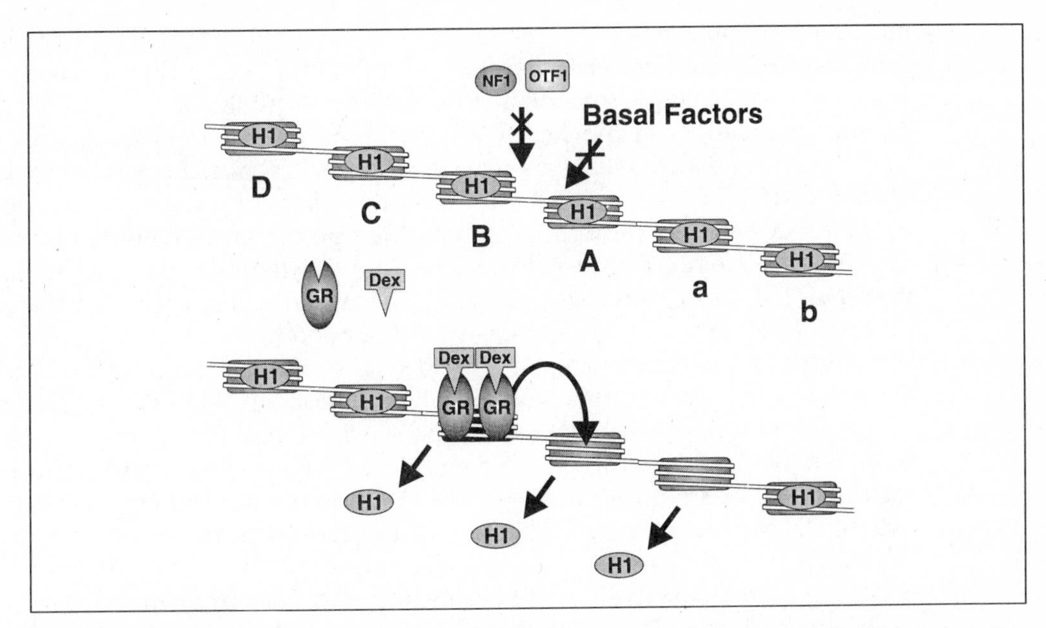

Fig. 4 Activation of the MMTV promoter is accompanied by loss of H1. Basal transcription factors, including NF1, OTF, and the basal initiation complex, are excluded from the promoter by some feature of chromatin structure. The hormone-induced transition in the nucleosome B region also involves the displacement of H1 from the immediate proximal promoter. This loss has been measured for the B–A–a–b region, including three linkers. Experiments at single linker resolution have not yet been completed.

inefficiency; raising the dose to increase the extent of cross-linking does too much damage to the DNA for subsequent analyses based on amplification of DNA sequence. Many alternative cross-linking strategies are available, however, and these approaches may ultimately provide critical insights into the nature of the 'open' and 'closed' states.

7. Final comments

7.1 Important questions

The *PHO5* and MMTV models have been very useful in evaluating the potential role of chromatin structure in gene activation. Results derived from these systems have provided some of the best evidence to date that chromatin transitions are mechanistically involved in gene regulation. Although considerable progress has been achieved, there is still much to be learned regarding the actual molecular mechanisms involved in the generation of altered chromatin states, and how these states modulate the binding of *trans*-acting factors and determine promoter activity.

The results obtained thus far highlight several key questions. How precisely are these systems organized at the nucleoprotein level? We know that nucleosomal location deviates markedly from random positioning. Do cores occupy unique rotational and translational positions, or do they populate a multiple but finite subset of possible positions? A recent study on core positioning over 5S DNA sequences in yeast demonstrated that essentially all potential translational frames consistent with a single rotational DNA path are occupied *in vivo* (63). Unexpected results have been obtained when synthetic optimal positioning sequences developed by Crothers and his colleagues (64) were placed *in vivo*. In both yeast (65) and *Drosophila* (Q. Lu and S. Elgin, unpublished data), these sequences appear to exclude nucleosomes rather than to provide optimal positioning. Similarly, these sequences are hypersensitive to DNase I when they are inserted in the *PHO5* promoter (W. Hörz, unpublished data). These findings suggest that the mechanisms for positioning *in vivo* are not well understood.

A second major issue concerns the mechanisms involved in factor binding, or exclusion, from nucleosomal templates. Some proteins, such as the glucocorticoid receptor, can bind to sites on nucleosomal surfaces, but others appear to be unable to interact with DNA wrapped on cores (49–51, 66–69). What are the rules? Many factors might be involved: distortion of DNA on the core surface, binding to one face of DNA v. 'wrap-around' binding, interaction with underlying histones, and others.

Another problem involves the exclusion of factors from the template during replication. Chromatin must be disassembled and reorganized during DNA synthesis; if factors are excluded from the nucleoprotein template, then how are these constraints maintained during replication?

Is there a chromatin disruption domain? The *trans*-activation domains of both Gal4 and Pho4 are required for chromatin disruption (10, 36). Several factors have been identified, which do not themselves activate transcription but rather set up a chromatin structure which allows other *trans*-acting factors to activate transcription (70–74). Pho4 and glucocorticoid receptor are thought of as bifunctional proteins which can fulfil both roles. It remains to be seen whether independent domains responsible for these two activities can be delineated.

The issue of greatest immediate import is the question of the modified, or 'open', chromatin state. Does this state indicate the complete, but transient, loss of a histone octamer? Other possibilities include partial loss, for example of a H2A–H2B dimer, or retention of all of the core histones but loss of H1 and an accompanying local decondensation of the polynucleosome array. What part does acetylation play in opening chromatin? Histone acetylation can influence the ability of factors to bind to nucleosomal sites (69). The binding of factors may also in turn influence histone acetylation patterns within a domain (75). Determination of the nature of the *in vivo* transitions will be invaluable in developing the appropriate *in vitro* chromatin assembly and modification systems that will be needed to address mechanistic questions.

These questions represent the primary issues to be addressed in our efforts to develop accurate models for the interaction of transcription factors with chromosomal templates. Although these questions are complex and technically challenging, a complete understanding of gene regulation from the native genetic template requires their answers.

7.2 Discussion

In discussion with other book authors, it was pointed out that many regulatory factors have multiple, adjaent upstream binding sites. This might facilitate binding and disruption of nucleosomes by (a) insuring that at least one site is on the 'outside' of the nucleosome and/or (b) allowing for cooperative binding. The nucleosome structure of the *PHO5* promoter suggests that nucleosomes are positioned so that one Pho4 binding site is extranucleosomal (UASp1). This site is normally required; deletion of this element makes the promoter inactive (31). Although this binding site seems to be relatively weak (at least *in vitro*), evidence cited above suggests that binding to this site most likely is the first step in activating the *PHO5* promoter. Therefore, it may be critical that this element be located in a nucleosome-free region. So far, there is no direct evidence of binding cooperativity between the binding sites for Pho4. In fact, if nucleosome -2 of the *PHO5* promoter is deleted (leaving only one Pho4 binding site), the gene can still be activated to about 40 per cent of the level reached by the wild type promoter (32), suggesting that binding to this isolated element is not greatly impaired. However, in the absence of binding cooperativity, two or more Pho4 molecules may none the less cooperate to disrupt chromatin structure in the domain.

Acknowledgements

This work was supported by grants to WH from the DFG (SFB 190) and Fonds der Chemischen Industrie. JS was supported by a grant from the National Science Foundation Program for Medium and Long Term Research at Foreign Centers of Excellence and by a National Science Foundation NATO Postdoctoral Fellowship.

References

1. Kornberg, R. D. (1974) Chromatin structure: a repeating unit of histones and DNA. *Science*, **184**, 868.
2. Elgin, S. C. (1988) The formation and function of DNase I hypersensitive sites in the process of gene activation. *J. Biol. Chem.*, **263**, 19259.
3. Felsenfeld, G. (1992) Chromatin as an essential part of the transcriptional mechanism. *Nature*, **355**, 219.
4. Kornberg, R. D. and Lorch, Y. (1991) Irresistible force meets immovable object–transcription and the nucleosome. *Cell*, **67**, 833.
5. Svaren, J. and Hörz, W. (1993) Histones, nucleosomes and transcription. *Curr. Opinion Genet. Dev.*, **3**, 219.
6. Reik, A., Schütz, G., and Stewart, A. F. (1991) Glucocorticoids are required for establishment and maintenance of an alteration in chromatin structure: induction leads to a reversible disruption of nucleosomes over an enhancer. *EMBO J.*, **10**, 2569.
7. Hirschhorn, J. N., Brown, S. A., Clark C. D., and Winston, F. (1992) Evidence that *SNF2/SWI2* and *SNF5* activate transcription in yeast by altering chromatin structure. *Genes and Dev.*, **6**, 2288.

8. Morgan, J. E. and Whitlock, J. P. (1992) Transcription-dependent and transcription-independent nucleosome disruption induced by dioxin. *Proc. Natl Acad. Sci. USA*, **89**, 11622.
9. Verdin, E., Paras, P. J., and Van Lint, C. (1993) Chromatin disruption in the promoter of human immunodeficiency virus type 1 during transcriptional activation. *EMBO J.*, **12**, 3249.
10. Axelrod, J. D., Reagan, M. S., and Majors, I. (1993) *GAL4* disrupts a repressing nucleosome during activation of *GAL1* transcription *in vivo*. *Genes and Dev.*, **7**, 857.
11. Almer, A. and Hörz, W. (1986) Nuclease hypersensitive regions with adjacent positioned nucleosomes mark the gene boundaries of the *PHO5/PHO3* locus in yeast. *EMBO J.*, **5**, 2681.
12. Almer, A., Rudolph, H., Hinnen, A., and Hörz, W. (1986) Removal of positioned nucleosomes from the yeast *PHO5* promoter upon *PHO5* induction releases additional upstream activating DNA elements. *EMBO J.*, **5**, 2689.
13. Richard Foy, H. and Hager, G. L. (1987) Sequence-specific positioning of nucleosomes over the steroid-inducible MMTV promoter. *EMBO J.*, **6**, 2321.
14. Cordingley, M. G., Riegel, A. T., and Hager, G. L. (1987) Steroid-dependent interaction of transcription factors with the inducible promoter of mouse mammary tumor virus *in vivo*. *Cell*, **48**, 261.
15. Hager, G. L. and Archer, T. K. (1991) The interaction of steroid receptors with chromatin. In *Structure and function on nuclear hormone receptors*. Parker, M. G. (ed.). Academic Press, London, p. 217.
16. Bresnick, E. H., Rories, C., and Hager, G. L. (1992) Evidence that nucleosomes on the mouse mammary tumor virus promoter adopt specific translational positions. *Nucleic Acids Res.*, **20**, 865.
17. Schmid, A., Fascher, K. D., and Hörz, W. (1992) Nucleosome disruption at the yeast *PHO5* promoter upon *PHO5* induction occurs in the absence of DNA replication. *Cell*, **71**, 853.
18. Rudolph, H. and Hinnen, A. (1987) The yeast *PHO5* promoter: phosphate-control elements and sequences mediating mRNA start-site selection. *Proc. Natl Acad. Sci. USA*, **84**, 1340.
19. Vogel, K., Hörz, W., and Hinnen, A. (1989) The two positively acting regulatory proteins *Pho2* and *Pho4* physically interact with *PHO5* upstream activation regions. *Mol. Cell. Biol.*, **9**, 2050.
20. Ogawa, N. and Oshima, Y. (1990) Functional domains of a positive regulatory protein, *Pho4*, for transcriptional control of the phosphatase regulon in *Saccharomyces cerevisiae*. *Mol. Cell. Biol.*, **10**, 2224.
21. Berben, G., Legrain, M., Gilliquet, V., and Hilger, F. (1990) The yeast regulatory gene *PHO4* encodes a helix-loop-helix motif. *Yeast*, **6**, 451.
22. Koren, R., LeVitre, J., and Bostian, K. A. (1986) Isolation of the positive-acting regulatory gene *PHO4* from *Saccharomyces cerevisiae*. *Gene*, **41**, 271.
23. Fisher, F., Jayaraman, P. S., and Goding, C. R. (1991) C-myc and the yeast transcription factor Pho4 share a common CACGTG-binding motif. *Oncogene*, **6**, 1099.
24. Beato, M. (1989) Gene regulation by steroid hormones. *Cell*, **56**, 335.
25. Yamamoto, K. R., Payvar, F., Firestone. G. L., Maler, B. A., Wrange, O., Carlstedt Duke, J., Gustafsson, J. A., and Chandler, V. I. (1983) Biological activity of cloned mammary tumor virus DNA fragments that bind purified glucocorticoid receptor protein *in vitro*. *Cold Spring Harb. Symp. Quant. Biol.*, **47** Pt 2, 977.

26. Lefebvre, P., Berard, D. S., Cordingley, M. G., and Hager, G. L. (1991) Two regions of the mouse mammary tumor virus long terminal repeat regulate the activity of its promoter in mammary cell lines. *Mol. Cell. Biol.*, **11,** 2529.
27. Yanagawa, S., Tanaka, H., and Ishimoto, A. (1991) Identification of a novel mammary cell line-specific enhancer element in the long terminal repeat of mouse mammary tumor virus, which interacts with its hormone-responsive element. *J. Virol.*, **65,** 526.
28. Mink, S., Hartig, E., Jennewein. P., Doppler, W., and Cato, A. C. (1992) A mammary cell-specific enhancer in mouse mammary tumor virus DNA is composed of multiple regulatory elements including binding sites for CTF/NFI and a novel transcription factor, mammary cell-activating factor. *Mol. Cell. Biol.*, **12,** 4906.
29. Schild, C., Claret, F. X., Wahli, W., and Wolffe, A. P. (1993) A nucleosome-dependent static loop potentiates estrogen-regulated transcription from the *Xenopus* vitellogenin B1 promoter *in vitro*. *EMBO J.*, **12,** 423.
30. Archer, T. K., Lefebvre, P., Wolford, R. G., and Hager, G. L. (1992) Transcription factor loading on the MMTV promoter: a bimodal mechanism for promoter activation. *Science*, **255,** 1573.
31. Fascher, K. D., Schmitz, J., and Hörz, W. (1993) Structural and functional requirements for the chromatin transition at the *PHO5* promoter in *Saccharomyces cerevisiae* upon *PHO5* activation. *J. Mol. Biol.*, **231,** 658.
32. Straka, C. and Hörz, W. (1991) A functional role for nucleosomes in the repression of a yeast promoter. *EMBO J.*, **10,** 361.
33. Venter, U., Svaren, J., Schmitz, J., Schmid A., and Hörz, W. (1994) A nucleosome precludes binding of the transcription factor Pho4 *in vivo* to a critical target site in the *PHO5* promoter. *EMBO J.*, **13,** 4848.
34. Han, M. and Grunstein, M. (1988) Nucleosome loss activates yeast downstream promoters *in vivo*. *Cell*, **55,** 1137.
35. Han, M., Kim, U. J., Kayne, P., and Grunstein, M. (1988) Depletion of histone H4 and nucleosomes activates the *PHO5* gene in *Saccharomyces cerevisiae*. *EMBO J.*, **7,** 2221.
36. Svaren, J., Schmitz, J., and Hörz, W. (1994). The transactivation domain of Pho4 is required for nucleosome disruption at the *PHO5* promoter. *EMBO J.*, **13**, 4856.
37. Archer, T. K., Lee, H. L., Cordingley, M. G., Mymryk, J. S., Fragoso, G., Berard, D. S., and Hager, G. L. (1994) Differential steroid hormone induction of transcription from the mouse mammary tumor virus promoter. *Mol. Endocrinol.*, **8,** 568.
38. Lee, H. L. and Archer, T. K (1994) Nucleosome-mediated disruption of transcription factor-chromatin initiation complexes at the mouse mammary tumor virus long terminal repeat in vivo. *Mol. Cell. Biol.*, **14,** 32.
39. Svaren, J. and Chalkley, R. (1990) The structure and assembly of active chromatin. *Trends Genet.*, **6,** 52.
40. Cavalli, G. and Thoma, F. (1993) Chromatin transitions during activation and repression of galactose-regulated genes in yeast. *EMBO J.*, **12,** 4603.
41. Smith, C. L., Archer, T. K., Hamlingreen, G., and Hager, G. L. (1993) Newly expressed progesterone receptor cannot activate stable, replicated mouse mammary tumor virus templates but acquires transactivation potential upon continuous expression. *Proc. Natl Acad. Sci. USA*, **90,** 11202.
42. Bürglin, T. R. (1988) The yeast regulatory gene *PHO2* encodes a homeo box. *Cell*, **53,** 339.
43. Sengstag, C. and Hinnen, A. (1987) The sequence of the *Saccharomyces cerevisiae* gene

PHO2 codes for a regulatory protein with unusual amino acid composition. *Nucleic Acids Res.*, **15,** 233.
44. Fascher, K. D., Schmitz, J., and Hörz, W. (1990) Role of trans-activating proteins in the generation of active chromatin at the *PHO5* promoter in *S. cerevisiae*. *EMBO J.*, **9,** 2523.
45. Sengstag, C. and Hinnen, A. (1988) A 28-bp segment of the *Saccharomyces cerevisiae PHO5* upstream activator sequence confers phosphate control to the *CYC1-lacZ* gene fusion. *Gene*, **67,** 223.
46. Brazas, R. M. and Stillman, D. J. (1993) Identification and purification of a protein that binds DNA cooperatively with the yeast Swi5 protein. *Mol. Cell. Biol.*, **13,** 5524.
47. Brazas, R. M. and Stillman, D. J. (1993) The Swi5 zinc-finger and Grf10 homeodomain proteins bind DNA cooperatively at the yeast HO promoter. *Proc. Natl Acad. Sci. USA*, **90,** 11237.
48. Kaffman, A., Herskowitz, I., Tjian, R., and O'Shea, E. K (1994) Phosphorylation of the transcription factor Pho4 by a cyclin-CDK complex, Pho80–Pho85. *Science*, **263,** 1153.
49. Perlmann, T. and Wrange, O. (1988) Specific glucocorticoid receptor binding to DNA reconstituted in a nucleosome. *EMBO J.*, **7,** 3073.
50. Piña, B., Brüggemeier, U., and Beato, M. (1990) Nucleosome positioning modulates accessibility of regulatory proteins to the mouse mammary tumor virus promoter. *Cell*, **60,** 719.
51. Archer, T. K., Cordingley, M. G., Wolford R. G., and Hager, G. L. (1991) Transcription factor access is mediated by accurately positioned nucleosomes on the mouse mammary tumor virus promoter. *Mol. Cell. Biol.*, **11,** 688.
52. Workman, J. L. and Kingston, R. E. (1992) Nucleosome core displacement *in vitro* via a metastable transcription factor nucleosome complex. *Science*, **258,** 1780.
53. Adams, C. C. and Workman, J. L. (1993) Nucleosome displacement in transcription. *Cell*, **72,** 305.
54. Morse, R. H. (1993) Nucleosome disruption by transcription factor binding in yeast. *Science*, **262,** 1563.
55. Chen, H., Li, B., and Workman, J. L. (1994) A histone-binding protein, nucleoplasmin, stimulates transcription factor binding to nucleosomes, and factor-induced nucleosome disassembly. *EMBO J.*, **13,** 380.
56. Tsukiyama, T., Becker, P. B., and Wu, C. (1994) ATP-dependent nucleosome disruption at a heat-shock promoter mediated by binding of GAGA transcription factor. *Nature*, **367,** 525.
57. Winston, F. and Carlson, M. (1992) Yeast *SNF/SWI* transcriptional activators and the *SPT/SIN* chromatin connection. *Trends Genet.*, **8,** 387.
58. Peterson, C. L. and Herskowitz, I. (1992) Characterization of the yeast *SWI1*, *SWI2*, and *SWI3* genes, which encode a global activator of transcription. *Cell*, **68,** 573.
59. Abrams, E., Neigeborn, L., and Carlson, M. (1986) Molecular analysis of *SNF2* and *SNF5*, genes required for expression of glucose-repressible genes in *Saccharomyces cerevisiae*. *Mol. Cell. Biol.*, **6,** 3643.
60. Yoshinaga, S. K., Peterson, C. L., Herskowitz I., and Yamamoto, K. R (1992) Roles of *SWI1*, *SWI2*, and *SWI3* proteins for transcriptional enhancement by steroid receptors. *Science*, **258,** 1598.
61. Muchardt, C. and Yaniv, M. (1993) A human homologue of *Saccharomyces cerevisiae SNF2/SWI2* and *Drosophila brm* genes potentiates transcriptional activation by the glucocorticoid receptor. *EMBO J.*, **12,** 4279.
62. Bresnick, E. H., Bustin, M., Marsaud, V. Richard Foy, H., and Hager, G. L. (1992) The

transcriptionally-active MMTV promoter is depleted of histone H1. *Nucleic Acids Res.*, **20,** 273.
63. Buttinelli, M., DiMauro, E., and Negri, R. (1993) Multiple nucleosome positioning with unique rotational setting for the *Saccharomyces cerevisiae* 5S rRNA gene *in vitro* and *in vivo*. *Proc. Natl Acad. Sci. USA*, **90,** 9315.
64. Shrader, T. E. and Crothers, D. M. (1989) Artificial nucleosome positioning sequences. *Proc. Natl Acad. Sci. USA*, **86,** 7418.
65. Tanaka, S., Zatchej, M., and Thoma, F. (1992) Artificial nucleosome positioning sequences tested in yeast minichromosomes: a strong rotational setting is not sufficient to position nucleosomes *in vivo*. *EMBO J.*, **11,** 1187.
66. Taylor, I. C., Workman, J. L., Schuetz, T. J., and Kingston, R. E. (1991) Facilitated binding of Gal4 and heat shock factor to nucleosomal templates: differential function of DNA-binding domains. *Genes and Dev.*, **5,** 1285.
67. Hayes, J. J. and Wolffe, A. P. (1992) Histones H2A-H2B inhibit the interaction of transcription factor IIIA with the *Xenopus borealis* somatic 5S RNA gene in a nucleosome. *Proc. Natl Acad. Sci. USA*, **89,** 1229.
68. Hayes, J. J. and Wolffe, A. P. (1992) The interaction of transcription factors with nucleosomal DNA. *Bioessays*, **14,** 597.
69. Lee, D. Y., Hayes, J. J., Pruss, D., and Wolffe, A. P. (1993) A positive role for histone acetylation in transcription factor access to nucleosomal DNA. *Cell*, **72,** 73.
70. Chasman, D. I., Lue, N. F., Buchman, A. R., LaPointe, J. W., Lorch, Y., and Kornberg, R. D. (1990) A yeast protein that influences the chromatin structure of UASG and functions as a powerful auxiliary gene activator. *Genes and Dev.*, **4,** 503.
71. Brandl, C. J. and Struhl, K. (1990) A nucleosome-positioning sequence is required for *GCN4* to activate transcription in the absence of a TATA element. *Mol. Cell. Biol.*, **10,** 4256.
72. Thomas, D., Jacquemin, I., and Surdin Kerjan, Y. (1992) MET4, a leucine zipper protein, and centromere-binding factor 1 are both required for transcriptional activation of sulfur metabolism in *Saccharomyces cerevisiae*. *Mol. Cell. Biol.*, **12,** 1719.
73. Croston, G. E., Kerrigan, L. A., Lira, L. M., Marshak, D. R., and Kadonaga, J. T. (1991) Sequence-specific antirepression of histone H1-mediated inhibition of basal RNA polymerase II transcription. *Science*, **251,** 643.
74. Lu, Q., Wallrath, L. L., Granok, H., and Elgin, S. C. R. (1993) (CT)n.(GA)n repeats and heat shock elements have distinct roles in chromatin structure and transcriptional activation of the *Drosophila-hsp26* gene. *Mol. Cell. Biol.*, **13,** 2802.
75. Turner, B. M. (1993) Decoding the nucleosome. *Cell*, **75,** 5.

6 Transcription on chromatin templates

DONAL LUSE and GARY FELSENFELD

1. Introduction

As it has become clear that the coding regions of transcriptionally active chromatin are packaged in nucleosomes (1, 2, 3), it has also become important to determine how such a histone-covered template can accommodate to the passage of RNA polymerase. For this purpose, it has proved useful to employ 'model' systems, in which templates reconstituted from specific DNA sequences and histone octamers are transcribed by either prokaryotic or eukaryotic polymerases. Experiments with prokaryotic enzymes such as the T7 or SP6 polymerases give us information about the intrinsic behaviour of histone octamers during transcription; they obviously cannot reveal possible contributions to the reaction that might be made by a eukaryotic polymerase, or by specific factors capable of modifying nucleosome stability that might be present in eukaryotic cells. Experiments with eukaryotic polymerases have the limitation that in using purified enzymes these specific factors may have been excluded. Despite these limitations, much valuable information about the properties of chromatin templates has been obtained with both kinds of polymerases.

2. Prokaryotic RNA polymerases

2.1 Experiments with high molecular weight templates

The earliest experiments with prokaryotic polymerases involved transcription of high molecular weight DNA (typically from T7 or SV40) complexed with core histones at a relatively high protein:DNA ratio (4, 5). These studies revealed that, although transcription rates were sometimes slower, long transcripts could be produced from octamer-covered templates, consistent with the idea that core histone octamers did not serve as an insurmountable obstacle to transcription. More recent experiments with better-defined templates and better analytic methods confirm and extend this conclusion. Thus, O'Neill *et al.* (6) have used T7 RNA polymerase to transcribe a chromatin template formed by reconstituting regularly spaced nucleosome cores on to a tandem array of 207 bp repeats from a sea urchin 5S RNA gene. Transcription is initiated from a T7 promoter included in the construct, and is found to proceed through the histone-covered template. However, the presence of nucleosome cores partially inhibits transcript elongation, and leads to the accumulation of shorter RNA products of discrete sizes corresponding to characteristic 'pause' sites, seen also on a naked DNA template. It is estimated that, under the conditions used, the maximum efficiency for transit through each nucleosome core is 85 per cent. Full-length transcripts are more efficiently produced by shorter templates that carry a smaller number of nucleosome cores but, even with a 2 kb template and ten cores in place, 20 per cent of the polymerases should completely traverse the DNA. Results consistent with this picture have also been obtained by Kirov *et al.* (7), who used T7 polymerase to transcribe a chromatin template formed from a 1400 bp DNA fragment containing the T7 promoter fused to the mouse α-globin gene. Transcription was shown to have no detectible effect on the template, suggesting that there had been no net disruption of the chromatin structure.

2.2 Possible mechanisms

None of the experiments described above permits straightforward conclusions to be drawn about the mechanisms involved. What happens to a histone octamer when polymerase passes through? Various possibilities can be envisaged (8, Fig. 1): (1) The octamer unfolds or transfers temporarily to the untranscribed strand, then reforms without moving from its original place on the DNA. (2) The octamer is displaced into solution. (3) The octamer slides ahead of the advancing polymerase but remains bound to the same DNA molecule (this is less likely on a template completely covered with histone octamers, but possible on a sparsely covered template (see Fig. 1)). (4) The octamer is displaced into solution, but is held within the polymer coil by electrostatic forces and quickly recaptured nearby on the same molecule. (5) The octamer, about to be displaced, is directly transferred elsewhere on the DNA by collision with a region of naked DNA.

Much recent effort has been devoted to distinguishing among these possibilities. Before discussing that work, however, it is useful to consider the nature of the interaction energies that stabilize histone octamers and nucleosome core particles, since the mechanisms proposed above require disruption or displacement of nucleosomes. For example, mechanism (1) involves the unfolding of a histone octamer. How might this occur? Although early ideas of octamer chemistry led to the suggestion that they might dissociate into heterotypic tetramers (containing one each of the four core histone species), subsequent studies make it clear that octamers dissociate to form a histone H3–H4 tetramer and two H2A–H2B dimers. From the measured free energies of association of free histones, it is possible to estimate the energetic cost of breaking the histone–histone contacts to unfold the nucleosome without allowing histones to dissociate from DNA. The net free energy includes the benefit of unwinding the DNA, normally bent into two tight, left-handed superhelical turns that wrap around the octamer. The estimated net minimum

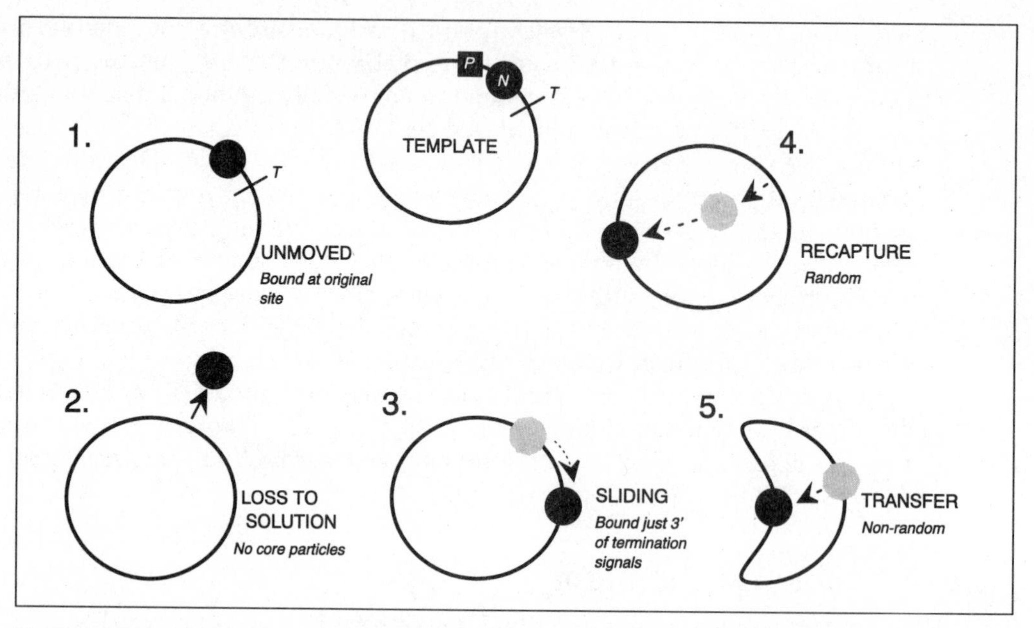

Fig. 1 Possible mechanisms of histone octamer accommodation to the passage of RNA polymerase during transcription. Code: P, promoter; N, nucleosome; T, termination signal. From (8), reproduced by permission of Cell Press.

cost is over 20 kcal/mol for two H2A–H2B dimers split from the octamer; unfolding the H3–H4 tetramer into separate H3–H4 dimers would cost at least another 15 kcal/mol (9, 10). These estimates include the energy gained from the unwinding of one superhelical turn when H2A–H2B unfolds, and of a second turn when H3–H4 unfolds.

The second mechanism described above involves complete displacement of the octamer from its site on DNA. Experimental estimates of the free energy of displacement vary considerably (11, 12, 13), but are generally in the range 10 to 50 kcal/mol of core particles in 0.1 M NaCl (which automatically includes the energy needed to bend the DNA). Because the interactions between DNA and histones are largely electrostatic, these values are expected to have a strong dependence on ionic strength. None the less under physiological conditions these energies are likely to be quite large. In comparison, mechanisms 3 and 5, which do not require the simultaneous disruption of all the electrostatic interactions, or mechanism 4, in which the octamer stays close to the DNA before recapture in such a way as to experience relatively unscreened electrostatic forces, do not involve any net change in free energy and are also likely to involve smaller activation energies. These mechanisms are therefore the most attractive *a priori*. It should be kept in mind that differences in binding free energy between different DNA sequences may be rather small. As has been shown by Shrader and Crothers (14), even optimally bent DNA contributes only −0.15 kcal/bend of favourable free energy to the free energy of octamer–DNA interaction, relative to random sequence DNA, under reversible binding conditions (15, 16). Although these energy differences may be enough to allow the observation of a 'preferred' position in a gel assay, they are in any event likely to be small compared with the overall energies of histone–DNA interaction.

In all these cases, a further contribution is involved if the DNA is topologically constrained and subject to supercoiling. A DNA linking number change of −1 is associated with the formation of a nucleosome core on a closed circular plasmid, followed by relaxation of the complex. Therefore histone octamers bind more tightly to negatively than to positively supercoiled plasmids. Taken in conjunction with the observation (17) that transcription can lead to the transient accumulation of positively supercoiled DNA ahead of the transcription complex, and negatively supercoiled DNA behind, this has led to the proposal that the act of transcription may disrupt histone octamers. Pfaffle *et al.* (18) have attempted to investigate this effect by using T7 RNA polymerase to transcribe a closed, circular template in the presence of topoisomerase I, which relaxes unconstrained supercoils. Under these conditions, a reduction in the superhelix density is observed, which the authors attribute to transient unfolding of the nucleosome cores as a result of positive superhelical stress (see below). It is unfortunately difficult to distinguish, in this complicated system, between unfolding of nucleosomes and their total displacement. There is strong physico-chemical evidence that positive supercoiling at superhelical densities as high as +0.07 is not sufficient to unfold or measurably distort a nucleosome core in 0.1 M NaCl solvents (19, 20). (It should be pointed out that these experiments made use of histone samples that were not highly acetylated; we do not know what effect acetylation would have on this behaviour.) On the other hand, considerable energy will be gained if a histone octamer bound to a positively supercoiled DNA segment can transfer to a negatively supercoiled segment (19).

2.3 Experiments with single nucleosomes

To distinguish among the mechanisms shown in Fig. 1 it is necessary to track the fate of single nucleosomes. In the absence of simple methods for tagging individual octamers, this requires that the template carry only a single nucleosome core. A number of investigations have used this approach. Lorch *et al.* (21) reconstituted a histone octamer on to a short DNA fragment, then ligated a promoter for SP6 polymerase to the complex, and transcribed with this enzyme. It was found under these conditions that transcription results in displacement of the histone octamer from the template, which can be detected both by the exposure of previously blocked restriction sites in the template DNA and by the appearance of protein-free DNA on electrophoretic gels. A different result was obtained by Losa and Brown (22), who carried out similar experiments by reconstituting nucleosome core particles on to 360 bp fragments carrying the SP6 promoter positioned to transcribe either the coding or the non-coding strand of the 5S ribosomal gene. Unlike Lorch *et al.*, Losa and Brown found that there was no change in the position of the octamer after transcription, as judged by DNase I footprinting. A later paper by Lorch *et al.* (23) attempted to reconcile the two sets of data by repeating experiments with the two templates under identical conditions, and came to the conclusion that the differences arose from some special sequence properties of 5S DNA, which resulted in retention of 50 to 75 per cent of the octamers on their original templates.

Subsequent attempts to determine the fate of histone octamers in detail have resolved some of these ambiguities. Clark and Felsenfeld (8) constructed a template in which a single nucleosome core was ligated into a plasmid so that there was an SP6 promoter upstream, and transcription terminators downstream of the insert. After transcription the template was digested with micrococcal nuclease, and the resulting nucleosome core DNA analysed to determine its sequence content. It was found that nearly all of the histone octamers had been displaced from their original binding site, but that they could be accounted for elsewhere on the plasmid. Although all sequences in the plasmid were represented, there was a 2.4-fold greater density of octamers positioned in the 902 bp region behind the promoter than in the half of the plasmid immediately downstream of the terminator. These results eliminate most of the models illustrated in Fig. 1: the observed movement of the octamers is inconsistent with model 1, and the observed recapture elsewhere on the plasmid is inconsistent with model 2. If sliding in front of the polymerase had occurred, most octamers would have been found just downstream of the termination signals. Since this did not happen, model 3 is also eliminated. Models 4 and 5 are thus the only mechanisms consistent with the observations.

To decide between these models, Studitsky *et al.* (24) returned to the study of a series of short linear templates coupled to the SP6 promoter. Reconstitution on these templates was found typically to result in several species of mononucleosome core particles, each with a discrete position of the octamer on the DNA. These species could be resolved and purified on electrophoretic gels; positions were identified by micrococcal nuclease and restriction endonuclease digestion. Similar methods were used to determine the positions of octamers after transcription. It was found that, in the absence of competitor DNA, histone octamers remained bound to their templates, but that in every case they had been shifted backward on the DNA sequence, in most cases by about 80 bp. Even in the presence of competitor, internal transfer was the preferred mechanism. Both in the absence and presence of competitor, the final positions of the octamer depended on its initial position. This appears to eliminate model 4, since it would be difficult to explain how the final position could depend on the initial position if there were an intermediate in which histones were free in solution. The data thus support mechanisms like model 5. A detailed model for transcription on short templates has been proposed (Fig. 2), in which the polymerase invades the nucleosome core, stripping the DNA from the surface. When transcription has proceeded sufficiently far, the 5′ end of the duplex is rebound to the octamer surface, producing a topologically closed domain. If the transcription complex is rotationally constrained, further advance of the polymerase will quickly generate large positive

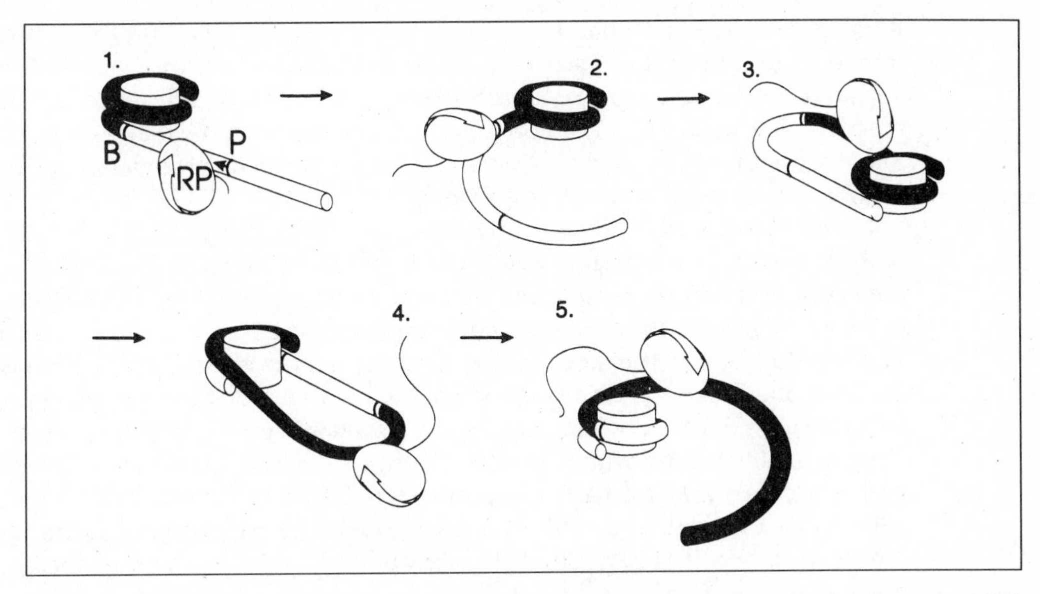

Fig. 2 Mechanism proposed for the transfer of histone octamers out of the path of a transcribing RNA polymerase. Code: P, promoter; RP, RNA polymerase; B, border of the nucleosome core; nucleosomal DNA is shaded. From (24), reproduced by permission of Cell Press.

superhelical stress ahead of the polymerase, and negative stress behind. The effect is to favour the transient displacement of DNA ahead of the advancing polymerase, allowing completion of the transcript. The octamer is depicted in Fig. 2 as remaining intact during the passage of the polymerase; support for this idea comes from the work of O'Neill *et al.* (25), who have shown that transcription of chromatin templates by T7 polymerase is not affected by the use of octamers that have been chemically cross-linked. It thus seems unlikely that octamer dissociation or partial unfolding is required during transcription by this polymerase. It should none the less be kept in mind that histone octamers are not stable when free in solution at physiological ionic strength. If the octamer were to be displaced completely from the template into the surrounding solvent (model 2, Fig. 1) it would not survive. On the other hand, it might well remain associated within the highly charged polymer coil in model 4 (Fig. 1), and would certainly do so in model 5 (Fig. 1), which corresponds to the observed mechanism. There is also direct evidence that hexamers and octamers retain their identity during transfer, which is consistent with the absence of dissociation (24).

In contrast to these results suggesting that octamers remain intact, Allfrey and his collaborators (26) have shown that transcriptionally active chromatin isolated from eukaryotic cells is preferentially retained on affinity columns able to react with sulphydryl groups. They suggest that transcription by eukaryotic polymerase might disrupt the histone octamer sufficiently to expose the histone H3 cysteine 110 sulphydryl group situated near the core particle dyad axis. As noted above, there is no evidence that a histone octamer can unfold into two heterotypic tetramers to expose the cysteine residues, and it is known that sulphydryl groups within a bound octamer are inaccessible to chemical modification. An unfolding reaction that exposed the sulphydryl groups might none the less be driven by histone modification, such as histone acetylation, or by the action of other unknown factors present in the eukaryotic nucleus.

The results described above show that, on short templates, the histone octamer is displaced from its starting position but retained by its original template DNA molecule. On longer templates, there is no reason why a more distant DNA segment might not move in to occupy the transiently vacant site on the octamer. This would result in transfer of the octamer to a more distant position on the DNA (in either the 3′ or 5′ direction), as is observed with the plasmid templates described above (8). It is less obvious how these mechanisms might function *in vivo* on templates fully loaded with octamers. During transcription, octamers might shuffle both backward and forward to vacant sites on the chromatin. In any case, the octamer is sufficiently versatile to accommodate the passing prokaryotic polymerase *in vitro*, and it is likely that this capability is exploited within the eukaryotic nucleus.

3. Eukaryotic RNA polymerases

3.1 The limitations of eukaryotic *in vitro* transcription approaches

In addition to work with model systems, a limited number of experiments have been done which address directly the ability of RNA polymerases II or III to elongate transcripts on nucleosomal templates. Since it is well established that the presence of nucleosomes on the promoter blocks initiation by RNA polymerases II (21, 27–28) and III (29–32), the basic strategy in these studies is to load RNA polymerase at the promoter before chromatin assembly. Polymerases are either held as pre-initiation complexes or allowed to transcribe until stable, early elongation complexes (RNAs of 10–20 nucleotides (nt)) are formed; in the latter case, NTPs are removed by gel filtration. The template DNA surrounding the transcription complex is then assembled into nucleosomes, either by direct exchange with existing chromatin or by incubation with histones and protein cofactors. Transcription is continued by adding NTPs.

There is an important limitation to the use of eukaryotic RNA polymerases in transcription of chromatin templates. Promoter-initiated *in vitro* transcription with RNA polymerases II or III usually results in RNA synthesis from only a few per cent of the templates (see, e.g., 33 and 34). Thus, the effect of transcription on the nucleosomes is very difficult to determine. It also becomes important to confirm that the small subset of the DNA which is transcriptionally active has been successfully reconstituted into chromatin. Finally, as noted above, there is the possibility that *in vitro* chromatin assembly methods might fail to duplicate faithfully some important aspects of *in vivo* chromatin structure. This last problem, which can occur regardless of which RNA polymerase is used for transcription, will be considered further in Section 4.

3.2 Transcript elongation on chromatin templates by RNA polymerase II *in vitro*

The first study that specifically addressed transcript elongation on nucleosomal templates by RNA polymerase II was that of Lorch *et al.* (21). In these experiments, analogous to those using prokaryotic polymerases described above, a short DNA fragment bearing a single nucleosome was ligated to another DNA segment containing the adenovirus major late promoter. The monosomes linked to upstream promoter regions were purified from free DNA by sucrose gradients. When these constructs and control (non-chromatin-assembled) DNAs were transcribed using partially purified rat liver transcription fractions, run-off RNAs of identical length were obtained in the presence or absence of the nucleosome. As expected, template usage was low in these experiments, so the fate of the nucleosomes on the transcribed templates could not be addressed. To overcome this difficulty, Lorch *et al.* added single-stranded extensions to the DNA of the individual nucleosomes and then incubated these constructs with pure RNA polymerase II and NTPs. It is well established that RNA polymerase II will efficiently initiate transcription at the ends of double-stranded DNAs bearing short 3′ single-stranded extensions (35). A gel shift analysis after 1 hour of transcription indicated that all of the nucleosomes had been disrupted; RNA polymerase II activity was required for this effect (since it was inhibited by α-amanitin). However, these results should be interpreted with caution, since transcription from template ends can lead to displacement of the non-template strand and the production of long RNA–DNA hybrids (35, 36). Furthermore, it is possible that circumventing the normal initiation step may make it impossible to recreate transcript elongation controls that rely on promoter-proximal loading of elongation factors.

The results of Lorch *et al.* could be taken to indicate that the nucleosome is not a significant barrier to the passage of RNA polymerase II. The question was revisited by Izban and Luse (37, 38), who measured the rate and efficiency of elongation during transcription of nucleosomal arrays by RNA polymerase II. In these studies, transcription complexes were assembled by incubating HeLa cell extracts with circular plasmid DNAs bearing the adenovirus major late promoter. The complexes were allowed to synthesize 15–20 base RNAs using a subset of the NTPs and were then purified by transient exposure to the detergent Sarkosyl followed by gel filtration. (Sarkosyl disrupts most protein–DNA interactions, including those of nucleosomes with DNA, but it does not inhibit the RNA polymerase II transcription complex once RNA synthesis has begun.) The purified transcription complexes were assembled into chromatin by incubation in *Xenopus* oocyte extracts. The plasmid templates were designed such that four nucleosomes were assembled at known locations beginning 50–70 bp downstream of transcription start. When the stalled RNA polymerases were released into elongation on these nucleosomal templates at 30 °C, nearly all of the polymerases paused in RNA synthesis somewhere within the first nucleosome; however, control complexes on which nucleosomes had not been assembled supported efficient elongation (Fig. 3; see 37). A small proportion of the polymerases continued transcription into the second nucleosome when reactions were performed at 37°C for 5 min (38). The majority of polymerases which paused during transcription of the nucleosomal templates had not terminated, since most polymerases continued transcription if the nucleosomes were disrupted by the addition of Sarkosyl or high salt (37).

One potential problem with the initial study (37) was that purification of the ternary complexes prior to chromatin assembly removed factors known to stimulate elongation by RNA polymerase II. The best characterized of these factors are TFIIF, which primarily increases the rate of elongation by RNA polymerase II (38–40), and SII, or TFIIS, which relieves pausing at DNA sequences which the polymerase has difficulty traversing but which does not increase the overall elongation rate (38, 40, 41). A third stimulatory fraction, TFIIX, can both increase transcription rate and reduce pausing (38, 40). TFIIX has not been extensively purified and it may contain several separate activities. Izban and Luse showed that the addition of TFIIF and SII to Sarkosyl-rinsed elongation complexes on DNA templates at 37°C restored *in vivo* elongation characteristics; that is, in the presence of these factors elongation proceeded at physiological rates (20–25 nt/s—see 42) and with minimal pausing during the synthesis of the initial 2 kb of transcript (38). When TFIIF, TFIIX, and SII were added individually or in combination to transcript elongation reactions on the nucleosomal templates, a definite increase in elongation efficiency was seen. In 5-minute reactions at 37°C with all three factors, nearly half of the polymerases successfully crossed one nucleosome and an easily detectable fraction succeeded in traversing two nucleosomes (Fig. 4, lane 8; see 38). However, while elongation was improved on the nucleosomal templates by the addition of saturating levels of factors, the rate of elongation on these templates was at best 1/10 of that seen on pure DNA templates in the presence of the elongation factors (38). Polymerases paused continually on the chromatin templates even in the presence of factors, quickly creating a broad distribution of transcript lengths within the com-

Fig. 3 Transcription of naked and nucleosomal templates by RNA polymerase II (see 37). The template is pML5–4NR, a plasmid DNA which contains four consecutive repeats of a nucleosome phasing signal cloned downstream of the adenovirus major late promoter. The position of the first nucleosome is indicated by the ellipsoid; the second repeat of the assembly sequence occurred roughly between positions +260 and +450, the third between roughly +450 and +640, etc. RNA polymerase was loaded on to the plasmid DNAs by incubation in HeLa cell nuclear extract and then advanced to position +15 with the addition of a subset of the NTPs, including radiolabelled CTP. The resulting ternary complexes were purified and the downstream DNA was reconstituted into nucleosomes by incubation with *Xenopus* oocyte extract (right-hand panel). Ternary complexes shown in the left-hand panel were incubated for the same period without *Xenopus* extract. After assembly or mock assembly, 1 mM non-labelled NTPs were added and the samples were incubated for the times indicated at 30°C. RNAs were purified and resolved by electrophoresis on a 10% polyacrylamide gel; the lengths of the DNA size markers are shown next to the right-hand panel.

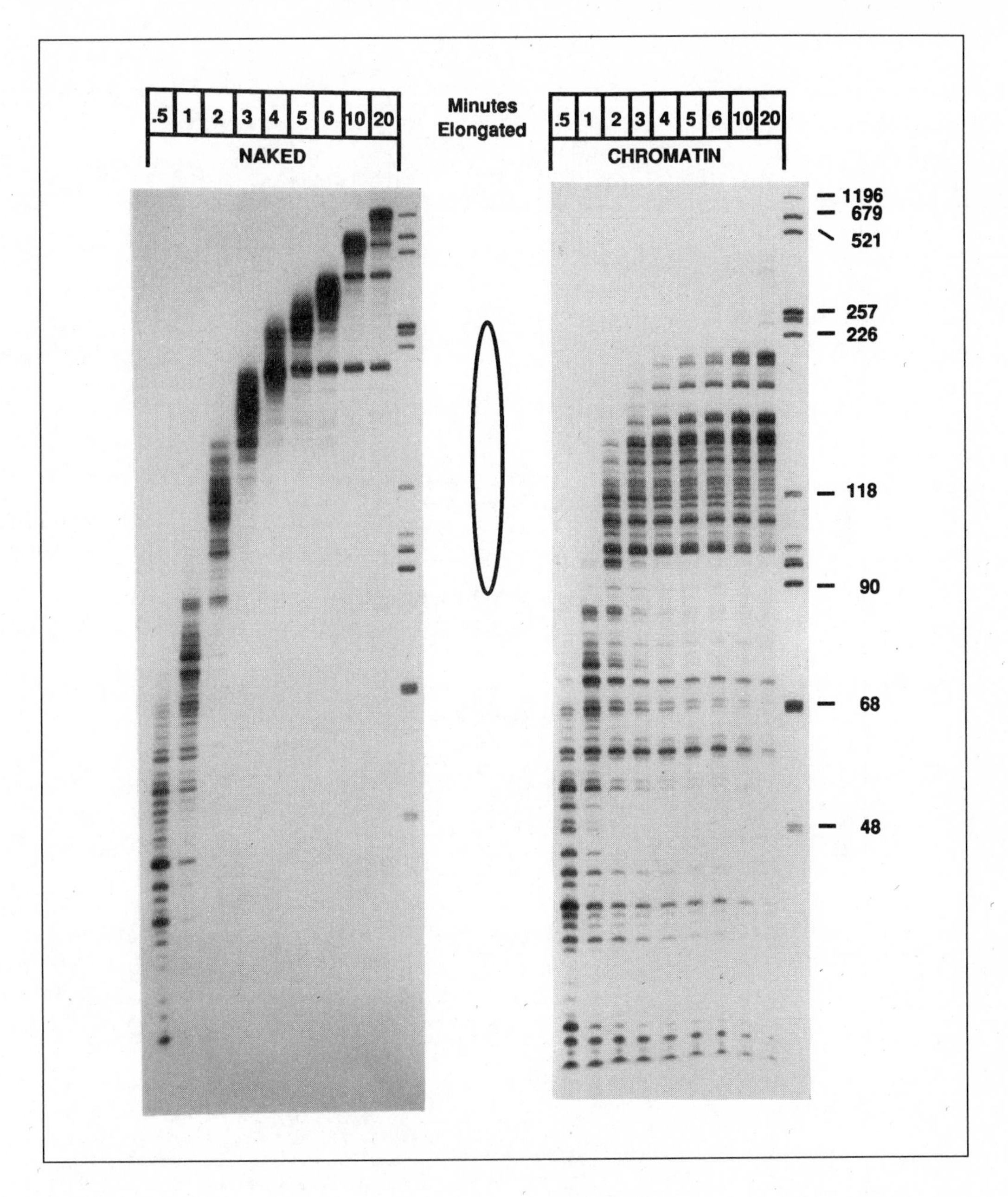

plexes, in contrast to the relatively narrow distribution of transcript lengths seen on pure DNA templates under the same conditions (Fig. 3).

The studies on transcription of nucleosomal arrays by RNA polymerase II (37, 38) do not contradict the earlier work on single nucleosome templates (21). Lorch *et al.* transcribed single nucleosomes with promoter-initiated polymerases for 20 min in the presence of partially fractionated rat initiation factors (21). These

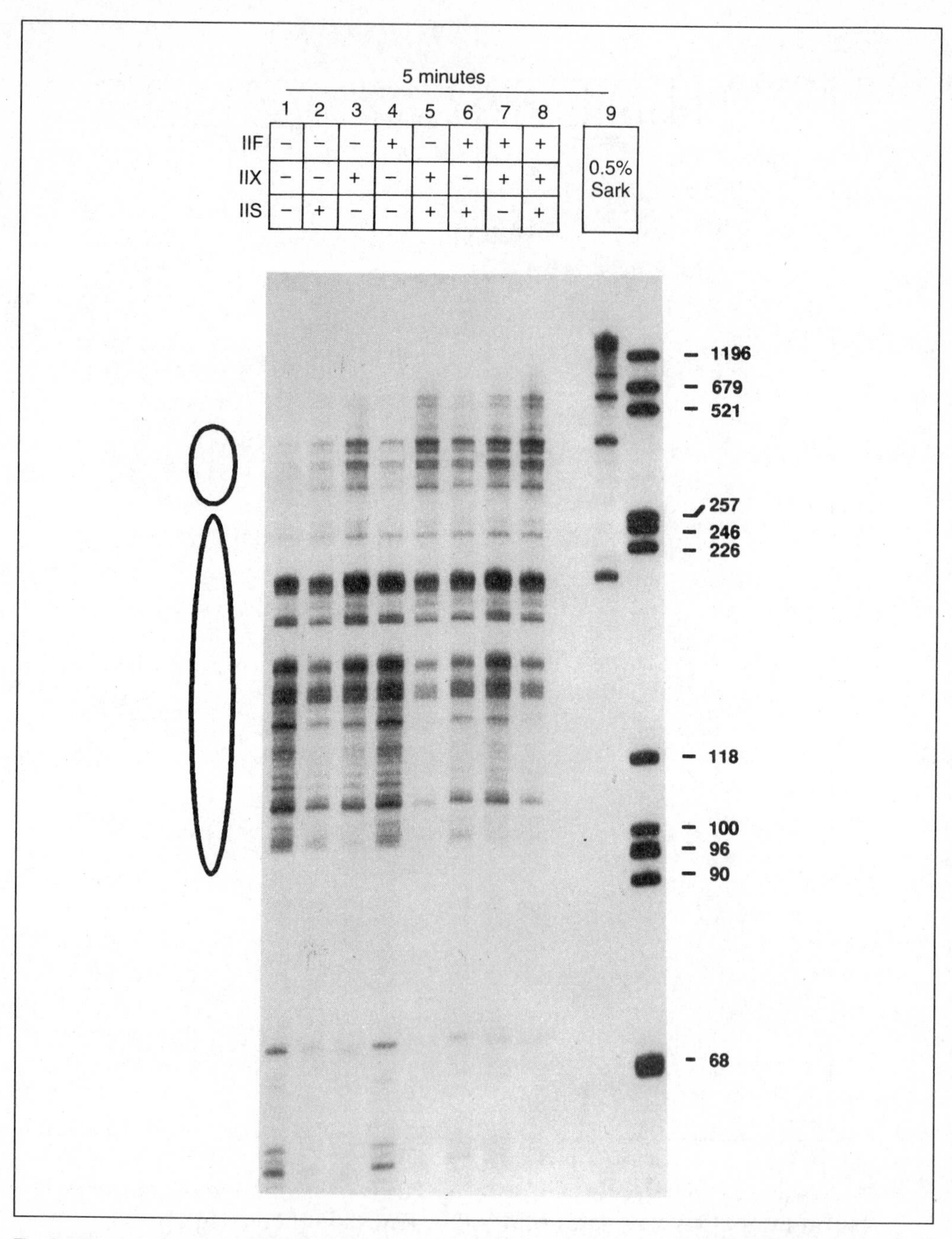

Fig. 4 Effect of elongation factors on the ability of RNA polymerase II to traverse nucleosomal templates (taken from panel B, Fig. 4, of 38; reproduced by permission of the American Society for Biochemistry and Molecular Biology). This experiment is essentially identical to that shown in Fig. 3, except that the elongation reactions were performed for 5 min at 37 °C and supplemented with transcript elongation factors or Sarkosyl as indicated. The positions of the first two nucleosomes are indicated by the ellipsoids.

fractions must have contained TFIIF (which is also a required initiation factor—recently reviewed in 43 and 44) and they probably contained TFIIX and SII as well. Thus, earlier results are not inconsistent with the observation that about half of the purified transcription complexes that were factor supplemented were able to cross a single nucleosome in 5 min (38). A small but easily detectable fraction of purified transcription complexes were able to traverse a nucleosome in only 1 min at 37 °C (38) even in the absence of elongation factors, which is consistent with the result of Lorch *et al.* that RNA polymerase II initiated from a single-stranded tail can efficiently transcribe through a single nucleosome in 1 h (21).

3.3 Transcript elongation on chromatin templates by RNA polymerase III *in vitro*

The ability of RNA polymerase III to negotiate nucleosomal templates *in vitro* has also been explored. The reader may question the utility of such experiments, since RNA polymerase III transcripts (primarily 5S rRNA and tRNA precursors) are shorter than the length of DNA within a single nucleosome. However, polymerase III provides a useful alternative to the bacteriophage RNA polymerases, since RNA polymerase III is a multi-subunit polymerase similar to RNA polymerase II (45) while phage polymerases are much smaller, single-subunit enzymes (see, e.g., 46). Felts *et al.* formed yeast RNA polymerase III pre-initiation complexes on 5S RNA maxigenes capable of giving rise to 240, 360, 600, or 1300 nt transcripts (31). The DNA surrounding these pre-initiation complexes was then assembled into chromatin using purified histones and nucleoplasmin. Felts *et al.* found that RNA polymerase III was able to synthesize maxigene transcripts as long as 600 nt with roughly the same efficiency on naked DNA and on chromatin templates bearing on average one nucleosome per 215 bp; however, production of the 1300 nt RNA from the chromatin templates was reduced to <1 per cent of the signal seen with naked DNA templates. The activity of the maxigenes was also tested *in vivo* (31). Production of the 1300 nt RNA was reduced relative to the shorter maxigene products but the inhibition was less severe than that seen *in vitro*. Thus, in this case RNA polymerase III was found to be capable of transcription through small numbers of nucleosomes, showing slightly greater effectiveness than (non factor-stimulated) RNA polymerase II in analogous studies.

Hansen and Wolffe (47) assembled varying numbers of nucleosomes on to a DNA fragment consisting of twelve tandem repeats of a 5S gene lacking a termination sequence. The promoters which were unoccupied by nucleosomes supported transcription initiation; the resulting RNA polymerase III elongation complexes continued transcription until they either ran off the fragment or were blocked by nucleosomes. It was found that, even when 8–10 nucleosomes had been assembled on these fragments, RNAs of 1000 bases or longer could be synthesized. Synthesis of long RNAs was less efficient, in comparison to reactions using pure DNA templates, on the templates with high levels of nucleosome assembly, in general agreement with the results of Felts *et al.* (31). However, Hansen and Wolffe also noted a significant effect of Mg^{2+} concentration in their studies. Efficient elongation on the nucleosomal arrays was observed only when Mg^{2+} and NTPs were present in equal concentration, giving an effective free Mg^{2+} concentration of about 0.1 mM. When Mg^{2+} was in excess over NTPs by 5 mM, essentially no transcript elongation was observed through the nucleosomal arrays. This inhibition was attributed to a Mg^{2+}-dependent chromatin compaction. (Note that reconstitution in this case was performed by salt dialysis using only core histones; thus the compaction should have resulted from nucleosome–nucleosome interactions and not histone H1—see 31.) The possibility that Mg^{2+}-induced chromatin structure changes may play a role in inhibiting transcript elongation could help to explain the early observation of Morse (48) that RNA polymerase III cannot elongate transcripts through even a single nucleosome. However, other authors have achieved moderate levels of transcript elongation (for example, in 31) without resorting to very low free Mg^{2+} levels.

4. Final comments

The overall lesson from all of the studies just discussed is that the nucleosome is a significant but not insurmountable barrier to transcript elongation by both RNA polymerases II and III *in vitro*. Since the *in vitro* system cannot reproduce fully an important property of RNA polymerase II *in vivo*, it is possible that auxiliary factors which assist RNA polymerase II in transcript elongation in the cell remain to be discovered. It is of course also possible that chromatin assembled in the test-tube is somehow lacking in features (such as histone modifications) that might be important for efficient passage of the RNA polymerase. In the following section we will speculate on new approaches that might allow a better reproduction *in vitro* of the behaviour of polymerase II in the cell nucleus.

4.1 Why is transcript elongation by RNA polymerase II on nucleosomal templates so inefficient, even in the presence of elongation factors?

Izban and Luse showed that most pausing sites which are observed when RNA polymerase transcribes a given stretch of DNA packaged in nucleosomes are also recognized, but for much shorter times, when the same DNA is transcribed in the absence of nucleosomes (37; see also Fig. 3). Thus, to a first approximation a difficult step in chain elongation on pure DNA has been made more difficult on chromatin templates. It seems unlikely that bond formation itself would be affected by the presence of nucleosomes, but translocation along the DNA by the polymerase could obviously be restricted on a chromatin template. Several lines of evidence suggest that RNA polymerase II may respond to this problem by falling into an arrested state. Arrested polymerases remain in ternary complex but cannot continue RNA synthesis, even in the presence of saturating levels of NTPs. These polymerases can, however, be restarted through the action of the SII elongation factor (49, 50). Reines and Mote (51) have shown that there is a high probability of RNA polymerase II falling into transcriptional arrest when it encounters *lac* repressor bound to the DNA. Several laboratories have also demonstrated that when RNA polymerase II approaches the end of a linear template (at which further downstream translocation is obviously impossible) the polymerase may fall into arrest (49, 52; Izban *et al.* submitted).

The results just noted are consistent with the idea that transcriptional arrest may result when RNA polymerase II attempts to transcribe through nucleosomes. If arrest is indeed occurring, why does SII have such a modest stimulatory effect on transcript elongation with the nucleosomal templates? It is important to note that SII functions in arrest relief by mediating a transcript cleavage activity. Arrested RNA polymerase II cannot restart unless a short segment is cleaved from the 3′ end of the nascent RNA (49, 50). Transcription resumes from this newly-generated 3′ end. The transcript cleavage reaction occurs once every 10–30 s (50), which is considerably slower than the normal transcript elongation rate (20–25 nt/s; see 42). Also, it appears that SII must cycle off the polymerase and then rebind for each round of transcript cleavage (53). If polymerases can arrest at many positions as they move through a nucleosome, which is consistent with the pattern of transcripts on nucleosomal templates (38), and arrest relief is slow, then this would account for the very slow overall elongation rate observed even in the presence of SII.

However, arrest cannot be the only fate for RNA polymerase II on the nucleosomal templates, since most polymerases paused within nucleosomes will chase if Sarkosyl or high salt are added in addition to NTPs. If the majority of polymerases on nucleosomal templates are not arrested and their rate of transcript elongation is not strongly stimulated by TFIIF, it is reasonable to assume that novel factors exist whose major function is mediating rapid elongation through nucleosomes. Such factors might work alone, but they could also function with SII, perhaps by helping to retain SII in a tight complex with the polymerase. This would eliminate the time required for SII to recycle between transcript cleavage events. This latter possibility is suggested by analogy with N-mediated anti-termination in bacteriophage lambda. In this case, a complex collection of protein factors (and transcript sequences) must all be tightly bound to the RNA polymerase to achieve antitermination (recently reviewed in 54).

4.2 How can chromatin assembled in the test tube be made to reflect more accurately the structure of actively transcribed chromatin in the cell?

If the chromatin templates assembled in the test-tube lack some feature which is important for efficient transcript elongation, then this problem will also have to be addressed before RNA polymerase II *in vitro* systems using reconstituted chromatin templates can achieve the expected rates and efficiencies of transcript elongation. A number of studies have indicated that hyperacetylated histones are preferentially associated with transcribed regions (reviewed in 55–57). This modification, which increases the ability of transcription factors to access their binding sites in nucleosomal DNA (see, e.g., 58) might also aid transcript elongation by facilitating template access by the RNA polymerase. Thus, it should be useful to test the effect of histone acetylation on elongation. Recent findings, discussed above, on the displacement of nucleosomes to upstream regions by elongating bacteriophage RNA polymerase raise the possibility that efficient transcript elongation may depend in part on the spacing of nucleosomes on the template. If efficient transcript elongation requires the exchange of nucleosomes from downstream to upstream of the RNA polymerase, then reconstituted templates containing long arrays of close-packed nucleosomes (which would lack an 'acceptor' for the nucleosome being invaded by the polymerase) may be unable to support efficient transcript elongation. Finally, it should be noted that *in vitro* approaches to chromatin transcription might be employed without *in vitro* reconstituted templates. Hansen and colleagues (59) have shown that SV40 minichromosomes will serve as templates for *in vitro* transcription by RNA polymerase II. In an initial study, these minichromosomes, but not naked DNA templates, gave the appropriate ratios of promoter usage among the various viral early and late promoters (59), which is consistent with the hope that the isolated minichromosomes retain at least some of the their native structure. Thus, *in vivo* assembled chromatin templates may also be useful in determining the biochemical requirements for the efficient passage of RNA polymerase II through nucleosomes.

4.3 Discussion

Several interesting points were raised in discussion with other authors. Allfrey and co-workers maintain that nucleosomes from active genes are preferentially retained on mercury columns, presumably because a normally buried SH group on histone H3 is revealed by unfolding of the octamer during the passage of RNA polymerase (26 and references therein). These results might be explained without invoking the unfolding of the octamer by supposing the nucleosomes from active genes are preferentially bound by nonhistone proteins with exposed SH groups. However, this seems unlikely to explain all of the binding to the mercury columns, for two reasons. First, Allfrey and colleagues have shown that a fraction of the material which binds to their mercury columns is released with a 0.5 M NaCl wash, as expected if nonhistone proteins which adhere to chromatin are causing column retention. (Note that the octamer itself will not dissociate until 0.8 M or higher salt levels.) However, the material retained after a 0.5 M salt wash elutes with 10 mM dithiothreitol, as expected if the core histones themselves were bound (26). Second, yeast normally lack a cysteine in histone H3; if yeast nucleosomes are applied to a mercury column, very little material sticks. However, when site-directed mutagenesis is used to change position 110 of yeast H3 to a cysteine, nucleosomes from active genes are preferentially retained on mercury columns (26). This is consistent with the idea that retention occurs through an SH group in the nucleosome octamer, and not through nonhistone proteins; however, it is also worth noting that in yeast, nucleosomes from genes which are potentially active, as well as genes which are actually being transcribed, may be retained on the column (26). This last result is difficult to explain simply by invoking a transient unfolding of the histone octamer as the RNA polymerase passes.

The experiments in which RNA polymerase II failed to traverse nucleosomes efficiently (37, 38) used plasmid DNA templates which were fully reconstituted into nucleosomes (that is, >1 nucleosome per 200 bp of DNA). In light of the results with bacteriophage polymerases (for example, Ref. 24), it might be possible to modify these experiments to increase the chance that RNA polymerase II can efficiently traverse nucleosomes. For example, if efficient elongation requires the displacement of at least some octamers to free DNA, it might be useful to produce arrays of nucleosomes on DNA fragments (as in Ref. 47) and then ligate these fragments downstream of other, non-nucleosomal DNA segments bearing RNA polymerase II promoters (as in Ref. 21). This would provide a potential upstream acceptor for migrating nucleosome cores. It is worth emphasizing that only a few per cent of these templates will be transcribed in a given experiment, which will make it very difficult to determine whether any nucleosomes did, in fact, move in response to transcription. To address this question it would be necessary to isolate the plasmid DNAs which were actually engaged in transcription. A useful tool for this purpose would be a RNA polymerase II-specific antibody (see Ref. 60).

The failure of RNA polymerase II to traverse nucleosomes (37, 38) could be attributed, not to inhibition by the nucleosomes alone, but to the action of histone H1 as well—perhaps through the creation of some higher-order structure in the reconstituted chromatin. Given the crude nature of both the chromatin reconstitution system and the nuclear extracts used to assemble transcription complexes, the authors of Ref. 37 felt that the possible involvement of histone H1 should be addressed. Their approach took advantage of the fact that the polymerase II transcription complex is stable to salt concentrations which remove H1 from chromatin. When reconstituted chromatin templates bearing RNA polymerase II transcription complexes were washed with 650 mM salt (only about 500 mM is needed to extract histone H1 from chromatin), the inhibition of elongation on these templates was unaffected (37). However, the situation *in vivo* is more complex, and the role of H1 remains an open question.

Acknowledgements

We thank fellow authors from this volume for many helpful comments on this manuscript. DSL gratefully acknowledges support from the National Institutes of Health (grant GM29487) for the work from his laboratory.

References

1. Elgin, S. C. R. (1988) The formation and function of DNase I hypersensitive sites in the process of gene activation. *J. Biol. Chem.*, **263,** 19259.
2. Grunstein, M. (1990) Histone function in transcription. *Annu. Rev. Cell Biol.*, **6,** 643.
3. Felsenfeld, G. (1992) Chromatin as an essential part of the transcriptional mechanism. *Nature*, **355,** 219.
4. Williamson, P. and Felsenfeld, G. (1978) Transcription of histone-covered T7 DNA by *Escherichia coli* RNA polymerase. *Biochemistry*, **17,** 5695.
5. Wasylyk, B., Thevenin, G., Oudet, P., and Chambon, P. (1979) Transcription of *in vitro* assembled chromatin by *Escherichia coli* RNA polymerase. *J. Mol. Biol.*, **128,** 411.
6. O'Neill, T. E., Roberge, M., and Bradbury, E. M. (1992) Nucleosome arrays inhibit both initiation and elongation of transcripts by bacteriophage T7 RNA polymerase. *J. Mol. Biol.*, **223,** 67.
7. Kirov, N., Tsaneva, I., Einbinder, E., and Tsanev, R. (1992) *In vitro* transcription through nucleosomes by T7 RNA polymerase. *EMBO J.*, **11,** 1941.
8. Clark, D. and Felsenfeld, G. (1992) A nucleosome core is transferred out of the path of a transcribing polymerase. *Cell*, **71,** 11.
9. Camerini-Otero, R. D. and Felsenfeld, G. (1977) Supercoiling energy and nucleosome formation: The role of the arginine-rich histone kernel. *Nucl. Acids Res.*, **4,** 1159.
10. van Holde, K. E. (1989) *Chromatin*. Springer-Verlag, New York.
11. McGhee, J. D. and Felsenfeld, G. (1980) The number of charge–charge interactions stabilizing the ends of nucleosome DNA. *Nucl. Acids Res.*, **8,** 2751.
12. Stein, A. (1979) DNA folding by histones: the kinetics of chromatin core particle reassembly and the interaction of nucleosomes with histones. *J. Mol. Biol.*, **130,** 103.
13. Ausio, J., Seger, D., and Eisenberg, H. (1984) Nucleosome core particle stability and conformational change. *J. Mol. Biol.*, **176,** 77.
14. Shrader, T. E. and Crothers, D. M. (1989) Artificial nucleosome positioning sequences. *Proc. Natl Acad. Sci. USA*, **86,** 7418.
15. Drew, H. R. (1991) Can one measure the free energy of binding of the histone octamer to different DNA sequences by salt-dependent reconstitution? *J. Mol. Biol.*, **219,** 391.
16. Meersseman, G., Pennings, S., and Bradbury, E. M. (1992) Mobile nucleosomes—a general behavior. *EMBO J.*, **8,** 2951.
17. Liu, L. F. and Wang, J. C. (1987) Supercoiling of the DNA template during transcription. *Proc. Natl Acad. Sci. USA*, **84,** 7024.
18. Pfaffle, P., Gerlach, V., Bunzel, L., and Jackson, V. (1990) *In vitro* evidence that transcription-induced stress causes nucleosome dissolution and regeneration. *J. Biol. Chem.*, **265,** 16830.
19. Clark, D. and Felsenfeld, G. (1991) Formation of nucleosomes on positively supercoiled DNA. *EMBO J.*, **10,** 387.
20. Clark, D. J., Ghirlando, R., Felsenfeld, G., and Eisenberg, H. (1993) Effect of positive supercoiling on DNA compaction by nucleosome cores. *J. Mol. Biol.*, **234,** 297.
21. Lorch, Y., LaPointe, J. W., and Kornberg, R. D. (1987) Nucleosomes inhibit the initiation of transcription but allow chain elongation with the displacement of histones. *Cell*, **49,** 203.
22. Losa, R. and Brown, D. D. (1987) A bacteriophage RNA polymerase transcribes *in vitro* through a nucleosome core without displacing it. *Cell*, **50,** 801.
23. Lorch, Y., LaPointe, J. W., and Kornberg, R. D. (1988) On the displacement of histones from DNA by transcription. *Cell*, **55,** 734.

24. Studitsky, V. M., Clark, D. J., and Felsenfeld, G. (1994) A histone octamer can step around a transcribing polymerase without leaving the template. *Cell*, **76**, 371.
25. O'Neill, T. E., Smith, J. G., and Bradbury, E. M. (1993) Histone octamer dissociation is not required for transcription elongation through arrays of nucleosome cores by phage T7 polymerase *in vitro*. *Proc. Natl Acad. Sci. USA*, **90**, 6203.
26. Chen, T. A., Smith, M. M., Le, S. Y., Sternglanz, R., and Allfrey, V. G. (1991) Nucleosome fractionation by mercury affinity chromatography. Contrasting distribution of transcriptionally active DNA sequences and acetylated histones in nucleosome fractions of wild-type yeast cells and cells expressing a histone H3 gene altered to encode a cysteine 110 residue. *J. Biol. Chem.*, **266**, 6489.
27. Knezetic, J. A., Jacob, G. A., and Luse, D. S. (1988) Assembly of RNA polymerase II preinitiation complexes before assembly of nucleosomes allows efficient initiation of transcription on nucleosomal templates. *Mol. Cell. Biol.*, **8**, 3114.
28. Workman, J. L. and Roeder, R. G. (1987). Binding of transcription factor TFIID to the major late promoter during *in vitro* nucleosome assembly potentiates subsequent initiation by RNA polymerase II. *Cell*, **51**, 613.
29. Gottesfeld, J. M. and Bloomer, L. S. (1982). Assembly of transcriptionally active 5S RNA gene chromatin *in vitro*. *Cell*, **28**, 781.
30. Shimamura, A., Tremethick, D., and Worcel, A. (1988) Characterization of the repressed 5S DNA minichromosomes assembled *in vitro* with a high-speed supernatant of *Xenopus laevis* oocytes. *Mol. Cell. Biol.*, **8**, 4257.
31. Felts, S. J., Weil, P. A., and Chalkley, R. (1990) Transcription factor requirements for *in vitro* formation of transcriptionally competent 5S rRNA gene chromatin. *Mol. Cell. Biol.*, **10**, 2390.
32. Hayes, J. J. and Wolffe, A. P. (1992). Histones H2A/H2B inhibit the interaction of transcription factor-IIIA with the *Xenopus borealis* somatic 5S RNA gene in a nucleosome. *Proc. Natl Acad. Sci. USA*, **89**, 1229.
33. Cai, H. and Luse, D. S. (1987) Variations in template protection by the RNA polymerase II transcription complex during the initiation process. *Mol. Cell. Biol.*, **7**, 3371.
34. Kovelman, R. and Roeder, R. G. (1990) Sarkosyl defines three intermediate steps in transcription initiation by RNA polymerase III: application to stimulation of transcription by E1A. *Genes and Dev.*, **4**, 646.
35. Kadesch, T. R. and Chamberlin, M. J. (1982) Studies of *in vitro* transcription by calf thymus RNA polymerase II using a novel duplex DNA template. *J. Biol. Chem.*, **257**, 5286.
36. Sluder, A. E., Price, D. H., and Greenleaf, A. L. (1988) Elongation by *Drosophila* RNA polymerase II. Transcription of 3′-extended DNA templates. *J. Biol. Chem.*, **263**, 9917.
37. Izban, M. G. and Luse, D. S. (1991). Transcription on nucleosomal templates by RNA polymerase II *in vitro*: inhibition of elongation with enhancement of sequence-specific pausing. *Genes and Dev.*, **5**, 683.
38. Izban, M. G. and Luse, D. S. (1992) Factor-stimulated RNA polymerase-II transcribes at physiological elongation rates on naked DNA but very poorly on chromatin templates. *J. Biol. Chem.*, **267**, 13647
39. Price, D. H., Sluder, A. E., and Greenleaf, A. L. (1989) Dynamic interaction between a *Drosophila* transcription factor and RNA polymerase II. *Mol. Cell. Biol.*, **9**, 1465.
40. Bengal, E., Flores, O., Krauskopf, A., Reinberg, D., and Aloni, Y. (1991) Role of the mammalian transcription factors IIF, IIS, and IIX during elongation by RNA polymerase II. *Mol. Cell. Biol.*, **11**, 1195.

41. Reines, D., Chamberlin, M. J., and Kane, C. M. (1989) Transcription elongation factor SII (TFIIS) enables RNA polymerase II to elongate through a block to transcription in a human gene *in vitro*. *J. Biol. Chem.*, **264,** 10799.
42. Ucker, D. S. and Yamamoto, K. R. (1984) Early events in the stimulation of mammary tumor virus RNA synthesis by glucocorticoids. *J. Biol. Chem.*, **259,** 7416.
43. Reinberg, D. and Zawel, L. (1993) Initiation of transcription by RNA polymerase II: a multi-step process. *Prog. Nucl. Acid. Res. Mol. Biol.*, **44,** 67.
44. Conaway, R. C. and Conaway, J. W. (1993) General initiation factors for RNA polymerase II. *Annu. Rev. Biochem.*, **62,** 161.
45. Woychik, N. A., Liao, S.-M., Kolodziej, P. A., and Young, R. A. (1990). Subunits shared by eukaryotic nuclear RNA polymerases. *Genes and Dev.*, **4,** 313.
46. Sousa, R., Chung, Y. J., Rose, J. P., and Wang, B. C. (1993) Crystal structure of bacteriophage-T7 RNA polymerase at 3.3-Angstrom resolution. *Nature*, **364,** 593.
47. Hansen, J. C. and Wolffe, A. P. (1992) Influence of chromatin folding on transcription initiation and elongation by RNA polymerase-III. *Biochemistry*, **31,** 7977.
48. Morse, R. H. (1989) Nucleosomes inhibit both transcriptional initiation and elongation by RNA polymerase III *in vitro*. *EMBO J.*, **8,** 2343.
49. Reines, D., Ghanouni, P., Li, Q. Q., and Mote, J. (1992) The RNA polymerase-II elongation complex—factor-dependent transcription elongation involves nascent RNA cleavage. *J. Biol. Chem.*, **267,** 15516.
50. Izban, M. G. and Luse, D. S. (1993). The increment of SII-facilitated transcript cleavage varies dramatically between elongation competent and incompetent RNA polymerase-II ternary complexes. *J. Biol. Chem.*, **268,** 12874.
51. Reines, D. and Mote, J. (1993) Elongation factor-SII-dependent transcription by RNA polymerase-II through a sequence-specific DNA-binding protein. *Proc. Natl Acad. Sci. USA*, **90,** 1917.
52. Wang, D. G. and Hawley, D. K. (1993). Identification of a 3′ → 5′ exonuclease activity associated with human RNA polymerase-II. *Proc. Natl Acad. Sci. USA*, **90,** 843.
53. Guo, H. and Price, D. H. (1993) Mechanism of DmS-II-mediated pause suppression by *Drosophila* RNA polymerase II. *J. Biol. Chem.*, **268,** 18762.
54. Greenblatt, J., Nodwell, J. R., and Mason, S. W. (1993) Transcriptional antitermination. *Nature*, **364,** 401.
55. Turner, B. M. (1991) Histone acetylation and control of gene expression. *J. Cell Sci.*, **99,** 13.
56. van Holde, K. (1993) Transcription—the omnipotent nucleosome. *Nature*, **362,** 111.
57. Turner, B. M. (1993) Decoding the nucleosome. *Cell*, **75,** 5.
58. Lee, D. Y., Hayes, J. J., Pruss, D., and Wolffe, A. P. (1993) A positive role for histone acetylation in transcription factor access to nucleosomal DNA. *Cell*, **72,** 73.
59. Batson, S. C., Sundseth, R., Heath, C. V., Samuels, M., and Hansen, U. (1992) *In vitro* initiation of transcription by RNA polymerase-II on *in vivo*-assembled chromatin templates. *Mol. Cell Biol.*, **12,** 1639.
60. Thompson, N. E., Aronson, D. B., and Burgess, R. R. (1990) Purification of eukaryotic RNA polymerase II by immunoaffinity chromatography. Elution of active enzyme with protein stabilizing agents from a polyol-responsive monoclonal antibody. *J. Biol. Chem.*, **265,** 7069.

7 Chromatin structure and epigenetic regulation in yeast

LORRAINE PILLUS and MICHAEL GRUNSTEIN

1. Introduction

Although the genetic information of each cell in a yeast colony is identical, the transcriptional patterns of individual cells within that colony may be very different. These variable, or epigenetic, transcriptional patterns arise at least in part from local differences in chromatin structure. The focus of this chapter is a discussion of how chromatin structure contributes to epigenetic transcriptional states. We begin with a brief description of how chromatin structure contributes to transcriptional regulation in multicellular eukaryotes, particularly the form of regulation known as position effect variegation (a topic considered in depth in Chapter 8). We then move on to consider in greater detail the role of chromatin structure in position effect transcriptional regulation in yeast.

2. Heterochromatin and position effects in multicellular eukaryotes

'Heterochromatin' is a cytogenetic description for chromosomal regions that remain condensed in interphase. These regions are known to contain highly repetitive DNA that is largely transcriptionally inactive, located primarily at centromeres and telomeres, and generally late replicating. Major progress in characterizing heterochromatin has come from its analysis in *Drosophila* (see Chapter 8). The dearth of transcription in heterochromatin is particularly interesting since that characteristic is not self-limited: chromosomal rearrangements in which euchromatic regions have become juxtaposed to heterochromatin result in spreading of transcriptional repression into euchromatic regions near the translocation junctions. The pattern of transcriptional activity near the breakpoints can vary in a clonal fashion, giving rise to variegation of markers that are near the endpoints of the translocations. These position effects have been investigated in detail using genetic approaches. Large numbers of both enhancers and suppressors have been identified in screens for modifiers of position effect variegation (PEV), and it has been hypothesized and demonstrated that some of these modifiers encode structural components of chromatin.

Other major structural components of *Drosophila* heterochromatin are the histones. When PEV is examined in strains with deletions leading to reduced histone gene copy number, one observes that the degree of repression by adjacent heterochromatin is decreased with decreased histone gene dosage (1, 2). Studies with immunological reagents specific for differentially acetylated lysine residues at the amino terminus of histone H4 have shown that different patterns of acetylation are associated with functionally distinct regions of heterochromatin and euchromatin (3, 4). Comparable analysis of modified histones in female mammals reveals a lack of acetylated histone H4 on the inactive X chromosome ((5), Fig. 1). Taken together, the results from *Drosophila* and mammals demonstrate that distinct patterns of H4 acetylation are associated with different chromosomal structural states defining both heterochromatin and transcriptionally active euchromatin. However, it remains to be determined whether different patterns of H4 acetylation are the cause or the consequence of altered transcription patterns in different classes of chromatin.

Position effects on transcription in mammals have also been observed. These effects are seen in transgenic animal studies (6), and have been implicated in human diseases with strong epigenetic or imprinting components, including several pediatric tumours (7, 8). Epigenetic regulation in mammals is discussed in detail in Chapter 10.

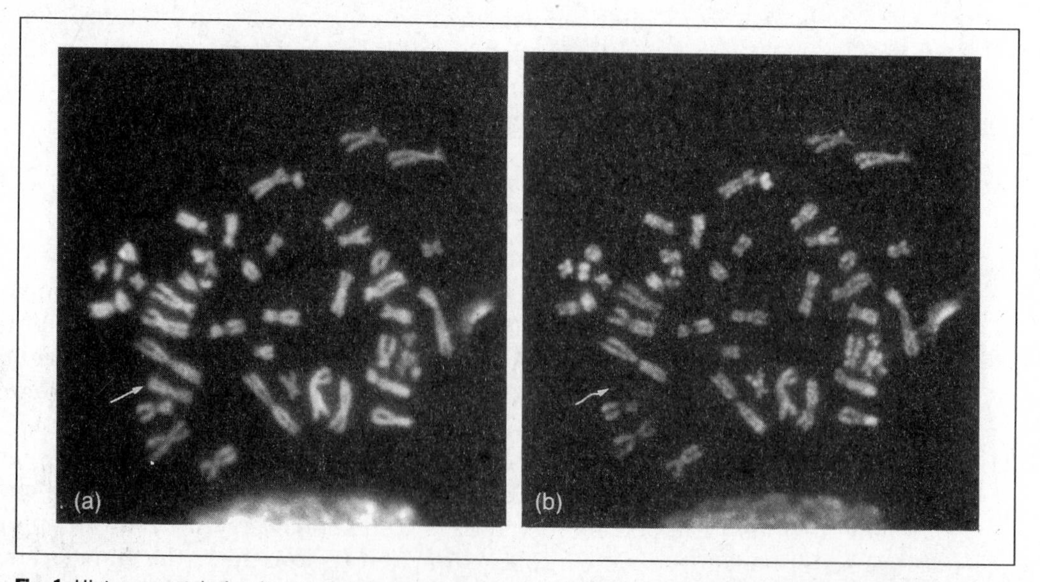

Fig. 1 Histone acetylation is correlated with transcription. In female mammals, the inactive X chromosome is preferentially underacetylated. Metaphase chromosomes from primary cultured human female fibroblasts have been stained with the DNA-binding fluorochrome Hoechst 33342 (a) and with rabbit antibodies to H4 acetylated at lysine$_8$ followed by FITC-conjugated antibodies to rabbit IgG (b). One X chromosome (arrow) is nearly unlabelled with antibodies. Figure provided courtesy of B. M. Turner and N. D. Belyaev.

In addition to the effects of a gene's chromosomal position on its regulation, the absolute position of that chromosomal region within the nucleus may also be crucial to its transcription. Heterochromatin is often observed at the periphery of the nucleus, whereas euchromatin is localized in more interior nuclear space. It will be important to determine how the nuclear membrane may participate in the assembly and/or condensation of chromatin domains. Insight into the interactions between nuclear membranes and chromatin comes from analysis of nuclear remodelling during mitosis. In multicellular eukaryotes, the nuclear envelope and chromatin dissociate early in mitosis. At the end of mitosis, the bilayer nuclear envelope is reformed by the interaction of fusing membranous vesicles with chromatin. Protein factors from both chromatin and the membrane vesicles participate in this nuclear reformation (9, 10). The cycle of assembly and disassembly is regulated by phosphorylation and dephosphorylation of a membrane-bound receptor; this appears to mediate release and binding of the chromatin to the nuclear membrane (11, 12). Heterochromatin binding with or near to the nuclear membrane may be accomplished through comparable activities.

3. Heterochromatin and position effects in yeast

Cytologically defined heterochromatin is not visible in chromosomes of the budding yeast *Saccharomyces*. In part, this may be because chromosomes of *Saccharomyces* are so small, some not significantly larger than the genomes of bacteriophage. Yet many of the features of heterochromatin and position effect characterized in *Drosophila* are directly analogous to those observed at the silent mating-type loci and telomere-proximal genes in yeast (13, 14): these regions of the yeast genome are replicated late in S-phase (15, 16), their associated transcriptional repression is region- rather than gene-specific (17–20), and the repression can spread over the chromosome (19, 21) and is subject to epigenetic and/or variegating patterns of inheritance (19, 22, 23). This spectrum of similarities suggests that many specific features of position effect transcriptional regulation are functionally conserved, even in the absence of cytologically obvious chromosomal condensation or strictly conserved higher order chromatin structures.

3.1 Position effects at silent mating type loci, telomeres, and centromeres

Position effects in yeast were initially identified through study of mating-type regulation (reviewed in 13, 24). The three cell types in *S. cerevisiae* are the **a** and α haploid cells and **a**/α diploids. Haploid cells of the opposite type mate with each other to form diploids after a cell–cell fusion event. Master regulatory genes are expressed from the centromere-proximal *MAT* locus (Fig. 2). Two sets of these genes (**a**), and α) exist: the particular set present and expressed at *MAT* defines both haploid cell identity and cell-type specific functions. In diploids, simultaneous

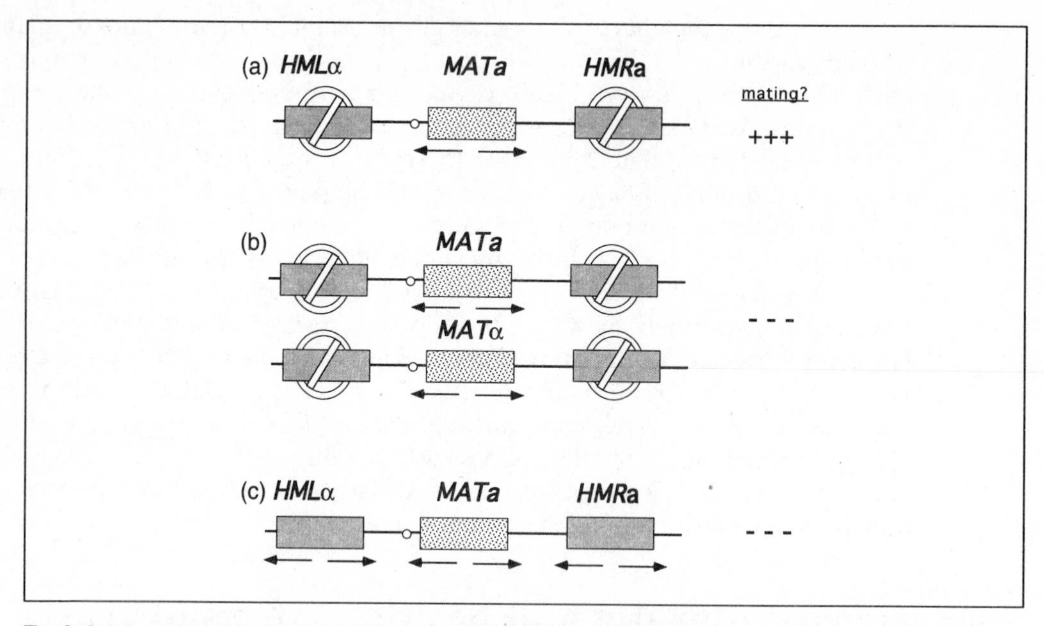

Fig. 2 Cell-type determination in *Saccharomyces*. (a) Wild-type haploid cells transcribe the *MAT* locus. The information present at *MAT*, **a** or α, determines whether a cell is of the **a** or α mating type, and whether it has the capacity to mate with other **a** or α haploid cells. Mating is possible if the information at *HML* and *HMR* is repressed. (b) Diploids express both *MATa* and *MAT*α information simultaneously, because they have two copies of chromosome III—one from which **a** information is transcribed, the other from which α information is transcribed. Diploids do not mate because of the simultaneous expression of both forms of cell-type information. (c) Mutant haploid cells, in which repression at *HML* and *HMR* has been disrupted, do not mate because, like diploids, there is simultaneous expression of both **a** and α information in a single cell.

expression of *MAT***a** and *MAT*α alleles represses haploid-specific cell functions, including the ability to mate. In addition to the transcriptionally active genes at the *MAT* locus, there are silent copies of the **a** and α genes at the *HML*α and *HMR***a** loci. These 'silent mating-type loci' (also called the *HM* loci) are 12 and 23 kb from the left and right telomeres of chromosome III (25). The silent mating-type loci must be repressed in haploids; if these loci are transcribed, the simultaneous expression of **a** and α information leads to sterility, the non-mating phenotype of **a**/α diploids.

Repression of each of the silent mating-type loci is regulated by two flanking silencer elements, termed E and I. These *silencers* (named by analogy to *enhancers*) repress transcription in an orientation and gene-independent manner. The silencers can function on plasmids, with varying degrees of efficiency, and at other regions in the genome (26, 27). The E and I silencers are composed of consensus sequences for autonomously replicating sequences (ARS), and Rap1 and Abf1 protein-binding sites (discussed below and reviewed in 13, 24). The significance of each element for silencer function has been revealed through both mutagenesis and demonstration that a 'synthetic silencer' comprising these three elements functions *in vivo* (28–30).

Position effects exist not merely for transcription at the *HM* loci; they also

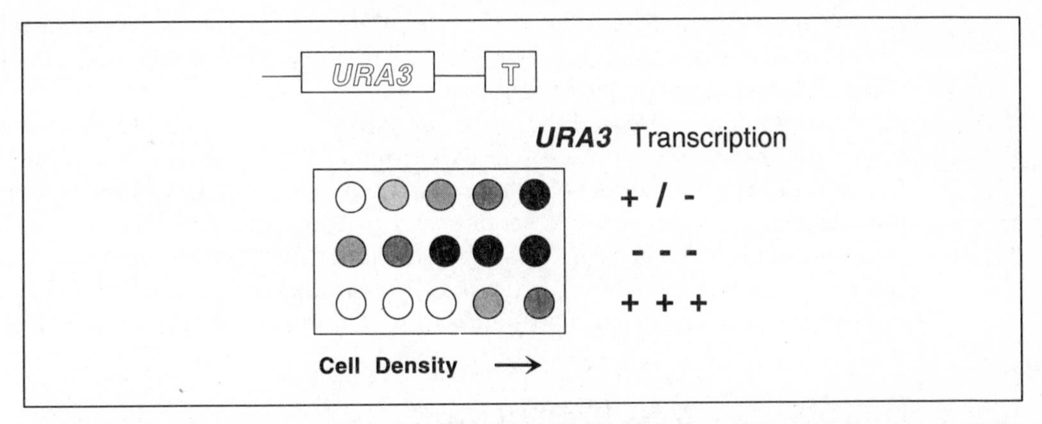

Fig. 3 Telomere-proximal genes are subject to position effects. An assay with the suicide-substrate 5-FOA reveals whether or not the *URA3* locus is transcribed. Cells are plated at increasing density on medium containing 5-FOA. If the *URA3* gene is near a telomere in wild-type cells, it will be variably expressed, giving rise to some growth (top row). If cells are unable to express the *URA3* gene, there will be abundant growth (middle row). If cells are mutated, so that repression is disrupted, the *URA3* gene will be continuously transcribed and there will be virtually no growth on 5-FOA (bottom row).

regulate transcription of genes at telomeres in a very similar manner. The telomeric repeats at the ends of yeast chromosomes contain an average of 350 bp of the C_{1-3} A repeat sequence (reviewed in 31). Adjacent to these repeats may be middle repetitive Y′ (5.2 or 6.7 kb) and X elements (0.3–3.7 kb). Genes experimentally introduced at the ends of chromosomes are repressed with properties similar to repression at *HM* (19, 20). Telomeric repression has been demonstrated by integration of several different genes, for which expression can readily be monitored, at several different chromosome ends. For example, expression of the integrated *URA3* gene is monitored by sensitivity of cells to 5-fluoroorotic acid (5-FOA). Transcription of *URA3* kills cells in the presence of 5-FOA due to its conversion to the toxic compound 5-fluorouracil (32). When *URA3* is present near the telomeric repeats it is strongly repressed, thereby allowing growth on 5-FOA. In contrast, when *URA3* is present at an internal chromosomal locus it is expressed, and cells are sensitive to 5-FOA (Fig. 3). Position effects can spread from the end of the chromosome for 3–4 kb. The absolute distance spread and degree of repression are influenced by many of the *trans*-acting factors that also regulate *HM* silencing (21). Moreover, the position effect at telomeres is completely dependent on the C_{1-3} A repeat, but not the X and Y′ repetitive sequences. These internal repeats may, however, serve to extend the region over which telomeric repression is propagated. Likewise, the C_{1-3} A repeat, although necessary, may not be sufficient for telomeric repression, since short stretches of the repeat do not repress when present at internal chromosomal sites (19, 21); however, it is possible that longer stretches (see Discussion, p. 140) may repress when present internally.

In *Schizosaccharomyces pombe*, transcriptional repression at analogous silent mating-type loci has been observed, and progress is being made in defining the *cis*- and *trans*-elements involved in this regulation (e.g., see 33, 34). It has also been demonstrated that position effects similar to those observed at the *HM* loci and telomeres operate at *S. pombe* centromeres (35). *S. pombe* centromeres are significantly larger than those of *S. cerevisiae*, and even include multiple genes for tRNAs (reviewed in 36, 37). To date, no experiments have been published to indicate centromeric position effects in *S. cerevisiae*, although it has been observed that transcription into centromeric sequences can compromise their function (38, 39). Thus although there is clear functional similarity between silencing in *S. cerevisiae* and the position effects observed at *S. pombe* centromeres, it is not yet clear to what extent the role of chromatin structure in mediating these effects is analogous.

3.2 Factors involved in silencing

Many *trans*-acting factors have been identified that are involved in yeast silencing (Fig. 4; reviewed in 13, 24). The four *SIR* (silent information regulator) genes encode proteins that function at the *HM* loci, apparently without direct DNA binding. Sir2, Sir3, and Sir4 proteins are also required for telomeric silencing: null mutations in any one of these genes results in a complete loss of repression. In contrast, Sir1p is not essential for silencing at *HM* and appears to have no function at the telomeres. Exactly how the Sir proteins function in silencing is not yet known. They are enriched in nuclear fractions, and Sir3 and Sir4 appear to localize in perinuclear foci when cells are examined by confocal immunofluorescence microscopy (40). The perinuclear localization of Sir3, Sir4, and Rap1 (see below) is reminiscent of the localization of heterochromatin in multicellular eukaryotes. In

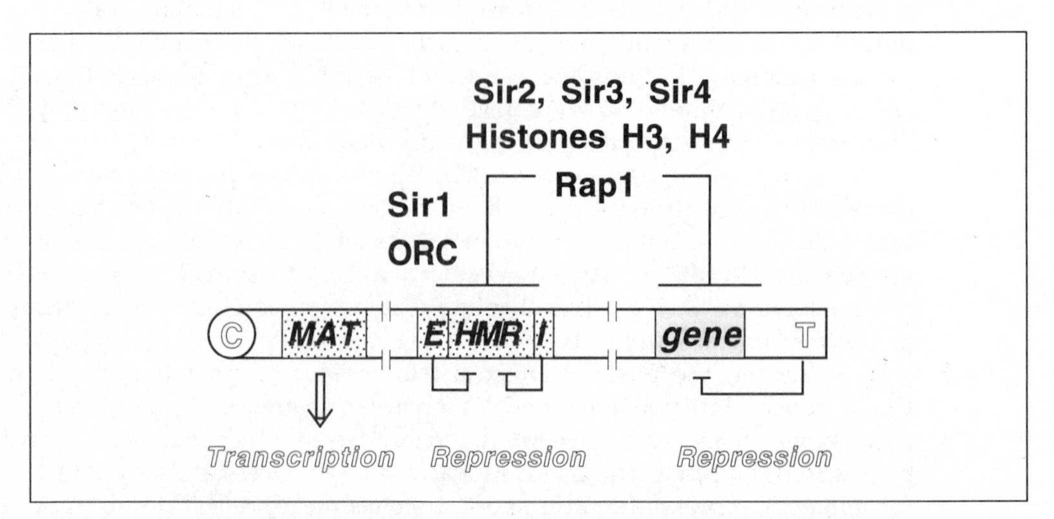

Fig. 4 Factors involved in yeast silencing. In this cartoon of the right arm of chromosome III, many factors are shown to act at both the silent mating-type locus and the telomere to mediate silencing. Some factors, like Sir1 and the ORC, may act only at internal loci and not at the telomeres. In addition to the factors shown in the diagram, a number of others may be involved. These include Nat1/Ard1 acetyltransferase, Abf1, Cdc7, Sum1, 2, 3, San1, and Rif1 (see text and (13) for a more complete discussion of these additional factors).

sir3 and *sir4* mutant strains, perinuclear localization is lost, suggesting that these proteins may have a role in maintaining this form of higher order nuclear structure (40).

In addition to the Sir proteins, the highly abundant, essential proteins Rap1 and Abf1 are involved in silencing. Rap1p binds sequences in the silencers (41, 42) and at the telomeric C_{1-3} A repeats (43). Rap1p localizes to the ends of pachytene-arrested meiotic chromosomes (44) and this localization suggests that the perinuclear foci described above may represent clustering of Rap1p-bound telomeres. Studies with mutated silencer elements also reveal a role for Rap1p in silencing *in vivo* (45–47). A C-terminal domain of Rap1p has been identified with specific functions in silencing (48) that may be distinguished from the roles of Rap1p in cell viability and transcriptional activation. Mutations in this domain cause a loss of repression at both *HM* loci and telomeres. Rap1p has been shown by the two-hybrid system to interact with Rif1p (Rap1 interacting factor), which is also required for repression at the *HM* loci (49). Evidence for the role of Abf1p (which shares extensive sequence similarity with Rap1p) is based largely on the decreased repression observed when its consensus binding site is deleted in the *HMR* synthetic silencer (30). There is no evidence demonstrating a role for Abf1p in telomeric repression. Other factors have been identified genetically (Fig. 3) to have a role in silencing, but it is not yet clear how these factors function. For example, *NAT1* and *ARD1* products, subunits of an N-terminal acetyltransferase activity (50, 51) may affect silencing strictly through co-translational modification of substrate protein(s) important for silencing.

Proteins of the origin recognition complex (ORC), a heteromeric complex that binds to chromosomal origins of replication (52) also appear to function in silencing (see below) (53–56). Histone H4 has a well-documented role in silencing, focusing attention on the direct participation of chromatin in this form of transcriptional regulation (57–61). Recent evidence suggests that H3 is also involved (62). The role of histones and regulation of their function are considered in greater detail below.

4. Yeast silencing has many characteristics of PEV

Silencing at yeast *HM* and telomere-proximal loci has striking similarities to the phenomenon of position effect variegation (PEV) in *Drosophila* (see Chapter 8). For example, silencing can influence transcription of a variety of polymerase II transcribed genes, as well as a polymerase III transcribed tRNA gene (17–19, 47). Also, silencing in yeast is inherited in an epigenetic manner. Repression of telomere-proximal genes can be monitored by looking at the expression of an integrated *ADE2* gene: in the absence of *ADE2* expression, yeast colonies are red, but when it is expressed they are white. At telomeric loci, the *ADE2* gene may be largely repressed, giving rise to mostly red colonies. However, within these colonies distinct white sectors are observed, indicating that in some cells of the colony, the *ADE2* gene has been switched into a transcriptionally activated state (19). Similar epigenetic transcriptional states at the *HM* loci are observed in *sir1* mutants (22) or in strains in which the silencers themselves have been partially mutated (23, 47). Thus, in these situations, cells are observed in two distinct populations: those that are transcriptionally active, and those that are repressed. The two distinct transcriptional states are both stable and heritable through many, many rounds of mitotic division. Only low levels of switching are observed between the two transcriptional states. Both of these on/off patterns, at the telomeres and at the *HM* loci, are similar to the patched patterns of eye pigmentation observed in *Drosophila* when the *white* gene is subject to PEV.

Another hallmark of PEV is that heterochromatic regions tend to undergo DNA replication late in S-phase. This pattern of late replication is also observed for the *HM* and telomeric loci in yeast (15, 16). Indeed, the replication timing of a specific origin of replication is directly influenced by its chromosomal location in yeast (63). A particular origin that normally replicates early in S-phase will replicate late in S-phase if it is translocated to a telomere-proximal region. These observations suggest that position effects are not limited to transcriptional regulation, but that the mechanism that inhibits access of transcription factors to particular regions of DNA may also limit access of replication factors. It is, of course, possible that late or delayed replication may function in the establishment of the silenced state. It has been shown that at least one of the ARS silencers functions as a required component of a chromosomal origin of replication (64). Furthermore, at least one round of replication is required to restore repression to a conditionally derepressed silent mating locus (65, 66). Other connections between replication and silencing are discussed below (Important questions).

5. A specialized histone-dependent chromatin structure is important for yeast silencing

5.1 Evidence for a specialized chromatin structure

A traditional method used in the analysis of chromatin structure is the digestion of chromatin with nucleases whose cleavage is influenced by the binding of proteins, particularly histones. Micrococcal nuclease and DNase I are used most commonly; both give characteristic cleavage patterns in the presence of nucleosomes. Data obtained from cleavage studies with both enzymes reveal that there are differences in the chromatin structure at the repressed *HM* loci compared with identical sequences at the transcribed *MAT* locus (67). Hypersensitive cleavage sites at the *MAT* locus are variably hypersensitive at the *HM* loci, whereas additional hypersensitive sites appear at the *HM* loci that do not appear at *MAT*. Derepression of the *HM* loci in a *sir* mutant background alters the cleavage pattern at the silent cassettes to resemble the pattern at *MAT*. These results indicate that repression is associated with a special chromatin structure.

Sequence-specific nucleases have also been used to demonstrate repression-specific alterations in chromatin structure at *HM* relative to the transcribed *MAT* locus (68–70). The silent loci in wild-type cells are resistant to DNA cleavage by the endogenously expressed HO endonuclease. Ordinarily, HO catalyses a double-stranded break at the *MAT* locus, one step in the switching of mating type in yeast cells. Although an identical HO recognition site is present at the silent mating-type loci, HO is unable to cleave these sites, except in *sir* mutant strains (68, 69). These effects have also been observed and extended in *in vitro* assays. In these experiments, nuclei were isolated such that the repressed chromatin state was preserved. It was shown that silencing results in the inaccessibility of *HMR* to HO and to many other sequence-specific protein–DNA interactions. Further, this structural inaccessibility is transcription independent: deleting the *HMR* promoters does not alter enzyme sensitivity in the region. The structurally distinct chromatin domains were shown to extend both between and beyond the silencer elements (70).

Unlike the precisely positioned nucleosomes observed across the silent mating-type loci, the simple telomeric sequence repeats at the extreme termini of yeast chromosomes appear to be devoid of nucleosomes (71). This contrasts with mammalian telomeres, which appear to be nucleosome bound (72). Yeast telomeric repeat sequences are bound by protein complexes, referred to as telosomes, that give distinct nuclease cleavage patterns when compared with regularly phased nucleosomes. However, nucleosomes do appear positioned across the X and Y′ repeat regions. Surprisingly, the nuclease cleavage pattern across a telomere-proximal *URA3* gene does not appear to differ from the pattern observed at its endogenous chromosomal locus (71). It is possible, however, that higher order chromatin structures may be present at these loci yet may be disrupted during the chromatin isolation procedures required for nuclease analysis, or that such higher order chromatin structures are not detected efficiently by these nucleases.

Analysis of chromatin by mercury affinity chromatography also distinguishes between transcriptionally active and inactive regions (73). As discussed in Chapter 6, analysis of chromatin from specially engineered yeast strains (carrying an H3 gene that includes a cysteine at aa 110) revealed that active regions of the genome are found in a column-bound fraction (indicating an open chromatin conformation) even if the gene was not being actively transcribed. Sequences from the silent mating-type loci were enriched in unbound fractions, indicating that these regions were in a closed, inaccessible conformation. If the fractionation was performed on chromatin isolated from *sir3* mutant strains, the silent mating-type loci sequences were shifted to the bound, accessible fraction (74). Although it is not yet clear precisely what structural differences exist between open/bound and closed/unbound regions, these data indicate that transcriptionally repressed regions possess structurally distinct forms of chromatin at the nucleosome level.

There is no evidence that methylation occurs naturally in yeast DNA; in its absence, introduction of genes for prokaryotic methyl transferases into yeast has proved to be a useful tool for analysing different chromatin structures (75, 76). Site-specific methylation is analysed by digestion of genomic DNA with different restriction enzyme isoschizomers that cleave either methylated or unmethylated recognition sequences. It has been shown that the repressed *HMR* locus is in a predominantly methylase-refractory state, whereas the *MAT* locus is accessible to methylation (75). Comparable analysis of *sir* mutants revealed increased accessibility at the silent mating-type loci. Likewise, a single methylation-sensitive site in a telomere-proximal *URA3* gene was inaccessible to modification when repressed, but became accessible after derepression (76). Inhibition of enzyme accessibility has also been observed at the silent mating-type loci for enzymes involved in UV radiation repair (77, 78).

Antibodies that specifically recognize acetylated forms of histone H4 have been utilized to determine whether acetylation of histones is involved in transcriptional regulation (79). In yeast it has been shown that the *HM* loci and subtelomeric Y′ sequences are unreactive to these antibodies in wild-type repressed strains, but are reactive in derepressed *sir* mutant strains (80). It was further demonstrated that this differential reactivity was unlikely to be due to transcription from the *HM* loci, because elimination of the promoters at the silent loci did not affect antibody reactivity in the *sir* mutants. Although it was concluded that the differential recognition was due to silenced regions being packaged with hypoacetylated histones, it remains a formal possibility that the failure of the antibodies to recognize silenced regions is due in part to a specialized chromatin structure that inhibits antibody accessibility, comparable to that observed for HO, restriction endonucleases, methyl transferases, and repair enzymes.

In sum, silenced regions of the yeast genome have chromatin structures that are different from their active counterparts. These differences in structure have been revealed by nucleases, thiol reagents, DNA-modifying enzymes, and antibodies against the acetylated H4 N-terminus. These differences may arise, in part, owing to gene activity, but probably also result from Sir proteins and other proteins that may participate in the formation of chromatin structures that prevent transcription factor and RNA polymerase access to promoter sequences.

5.2 Heterochromatin structure is dependent on histones H4 and H3

In addition to the structural analysis of yeast chromatin, genetic analysis has allowed detailed study of the histone proteins themselves. Such studies have revealed that the histones, especially the N-terminal tails of H4 and H3, have important roles in the formation of repressive chromatin.

5.2.1 Histone H4

Histone molecules are composed of two primary domains: a hydrophilic N-terminal domain, and a globular hydrophobic domain (see Chapter 1). Histone H4 residues 1–29 form the hydrophilic N-terminus that extends from the nucleosomal core and is relatively unstructured (reviewed in 81–83). Surprisingly, this region, which is highly evolutionarily conserved and contains conserved sites for acetylation (at positions 5, 8, 12, and 16) and phosphorylation (at position 18) is not essential for viability (57). Residues 33–102 form the hydrophobic core, which contains a number of α-helical regions. This domain, essential for viability (57), functions in histone–histone interactions and the histone–DNA interactions necessary for assembly and stability of nucleosomes (81–83).

Deletion and point mutation analysis of histone H4 has revealed the presence of two domains in the N-terminus that are required for repression at the silent mating-type loci (61). The R1 domain is a basic region spanning residues 16–19 (58, 84) and the R2 domain encompasses the non-basic residues 21–29 (61). Substitutions in either domain lead to derepression at *HML* and at telomeres (Fig. 5 and (20, 62)). The R1 domain appears most sensitive to mutations that alter charge. Conservative substitutions that maintain the positive charge produce significantly weaker silencing defects than do non-conservative neutrally charged substitutions. In contrast, the R2 domain is strongly affected by proline substitutions, which suggests that R2 may exist with other H4 sequences in a helix whose function is disrupted by these presumptive helix-breaking residues. Mating defects caused by substitutions in R2 can be suppressed by *sir3* mutants that also suppress mutations in the R1 domain (58, 61). Although the H4 N-terminus is thought to be unstructured, R1 and R2 can theoretically form an amphipathic α-helix that might exist in the presence of an interacting protein (perhaps Sir3p) or in the presence of bound DNA (61).

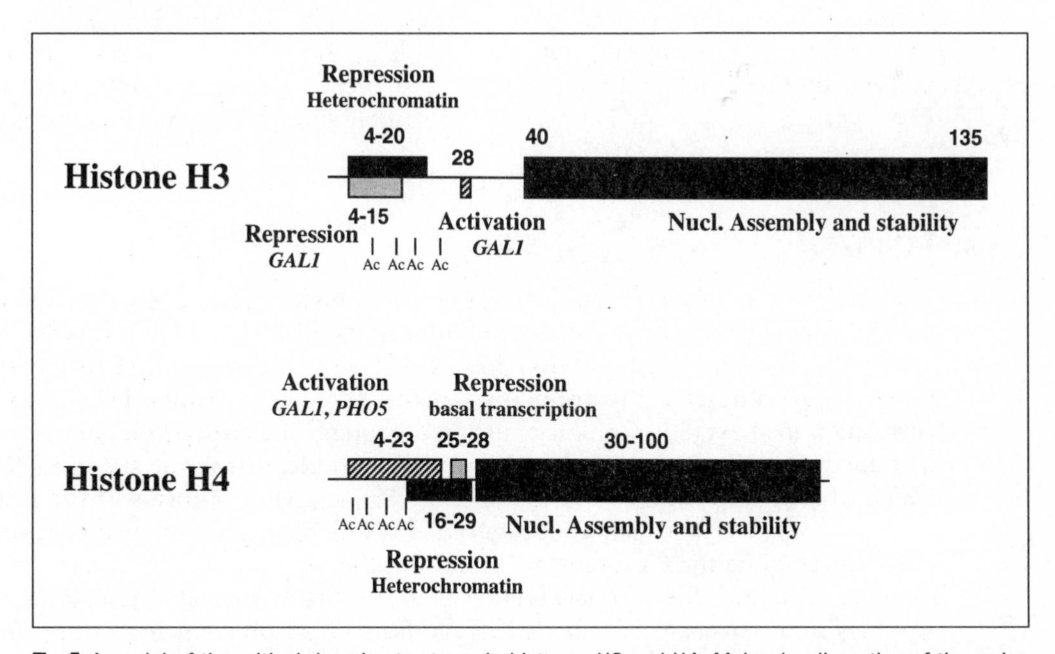

Fig. 5 A model of the critical domain structures in histones H3 and H4. Molecular dissection of the amino terminal regions of both proteins reveals residues important for silencing ('repression heterochromatin') as well as those that function in other forms of transcriptional repression and activation (see text and (62, 89) for details).

5.2.2 Histone H3

Histones H3 and H4 share a number of features that suggest that H3 might also be involved in transcriptional repression in yeast. Both proteins are nearly invariant between plants and animals, and both are modified similarly by acetylation. H3 and H4 form the tetrameric subunit upon which H2A and H2B dimers assemble to form the complete octameric nucleosome. *In vitro* reconstitution experiments suggest that the H3–H4 tetramer has structural properties similar to the complete nucleosome octamer, including the ability to position itself over a defined DNA sequence and to block transcription factor access at a promoter (see Chapter 2 and 85).

In contrast to H4, deletion of the H3 N-terminus produced only minor derepression at the silent mating-type loci (86). Deletion of residues 4–35 decreased mating efficiency of *MAT***a** strains about eight-fold (measuring derepression of *HML*). Deletion of residues 4–40 decreased mating efficiency about a hundred-fold. (Derepression due to mutations in H4 typically results in decreased mating efficiency of approximately six orders of magnitude.) These relatively minor mating defects suggested a negligible role for H3 in repression at the silent mating-type loci. Analysis of telomeric repression, however, revealed a significant role for H3. Utilizing the *URA3* 5-FOA assay (see Fig. 3), it was observed that N-terminal H3 deletions result in marked sensitivity to 5-FOA (62), indicating significant derepression of *URA3* at the telomere. A minimal region of residues 4–20 has been defined that is required for telomeric repression (Fig. 5). Thus, H3 has a clear role in telomeric repression and may have a role at the *HM* loci that has so far been masked by the functional redundancy of silencing elements at *HML* and *HMR*. However, analyses of the H3 mutants in combination with other mutations show clear effects on silencing (62).

5.3 Histone acetylation in silencing

Acetylation of amino terminal lysine residues in histones has been correlated with transcriptional activity in *Drosophila* and mammals (3, 87–89): high levels are associated with transcriptional activity, low levels with transcriptional inactivity. As noted above, studies using antibodies against acetylated histone H4 support this correlation in yeast (80). Genetic analyses also suggest a critical role for acetylation in transcriptional control, and a particularly significant role in silencing for the lysine residue at position 16 in histone H4 (58–60). Mutant forms of the histones have been created in which amino terminal lysine residues have been substituted with either glutamine or glycine to mimic the hyperacetylated state, or with arginine to mimic the hypoacetylated state. Substitutions at acetylatable lysine residues at positions 5, 8, and/or 12 have little effect on silencing at the *HM* loci or telomeres (58, 62). In contrast, strong derepression is observed if $lysine_{16}$ is substituted by glycine or glutamine. Little effect is observed, however, when

$lysine_{16}$ is changed to arginine. All three of these substitutions cause increased sensitivity to 5-FOA when telomere-proximal *URA3* expression is measured. A positive charge at position 16 thus correlates well with transcriptional repression, and acetylation at this site may lead to derepression. Taken together, these results point to the importance of $lysine_{16}$ in H4. It should be noted that it is not known what structural consequences these substitutions may have on the N-terminus—consequences that may be independent of direct charge-associated effects.

Analysis of histone H3 reveals somewhat different results with lysine substitutions to those seen with H4. When comparable substitutions were created it was observed that a decrease in repression was correlated with an increase in the number of mutated lysine residues (62). Although single lysine substitutions had little effect, multiple mutations (at amino acid positions 9, 14, and 18, or 9, 14, 18, and 23) caused increased levels of expression of the telomere-proximal *URA3* gene. As in the H4 experiments, non-conservative substitutions to glycine yielded higher derepression than corresponding substitutions to arginine. This again emphasizes that hypoacetylation of histones may be required for the repressed state.

Although no histone acetylases or deacetylases have yet been identified in yeast, it has been hypothesized that Sir2p may function in histone deacetylation. Overexpression of Sir2p results in decreased acetylation of several types of histones (80). It will be important to determine whether Sir2p is an actual deacetylase, or functions to inhibit histone acetylation or acts indirectly otherwise.

Histone acetylation/deacetylation provides one mechanistic explanation for the role of the amino terminal tails in the formation of heterochromatin. As DNA is wrapped around the histone octamer, it undergoes a sharp bend or kink at a defined distance from the dyad axis. H4 amino acid $histidine_{18}$ may be cross-linked nearby, suggesting that the R1 domain may be involved in stabilizing this bend (90). This part of the H4 tail is also necessary for nucleosome positioning by the $\alpha 2$ repressor (91); it is possible that modification of residues close to this region may dissociate or otherwise weaken electrostatic interactions between the H4 tail and nucleosomal DNA, thus altering nucleosome positioning. It is also possible that alteration of these potential electrostatic interactions may influence the degree to which the DNA is wrapped around the nucleosome, which might in turn affect the ability of the 10 nm chromatin fibre ultimately to condense into higher order structures (see Chapter 1).

6. Long-range position effects

The position effects in yeast described thus far, although defining regional rather than gene-specific control, extend only over distances of several kilobases. This is relatively short when compared with position effects in *Drosophila*, which may extend as far as multiple megabases. Several issues may be relevant to this apparent difference, including limitation of silencing factors and special contributions that the telomere may make to silencing in yeast.

6.1 Sir3p can promote long-range silencing and may be limiting in cells

One possibility for the apparent difference in extent of silencing between yeast and flies is that silencing factors in yeast may quantitatively limit the amount of silenced chromatin assembled in the cell. Evidence supporting this idea comes from the observation that if the gene dosage of *SIR3* is increased, the distance over which telomeres can impose position effects is also increased (21). Telomeric position effects typically act up to 3–4 kb from the end of the chromosome. If *SIR3* gene dosage is increased with high copy plasmids, position effects may be detected more than 20 kb away from the end of the chromosome. This suggests that Sir3p may not only be limiting for silencing in cells, but that it may also have a direct structural role in the formation of repressive chromatin. This idea is consistent with the immunofluorescence localization studies described above (40). It will be important to determine whether the increase in repression observed with increased *SIR3* gene dosage reflects an increase in the continuous assembly and/or involves propagation of repressed chromatin from the telomere along the chromosome. This question is motivated in part by the observation that the *SIR3* gene maps to a telomere-proximal position on chromosome XII. This location may indicate that repression of *SIR3* transcription itself provides a form of feedback regulation that may limit the extent of heterochromatic repression at the ends of yeast chromosomes (21).

6.2 Telomere proximity to *HMR* may have a role in long-range silencing

It has recently been discovered that some position effects may occur over long ranges in yeast under normal conditions. This has been revealed in studies that focus on *HMR* and its telomere 23 kb away (27). The silent mating-type locus *HMR* appears to have the strongest silencer in the chromosome. In particular, it is resistant to a variety of derepressing mutations, including those in the N-terminus of H4 described above. However, one factor that does influence the extent of repression at *HMR* is the proximity of the telomere on chromosome III. If *HMR* is moved far from the telomere, it is derepressed by the same H4 mutations that caused no phenotype in its normal location. Thus proximity to the telomere may make *HMR* a stronger silencer. Interestingly, this effect is not specific to the chromosome III telomere; if *HMR* is integrated adjacent to a telomere on chromosome VII, its silencing is entirely immune to the H4 mutations. *HML* is apparently not strongly affected by its chromosomal context; thus the telomeric effect observed appears specific for *HMR*. Although the mechanism of the effect of the telomere on *HMR* is not yet understood, these results do indicate that some forms of position effect in yeast may extend over greater distances than those originally monitored with telomere proximal gene constructs.

6.3 Long-range effects in *S. pombe*

The position effects observed at the *S. pombe* silent mating-type loci and centromeres are somewhat more extensive than those observed under most conditions in *S. cerevisiae*. Whether this is a function of greater steady-state levels of silencing proteins, different chromosomal and chromatin organization, or some other factor(s) is not yet known (35).

7. Final comments

Although significant progress has been made in identifying many of the molecules important in yeast silencing, we do not yet have a detailed mechanistic understanding of this form of repression. Among the important questions to consider are: which proteins are present in the repressed chromatin? How do they become assembled in that state? The answers to these questions will certainly involve the biochemical definition of protein–protein interactions in silenced complexes, as well as a dissection of the processes of establishment and maintenance of repressed chromatin. In particular, recent progress made in understanding the control of DNA replication may lead to new insights into mechanistic connections between replication and silencing.

7.1 Which proteins interact to form repressed chromatin?

The proteins known to function in silencing (Fig. 4) were identified through a combination of genetic and biochemical studies. Ideas about how repressed chromatin may be assembled are based on some of the known interactions between these silencing factors. The use of the two-hybrid system has provided additional support for some potential silencing factor interactions. For example, using the two-hybrid system, Rap1p appears to interact with both Sir3p and Sir4p. Sir3p and Sir4p also appear capable of interacting with one another (D. Shore and S. Gasser, unpublished results). These results complement previous genetic data suggesting Sir3p–Sir4p interactions (92, 93). The identification of suppressor mutations in *SIR3* that restore repression in H4 mutant strains also suggests potential physical interaction between these two proteins (58). New data suggest that H3 and H4 interact directly with both Sir3p and Sir4p *in vitro* (A. Hecht and M. Grunstein, unpublished data).

Some morphological data also address questions about protein–protein interactions. As discussed above (Sections 2 and 3) it appears that repressed regions of the genome may interact with the nuclear periphery. In multicellular organisms, heterochromatic regions are found almost exclusively at or near the nuclear membrane. Yeast telomeres also appear to localize to distinct regions of the nuclear periphery in clusters (44). Furthermore, this peripheral localization is dependent on *SIR3*, *SIR4*, H3, and H4 function (40; A. Hecht, M. Grunstein, and S. Gasser, unpublished results). It is possible that Sir4p may play a direct role in the association of the telomeres with the nuclear membrane. This idea is based, at least in part, on the predicted coiled-coil structure of the C-terminus of Sir4p, its ability to self-associate through the C-terminus, and its sequence similarity to human nuclear lamin proteins (94, 95).

An attractive idea is that many of the silencing proteins interact to form a multi-subunit complex that is tethered to the nuclear periphery via association with the Sir4 protein or a Sir3p–Sir4p complex (Fig. 6). Rap1p, which binds to many regions in the genome, including binding sites at the silencers and the telomeric repeat sequences, might serve to recruit Sir3p and Sir4p, or in some fashion seed assembly of silent chromatin. Rap1p might then function by first recruiting and then distributing Sir3p and Sir4p to adjacently bound nucleosomes. Sir3p might then facilitate chromatin compaction that is dependent on local regions of hypoacetylated histones to allow tight association of their N-termini with adjacent DNA. Other reversible weak interactions may be critically important for the overall assembly and integrity of the silencing complex. These entire nucleosome–Sir3p–Sir4p–Rap1p complexes might be anchored at the nuclear periphery by Sir4p. Because Sir4p is phosphorylated (96) it is possible that regulation of this modification may influence both regulation of complex assembly and positioning. Distinct roles for other proteins known to function in silencing are more vague. For example, Sir2p may function critically at some level in histone acetylation, or it may interact more directly in the silencing protein complex.

It seems likely that silencing is mediated by a variety of complex interactions

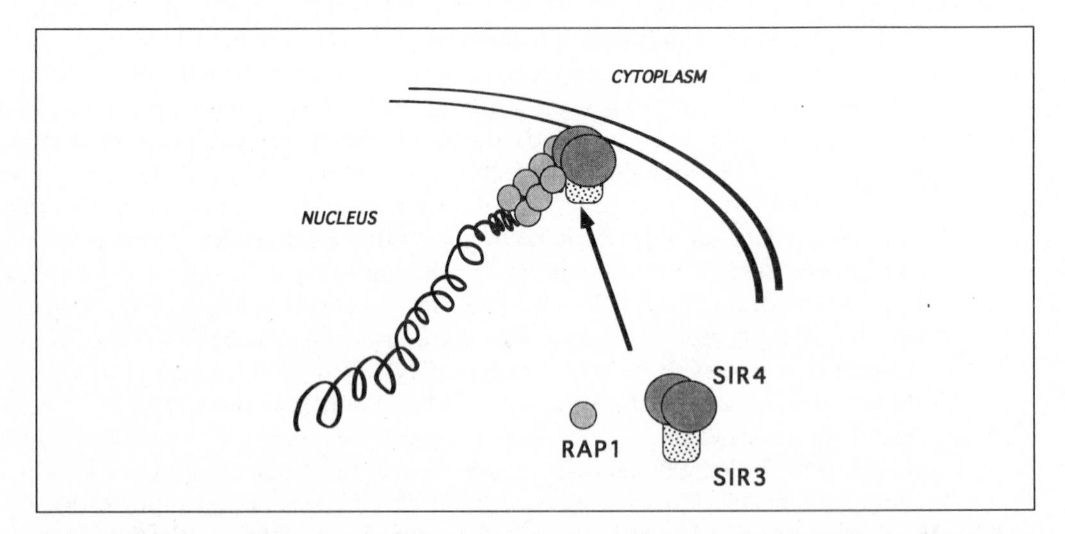

Fig. 6 A model for the assembly of silenced chromatin in the nucleus. The dark line represents nucleosome-bound chromosomal DNA. Proteins such as Sir3, Sir4, and Rap1 may interact with each other to facilitate both DNA binding and assembly of repressed complexes that are specifically localized to the periphery of the nucleus. Other specific protein–protein interactions are certainly involved, including those involving N-terminal tails of the histone H3 and H4 molecules in nucleosomes. Likewise, there may be other distinct assembly factors (perhaps Sir1) that function specifically in the formation of the repressed chromatin.

between Sir proteins, nucleosomes, and the nuclear periphery. These interactions ultimately lead to the formation of a condensed chromatin structure that prevents access of many DNA-binding proteins, including RNA polymerase and accessory factors. The post-translational modification or quantitative limitation of proteins provide potentially important mechanisms for regulating both the state and extent of repression.

7.2 Which factors establish silencing? Does DNA replication function in defining silenced chromatin?

The model suggested above describes molecular interactions that may exist in stably silenced complexes at the telomeres and silent mating-type loci. How is this complex assembled? Are there special assembly factors that recruit proteins to the complex and are then no longer necessary for its maintenance? Are there cell-cycle-specific signals necessary for the assembly and/or maintenance of silenced complexes?

Of the proteins known to be involved in silencing (Fig. 4), Sir1p and the proteins of ORC may have special functions in assembling repressed complexes. Single-cell experiments defined a clear role for *SIR1* in the establishment of silencing at *HML* and showed that there was absolutely no requirement for Sir1p in the maintenance of repressed complexes after they were formed (22). Sir1p is almost certainly not the only protein involved in establishment and, in fact, region-specific establishment functions may exist, since Sir1p appears to have no major role in telomeric repression (20). Exactly how Sir1p and other proteins may function to establish silencing is not yet known, but some process akin to the facilitated assembly seen in phage morphogenesis or chaperone function is attractive. Support for this idea comes from experiments in which *SIR1* is fused to a *GAL4* DNA-binding domain and the target for that domain is substituted for chromosomal silencer elements (95). In this case, there is a unique dependence for the *SIR1* fusion product in silencing. This suggests that *SIR1* has the capacity to recruit other silencing proteins to a targeted DNA domain. Likewise, targeting *SIR1* to the telomeric domains improves repression, perhaps by improving the efficiency of complex assembly. An alternative possibility is that *in vivo* other silencer-bound proteins recruit Sir1p to repressed complexes, but this seems less likely since complexes may be stably repressed in the absence of Sir1p for up to 20 generations (22).

If Sir1p and other proteins facilitate assembly of the silenced chromatin complexes, when do the assembly reactions occur? One attractive possibility is that DNA replication either allows or promotes assembly. This idea derives support from a series of experiments. As noted above (Section 4), experiments with a conditionally derepressed silent mating locus revealed that at least one passage through S-phase is necessary to restore repression, although repression can be disrupted at any point in the cell cycle (65). Each of the silencer elements contain ARS consensus sequences, at least one of which functions as a chromosomal origin of replication and whose function is necessary in a synthetic silencer (30, 64). The six-subunit ORC (52) not only binds the ARS elements in silencers, but also is necessary for repression. Mutations in the 80 kDa subunit Orc2p derepress the silent mating-type loci and also cause structural defects in both the complex and DNA replication (53, 54, 56).

The molecular connection between replication and silencing is intriguing, yet many hard questions remain. For example, is the function of ORC in silencing dependent or independent of replication? It is possible that ORC has independent functions in both silencing and replication. Perhaps there are subunit exchanges that promote replication activity, and this activity is distinct from silencing. Alternatively, it is possible that firing of an origin of replication, even infrequently, may be necessary to allow remodelling of chromatin and assembly of transcriptionally repressed domains. After formation, the repressed complex might be maintained with high fidelity through many rounds of cell division as a result of a self-templated assembly. Only if the repressed complex becomes damaged or otherwise destabilized would there again be a local requirement for replication initiation.

7.3 Discussion

Several interesting points were raised in discussion with other authors. There appears to be a clear hierarchy of silencing—*HMR* is more strongly silenced than *HML* than are the telomeres. This should be reflected in any model of silencing. The silent mating type loci appear to have the greatest number of genetically redundant elements, ensuring the stability of their silencing. Some of the silencing proteins may be stoichiometrically limiting in the cell, enabling regions with higher affinity for these proteins to be more readily (or more stably) silenced. Recent studies have shown that long, but not short, clusters of telomere repeats can cause repression at internal sites (97), suggesting that cooperative interactions may be important in silencing.

The relationship between the *SPT/SIN* class of genes and the *SNF/SWI* class of genes in yeast is reminiscent of the antipodal effects of the Pc-G and trx-G proteins in *Drosophila* (Chapter 8). One might be tempted to consider whether or not these gene products play a role in epigenetic regulation in yeast. The Snf/Swi proteins are required for transcriptional activation of many diversely regulated genes in yeast and appear to function by somehow overcoming an inactive/repressed state caused by the Spt/Sin proteins (reviewed in Ref. 98; see Chapter 5). These proteins do appear to have 'antipodal' effects on the genes subject to their regulation; however, no underlying epigenetic aspect of their activity has yet been described. Such epigenetic effects might be observed if experiments were performed with the sensitivity to evaluate changes in transcriptional states in single cells, such as those observed in embryonic development in *Drososphila*, or in *HM* and telomeric transcription in yeast.

7.4 Conclusions

These models for both the establishment or assembly of repressed complexes, and the identity of components within these complexes, are based largely on molecular interactions that have not yet been demonstrated in their functional contexts. The continuing experimental challenge is the further analysis of both relevant mutants and *in vivo* and *in vitro* assays for chromatin and nuclear membrane interactions, DNA replication, and the assembly of multimeric repressed chromatin complexes.

The combined genetic and molecular data that exist support these ideas and clearly indicate that silencing in yeast is mediated by a specialized histone-dependent chromatin structure. This chromatin structure bears many features comparable to cytologically defined heterochromatin in multicellular eukaryotes and, as in those organisms, is fundamental to epigenetic transcriptional states that are critical both in single cells and in the controlled expression of developmental programmes.

Acknowledgements

We thank the members of our laboratories for their contributions and E. M. Stone and our fellow authors-readers for their comments on the text. We also thank B. Turner and N. Belyaec for kindly providing Fig. 1. L. P. is a Pew Scholar in the Biomedical Sciences and a National Science Foundation New Young Investigator, and gratefully acknowledges support from The Pew Charitable Trusts and the National Science Foundation. M. G. acknowledges support from the National Institutes of Health.

References

1. Moore, G. D., Procunier, J. D., Cross, D. P., and Grigliatti, T. A. (1979) Histone gene deficiencies and position effect variegation in *Drosophila*. *Nature*, **282**, 312.
2. Moore, G. D., Sinclair, D. A., and Grigliatti, T. A. (1983) Histone gene multiplicity and position effect variegation in *Drosophila melanogaster*. *Genetics*, **105**, 327.
3. Turner, B. M., Birley, A. J., and Jayne, L. (1992) Histone H4 isoforms acetylated at specific lysine residues define individual chromosomes and chromatin domains in *Drosophila* polytene nuclei. *Cell*, **69**, 375.
4. Bone, J. R., Lavendar, J., Richman, R., Palmer, M. J., Turner, B. M., and Kuroda, M. I. (1994) Acetylated histone H4 on the male X chromosome is associated with dosage compensation in *Drosophila*. *Genes and Dev.*, **8**, 96.
5. Jeppesen, P. and Turner, B. M. (1993) The inactive X chromosome in female mammals is distinguished by a lack of histone H4 acetylation, a cytogenetic marker for gene expression. *Cell*, **74**, 281.
6. Wilson, C., Bellen, H., and Gehring, W. (1990) Position effects on eukaryotic gene expression. *Annu. Rev. Cell. Biol.*, **6**, 679.
7. Reik, W. (1989) Genomic imprinting and genetic disorders in man. *Trends Genet.*, **5**, 331.

8. Sapienza, C. and Peterson, K. (1993) Imprinting the genome: imprinted genes, imprinting genes, and a hypothesis for their interaction. *Annu. Rev. Genet.*, **27**, 7.
9. Wilson, K. and Newport, J. (1988) A trypsin-sensitive receptor on membrane vesicles is required for nuclear envelope formation *in vitro*. *J. Cell Biol.*, **107**, 57.
10. Newport, J. and Dunphy, W. (1992) Characterization of the membrane binding and fusion events during nuclear envelope assembly using purified components. *J. Cell Biol.*, **116**, 295.
11. Pfaller, R., Smythe, C. and Newport, J. (1991) Assembly/disassembly of the nuclear envelope membrane: cell-cycle dependent binding of nuclear membrane vesicles to chromatin *in vitro*. *Cell*, **65**, 209.
12. Fosner, R. and Gerace, L. (1993) Integral membrane proteins of the nuclear envelope interact with lamins and chromosomes, and binding is modulated by mitotic phosphorylation. *Cell*, **73**, 1287.
13. Laurenson, P. and Rine, J. (1992) Silencers, silencing, and heritable transcriptional states. *Microb. Rev.*, **56**, 543.
14. Sandell, L. J. and Zakian, V. A. (1992) Telomeric position effect in yeast. *Trends Cell Biol.*, **2**, 10.
15. McCarroll, R. M. and Fangman, W. L. (1988) Time of replication of yeast centromeres and telomeres. *Cell*, **54**, 505.
16. Reynolds, A. E., McCarroll, R. M., Newlon, C. S., and Fangman, C. S. (1989) Time of replication of ARS elements along yeast chromosome III. *Mol. Cell. Biol.*, **9**, 4488.
17. Brand, A. H., Breeden, L., Abraham, J., Sternglanz, R., and Nasmyth, K. (1985) Characterization of a 'silencer' in yeast: a DNA sequence with properties opposite to those of a transcriptional enhancer. *Cell*, **41**, 41.
18. Schnell, R. and Rine, J. (1986) A position effect on the expression of a tRNA gene mediated by the *SIR* genes of *Saccharomyces cerevisiae*. *Mol. Cell. Biol.*, **6**, 494.
19. Gottschling, D. E., Aparicio, O. M., Billington, B. L., and Zakian, V. A. (1990) Position effect at *S. cerevisiae* telomeres: reversible repression of Pol II transcription. *Cell*, **63**, 751.
20. Aparicio, O. M., Billington, B. L., and Gottschling, D. E. (1991) Modifiers of position effect are shared between telomeric and silent mating-type loci in *S. cerevisiae*. *Cell*, **66**, 1279.
21. Renauld, H., Aparicio, O. M., Zierath, P. D., Billington, B. L., Chablani, S. K., and Gottschling, D. E. (1993) Silent domains are assembled continuously from the telomere and are defined by promoter distance and strength, and *SIR3* dosage. *Genes and Dev.*, **7**, 1133.
22. Pillus, L. and Rine, J. (1989) Epigenetic inheritance of transcriptional states in *S. cerevisiae*. *Cell*, **59**, 637.
23. Mahoney, D., Marquardt, R., Shei, G., Rose, A., and Broach, J. (1991) Mutations in the *HML E* silencer of *Saccharomyces cerevisiae* yield metastable inheritance of transcriptional repression. *Genes and Dev.*, **5**, 605.
24. Herskowitz, I., Rine, J., and Strathern, J. N. (1992) Mating-type determination and mating-type interconversion. In *The molecular and cellular biology of the yeast Saccharomyces*. Jones, E. W., Pringle, J. R., and Broach, J. R. (ed.). Cold Spring Harbor Laboratory Press, Cold Spring Harbor, New York, p. 583.
25. Oliver, S. G. *et al.* (1992) The complete DNA sequence of yeast chromosome III. *Nature*, **357**, 38.
26. Lee, S. and Gross, D. S. (1993) Conditional silencing: the *HMRE* mating-type silencer exerts a rapidly reversible position effect on the yeast *HSP82* heat shock gene. *Mol. Cell. Biol.*, **13**, 727.

27. Thompson, J. S., Johnson, L. M., and Grunstein, M. (1994) Specific repression of the yeast silent mating locus *HMR* by an adjacent telomere. *Mol. Cell. Biol.*, **14,** 446.
28. Brand, A. H., Micklem, G., and Nasmyth, K. (1987) A yeast silencer contains sequences that can promote autonomous plasmid replication and transcriptional activation. *Cell,* **51,** 709.
29. Kimmerly, W., Buchman, A., Kornberg, R., and Rine, J. (1988) Roles of two DNA-binding factors in replication, segregation and transcriptional repression mediated by a yeast silencer. *EMBO J.*, **7,** 2241.
30. McNally, F. J. and Rine, J. (1991) A synthetic silencer mediates *SIR*-dependent functions in *Saccharomyces cerevisiae*. *Mol. Cell. Biol.*, **11,** 5648.
31. Zakian, V. A. (1989) Structure and function of telomeres. *Annu. Rev. Genet.*, **23,** 579.
32. Boeke, J. D., Trueheart, J., Natsoulis, G., and Fink, G. R. (1987) 5-Fluoroorotic acid as a selective agent in yeast molecular genetics. *Meth. Enzymol.*, **154,** 164.
33. Thon, G. and Klar, A. J. S. (1992) The *clr1* locus regulates the expression of the cryptic mating-type loci of fission yeast. *Genetics,* **131,** 287.
34. Ekwall, K. and Ruusala, T. (1994) Mutations in *rik1, clr2, clr3* and *clr4* asymmetrically derepress the silent mating-type loci in fission yeast. *Genetics,* **136,** 53.
35. Allshire, R. C., Javerzat, J.-P., Redhead, N. J., and Cranston, G. (1994) Position effect variegation at fission yeast centromeres. *Cell,* **76,** 157.
36. Schulman, I. and Bloom, K. (1991) Centromeres: an integrated protein/DNA complex required for chromosome movement. *Annu. Rev. Cell Biol.*, **7,** 311.
37. Price, C. (1992) Centromeres and telomeres. *Curr. Opinion Cell Biol.*, **4,** 379.
38. Panzeri, L., Groth-Clausen, I., Shepard, J., Stotz, A., and Phillippsen, P. (1984) Centromeric DNA in yeast. *Chromosomes Today,* **8,** 46.
39. Snyder, M., Sapolsky, R. J., and Davis, R. W. (1988) Transcription interferes with elements important for chromosome maintenance in *Saccharomyces cerevisiae*. *Mol. Cell Biol.*, **8,** 2184.
40. Palladino, F., Laroche, T., Gilson, E., Axelrod, A., Pillus, L., and Gasser, S. (1993) SIR3 and SIR4 proteins are required for the positioning and integrity of yeast telomeres. *Cell,* **75,** 543.
41. Buchman, A. R., Kimmerly, W. J., Rine, J., and Kornberg, R. D. (1988) Two DNA-binding factors recognize specific sequences at silencers, upstream activating sequences, autonomously replicating sequences, and telomeres in *Saccharomyces cerevisiae*. *Mol. Cell. Biol.*, **8,** 210.
42. Shore, D., Stillman, D., Brand, A., and Nasmyth, K. (1987) Identification of silencer binding proteins from yeast; possible role in *SIR* control and replication. *EMBO J.*, **6,** 461.
43. Longtine, M. S., Wilson, N. M., Petracek, M. E., and Berman, J. (1989) A yeast telomere binding activity binds to two related telomere sequence motifs and is indistinguishable from RAP1. *Curr. Genet.*, **16,** 225.
44. Klein, F., Laroche, T., Cardenas, M. E., Hofmann, J. F.-X., Schwartz, D., and Gasser, S. M. (1992) Localization of RAP1 and topoisomerase II in nuclei and meiotic chromosomes of yeast. *J. Cell Biol.*, **117,** 935.
45. Kyrion, G., Boyakye, K., and Lustig, A. (1992) C-terminal truncation of RAP1 results in the deregulation of telomere size, stability, and function in *Saccharomyces cerevisiae*. *Mol. Cell. Biol.*, **12,** 5159.
46. Sussel, L. and Shore, D. (1991) Separation of transcriptional activation and silencing functions of *RAP1*: isolation of viable mutants affecting both silencing and telomere length. *Proc. Natl Acad. Sci. USA,* **88,** 7749.

47. Sussel, L., Vannier, D., and Shore, D. (1993) Epigenetic switching of transcriptional states: *cis-* and *trans*-acting factors affecting establishment of silencing at the *HMR* locus in *Saccharomyces cerevisiae*. *Mol. Cell. Biol.*, **13,** 3919.
48. Hardy, C. F. J., Balderes, D., and Shore, D. (1992) Dissection of a carboxy-terminal region of the yeast regulatory protein Rap1 with effects on both transcriptional activation and silencing. *Mol. Cell. Biol.*, **12,** 1209.
49. Hardy, C. F. J., Sussel, L., and Shore, D. (1992) A RAP1-interacting protein involved in transcriptional silencing and telomere length regulation. *Genes and Dev.*, **6,** 801.
50. Mullen, J. R., Kayne, P. S., Moerschell, R. P., Tsunasawa, S., Gribskov, M., Colavito-Shapanski, M., Grunstein, M., Sherman, F., and Sternglanz, R. (1989) Identification and characterization of genes and mutants for an N-terminal acetyltransferase from yeast. *EMBO J.*, **8,** 2067.
51. Park, E.-C. and Szostak, J. W. (1992) ARD1 and NAT1 proteins form a complex that has N-terminal acetyltransferase activity. *EMBO J.*, **11,** 2087.
52. Bell, S. P. and Stillman, B. (1992) ATP-dependent recognition of eukaryotic origins of DNA replication by a multiprotein complex. *Nature*, **357,** 128.
53. Foss, M., McNally, F. J., Laurenson, P., and Rine, J. R. (1993) Origin recognition complex (ORC) in transcriptional silencing and DNA replication in *S. cerevisiae*. *Science*, **262,** 1838.
54. Bell, S. P., Kobayashi, R., and Stillman, B. (1993) Yeast origin recognition complex functions in transcription silencing and DNA replication. *Science*, **262,** 1844.
55. Li, J. J. and Herskowitz, I. (1993) Isolation of ORC6, a component of the yeast Origin Recognition Complex by a one-hybrid screen. *Science*, **262,** 1870.
56. Micklem, G., Rowley, A., Harwood, J., Nasmyth, K., and Diffley, J. F. X. (1993) Yeast origin recognition complex is involved in DNA replication and transcriptional silencing. *Nature*, **366,** 87.
57. Kayne, P. S., Kim, U.-J., Han, M., Mullen, J. R., Yoshizaki, F., and Grunstein, M. (1988) Extremely conserved histone H4 N-terminus is dispensible for growth but essential for repressing the silent mating type loci in yeast. *Cell*, **55,** 27.
58. Johnson, L. M., Kayne, P. S., Kahn, E. S., and Grunstein, M. (1990) Genetic evidence for an interaction between SIR3 and histone H4 in the repression of the silent mating loci in *Saccharomyces cerevisiae*. *Proc. Natl Acad. Sci. USA*, **87,** 6286.
59. Megee, P. C., Morgan, B. A., Mittman, B. A., and Smith, M. M. (1990) Genetic analysis of histone H4: essential role of lysines subject to reversible acetylation. *Science*, **247,** 841.
60. Park, E. and Szostak, J. W. (1990) Point mutations in the yeast histone H4 gene prevent silencing of the silent mating type locus *HML*. *Mol. Cell. Biol.*, **10,** 4932.
61. Johnson, L. M., Fisher-Adams, G., and Grunstein, M. (1992) Identification of a non-basic domain in the histone H4 N-terminus required for repression of the yeast silent mating loci. *EMBO J.*, **11,** 2201.
62. Thompson, J. S., Ling, X., and Grunstein, M. (1994) The histone H3 N-terminus is required for both telomeric and silent mating locus repression in yeast. *Nature*, **369,** 245.
63. Ferguson, B. M. and Fangman, W. L. (1992) A position effect on the time of replication origin activation in yeast. *Cell*, **68,** 333.
64. Rivier, D. and Rine, J. (1992) An origin of DNA replication and a transcription silencer require a common element. *Science*, **256,** 659.
65. Miller, A. M. and Nasmyth, K. A. (1984) Role of DNA replication in the repression of silent mating type loci in yeast. *Nature*, **312,** 247.
66. Miller, A. M., Sternglanz, R., and Nasmyth, K. (1985) The role of DNA replication in

the repression of the yeast mating-type silent loci. *Cold Spring Harbor Symp. Quant. Biol.*, **49,** 105.

67. Nasmyth, K. (1982) The regulation of yeast mating-type chromatin structure by SIR: an action at a distance affecting both transcription and transposition. *Cell*, **30,** 567.
68. Strathern, J. N., Klar, A. J. S., Hicks, J. B., Abraham, J. A., Ivy, J. M., Nasmyth, K. A., and McGill, C. (1982) Homothallic switching of yeast mating type cassettes is initiated by a double-stranded cut in the *MAT* locus. *Cell*, **31,** 183.
69. Klar, A. J. S., Strathern, J. N., and Abraham, J. A. (1984) Involvement of double-strand chromosomal breaks for mating-type switching in *Saccharomyces cerevisiae*. *Cold Spring Harbor Symp. Quant. Biol.*, **49,** 77.
70. Loo, S. and Rine, J. (1994) Silencers and domains of generalized repression on a eukaryotic chromosome. *Science*, **264**, 1768.
71. Wright, J. H., Gottschling, D. E., and Zakian, V. A. (1992) *Saccharomyces* telomeres assume a non-nucleosomal chromatin structure. *Genes and Dev.*, **6,** 197.
72. Makarov, V. L., Lejnine, S., Bedoyan, J., and Langmore, J. P. (1993) Nucleosomal organization of telomere-specific chromatin in rat. *Cell*, **73,** 775.
73. Chen, T. A., Smith, M. M., Le, S., Sternglanz, R., and Allfrey, V. G. (1991) Nucleosome fractionation by mercury-affinity chromatography. Contrasting distribution of transcriptionally active DNA sequences and acetylated histones in nucleosome fractions of wild-type yeast cells and cells expressing a histone H3 gene altered to encode a cysteine-110 residue. *J. Biol. Chem.*, **266,** 6489.
74. Chen-Cleland, T. A., Smith, M. M., Le, S., Sternglanz, R., and Allfrey, V. G. (1993) Nucleosome structural changes during derepression of silent mating-type loci in yeast. *J. Biol. Chem.*, **268,** 1118.
75. Singh, J. and Klar, A. S. (1992) Active genes in budding yeast display enhanced *in vivo* accessibility to foreign DNA methylases: a novel *in vivo* probe for chromatin structure of yeast. *Genes and Dev.*, **6,** 186.
76. Gottschling, D. E. (1992) Telomere proximal DNA in *S. cerevisiae* is refractory to methyltransferase activity *in vivo*. *Proc. Natl Acad. Sci. USA*, **89,** 4062.
77. Terleth, C., van Sluis, C. A., and van de Putte, P. (1989) Differential repair of UV damage in *Saccharomyces cerevisiae*. *Nucleic Acids Res.*, **17,** 4433.
78. Terleth, C., Schenk, P., Poot, R., Brouwer, J., and van de Putte, P. (1992) Differential repair of UV damage in *rad* mutants of *Saccharomyces cerevisiae*: a possible function of G2 arrest upon UV irradiation. *Mol. Cell. Biol.*, **10,** 4678.
79. Lin, R., Leone, J. W., Cook, R. J., and Allis, C. D. (1989) Antibodies specific to acetylated histones document the existence of deposition- and transcription-related histone acetylation in *Tetrahymena*. *J. Cell Biol.*, **108,** 1577.
80. Braunstein, M., Rose, A. B., Holmes, S. G., Allis, C. D., and Broach, J. R. (1993) Transcriptional silencing in yeast is associated with reduced nucleosome acetylation. *Genes and Dev.*, **7,** 592.
81. van Holde, K. E. (1989) *Chromatin*. Springer-Verlag, New York.
82. Smith, M. M. (1991) Histone structure and function. *Curr. Opinion Cell Biol.*, **3,** 429.
83. Wolffe, A. (1992) *Chromatin structure and function*. Academic Press, New York.
84. Johnson, E. S., Gonda, D. K., and Varshavsky, A. (1990) *Cis-trans* recognition and subunit specific degradation for short-lived proteins. *Nature*, **346,** 287.
85. Lee, D., Hayes, J., Pruss, D., and Wolffe, A. (1993) A positive role for histone acetylation in transcription factor access to nucleosomal DNA. *Cell*, **72,** 73.

86. Mann, R. K. and Grunstein, M. (1992) Histone H3 N-terminal mutations allow hyperactivation of the yeast *GAL1* gene *in vivo*. *EMBO J.*, **11,** 3297.
87. Allfrey, V. (1977) In *Chromatin and chromosome structure*. Li, H. J. and Eckhardt, R. A. (ed.). Academic Press, New York, pp. 167–91.
88. Hebbes, T. R., Thome, A. W., and Crane-Robinson, C. (1988) A direct link between core histone acetylation and transcriptionally active chromatin. *EMBO J.*, **7,** 1395.
89. Grunstein, M. (1990) Histone function in transcription. *Annu. Rev. Cell Biol.*, **6,** 643.
90. Ebralidse, K. K., Grachev, S. A., and Mirzabekov, A. D. (1988) A highly basic histone H4 domain bound to the sharply bent region of nucleosomal DNA. *Nature*, **331,** 365.
91. Roth, S., Shimizu, M., Johnson, L., and Grunstein, M. (1992) Stable nucleosome positioning and complete repression by the α repressor are disrupted by amino-terminal mutations in histone H4. *Genes and Dev.*, **6,** 411.
92. Ivy, J. M., Klar, A. J. S., and Hicks, J. B. (1986) Cloning and characterization of four *SIR* genes of *Saccharomyces cerevisiae*. *Mol. Cell. Biol.*, **6,** 688.
93. Marshall, M., Mahoney, D., Rose, A., Hicks, J. B., and Broach, J. R. (1987) Functional domains of *SIR4*, a gene required for position effect regulation in *Saccharomyces cerevisiae*. *Mol. Cell. Biol.*, **7,** 4441.
94. Diffley, J. F. X. and Stillman, B. (1989) Similarity between the transcriptional silencer binding proteins ABF1 and RAP1. *Science*, **246,** 1034.
95. Chien, C.-T., Bartel, P. L., Sternglanz, R., and Fields, S. (1991) The two-hybrid system: a method to identify and clone genes for proteins that interact with a protein of interest. *Proc. Natl Acad. Sci.*, **88,** 9578.
96. Kimmerly, W. J. (1988) *Cis*- and *trans*-acting regulators of the silent mating type loci of *Saccharomyces cerevisiae*. PhD thesis, University of California, Berkeley.
97. Stavenhagen, J. B. and Zakian, V. A. (1994) Internal tracts of telomeric DNA act as silencers in *Saccharomyces cerevisiae*. *Genes and Dev.*, **8,** 1411.
98. Winston, F. and Carlson, M. (1992) Yeast SNF/SWI transcriptional activators and the SPT/SIN connection. *Trends Genet.*, **8,** 387.

8 Epigenetic regulation in *Drosophila*: a conspiracy of silence

JOEL C. EISSENBERG, SARAH C. R. ELGIN, and
RENATO PARO

1. Chromatin structure and gene silencing

Understanding the mechanisms of epigenetic regulation, by which cell-specific gene silencing is established and stably maintained, is a major unresolved challenge. Gene expression in eukaryotes relies upon the accessibility of the DNA template to RNA polymerase and a variety of accessory proteins. It is clear that the accessibility of regulatory elements can be controlled by the chromatin structure at a given locus: in spite of the availability in a given cell of all of the requisite DNA-binding proteins, a gene can remain silent, and this silenced state can be propagated faithfully through mitosis. This might be accomplished by alternative modes of packaging DNA into a nucleosome array, but almost certainly involves other types of packaging as well.

Nucleosome depletion experiments in yeast provide the most direct evidence for a role for positioned nucleosomes in the repression of certain genes, and a growing number of examples in multicellular eukaryotes suggests that transcriptional repression by stable protein complexes including positioned nucleosomes may be a widely employed mechanism (see Chapter 5). Beyond the level of individual nucleosomal complexes a detailed description of higher order chromatin structure and of the role of higher order structural transitions in gene regulation is lacking. In *Drosophila*, the inference that transcriptional repression can be associated with higher level chromatin compaction rests primarily on the distinctive cytological appearance of heterochromatin; we still await a detailed description of the topology and composition of heterochromatin, and an assessment of the effect of heterochromatin packaging on the nucleosome array. *Drosophila* heterochromatin is rich in simple-sequence satellite and middle repetitious DNA. While the satellite DNA has been shown to be packaged in nucleosomes (1), the lack of specific markers has made mapping of the nucleosomal organization and evaluation of its functional significance difficult. In the examples of epigenetic regulation to be discussed below—heterochromatic position effect and homeotic gene repression—the involvement of large chromosomal regions argues for a mechanism beyond single-nucleosome promoter extinction.

In plants and vertebrates, transcriptional repression is frequently associated with hypermethylation of the repressed gene DNA and/or associated regulatory sequences. The functional significance of DNA methylation in gene repression (e.g., is it a cause or a consequence?) is not resolved (see Chapter 10). Since there is little or no detectable DNA methylation in *Drosophila* (2), the mitotically stable repression of genes in *Drosophila* is presumably the consequence of a shift in DNA–protein interactions *per se*. The ways in which chromosomal proteins conspire to silence genes—the members of this conspiracy, their hierarchy of interactions—are amenable to genetic and biochemical investigation, and are now beginning to be unravelled.

2. Heterochromatic position effect variegation

2.1 Gene silencing associated with chromosomal position

An early recognition of the differential packaging of chromatin in eukaryotic cells was made by Heitz (3), who observed a fraction of the nuclear material that failed to decondense after mitosis which he termed 'heterochromatin'. Subsequent studies have revealed that heterochromatin is rich in repetitive DNA, relatively poor in classical genes, late replicating in the cell cycle, and minimally transcribed (4). Genetic evidence suggests that heterochromatin formation interferes with expression of normally euchromatic genes; for example, all but one X chromosome in mammalian cells appear to be generally transcriptionally inactive and heterochromatic (5). Genes relocated to heterochromatin through rearrangement or transposition experience an abnormal, position-dependent mosaic silencing, termed 'position effect variegation' (PEV), which can serve as a model for mitotically stable gene repression (4, 6).

PEV was initially observed as a result of chromosomal rearrangements (inversions, translocations), generally a product of X-irradiation, which placed euchromatic genes close to a heterochromatic breakpoint. This position-dependent inactivation suggested a *cis*-acting silencing activity of heterochromatin acting across the rearrangement breakpoint. The acquisition of heterochromatic structure by normally euchromatic material at such breakpoints is visibly manifest in the giant polytene chromosomes of the *Drosophila* larval salivary glands. In these nuclei, the euchromatic DNA is amplified a thousand-fold; the chromatin strands remain tightly synapsed to give a highly ordered, stereotypical banded appearance. In contrast, the pericentric heterochromatin in these nuclei is underrepresented, densely staining, and disordered in appearance; the heterochromatin surrounding the centromeres of all of the chromosomes aggregates to form a single, cytologically distinct, compact focus (the 'chromocentre') from which the banded euchromatic arms are seen to radiate (see Fig. 1a). In strains of flies carrying variegating chromosome rearrangements, the regular banded appearance of the euchromatin in the vicinity of the breakpoint is lost in a subset of nuclei, and the region takes on the cytological appearance of heterochromatin (7–10). Thus, the mosaic nature of the silencing of variegating loci has a cytological correlate in differential chromatin compaction. Although the decision as to whether the rearranged gene is to be on or off appears to be random, once made it is maintained through multiple rounds of mitosis. This mitotic stability results in patches of cells with the gene off and other patches with it on. Generally an adult phenotype is scored, with clones of 'on' cells ranging from one to several hundred cells in size.

In several instances, the DNA sequences of loci subject to variegation appear to be underrepresented in polytene tissue (11, 12). Such underrepresentation could be the consequence of a failure to complete replication of the heterochromatic DNA during polytenization (11), or the consequence of sequence elimination (12). Either mechanism could, in principle, account for the reduced or undetectable expression of genes experiencing PEV in polytene tissues. However, copy number effects cannot account for other known examples of variegation that have been studied (7, 13). It is not clear whether the sequence underrepresentation that has been observed is a cause or merely concomitant with heterochromatin formation.

Virtually every locus that has been examined in an appropriate rearrangement has been found to variegate, and rearrangements involving the pericentric heterochromatin of the X chromosome or any of the autosomes can lead to PEV. In the cases studied, variegation has been associated with reduced transcript accumulation for the affected gene (7, 9, 13). The molecular basis for gene inactivation by PEV must involve a fundamental aspect of transcription common to genes which otherwise differ in their temporal and spatial regulation. One can easily imagine that the densely compacted quality of heterochromatin is sufficient to occlude sites of DNA–protein interaction necessary for gene activation, e.g., preventing assembly of the transcription initiation complex in promoter regions. Alternatively, association with heterochromatin, which is generally observed clumped in a single mass at the nuclear periphery, may place the gene in a 'compartment' inaccessible to transcription factors.

Although heterochromatin appears to be an inhospitable environment for the expression of euchromatic genes, there are a number of loci that behave as classical Mendelian genes that map to heterochromatic regions (14). Interestingly, several of these genes have been found to variegate in rearrangements placing them next to a *euchromatic* breakpoint (15, 16), suggesting that certain genes may not only tolerate a heterochromatic environment, but in fact have come to depend on this environment for proper expression (14). The degree of variegation experienced by a heterochromatic locus depends upon how far out along the euchromatic chromosome arm the gene is displaced, with more distal breakpoints giving more extreme variegation (15–17). This suggests that proximity to the pericentric heterochromatin has an effect; stable pairing has been suggested to play a role.

The importance of chromosome pairing to heterochromatin formation and/or

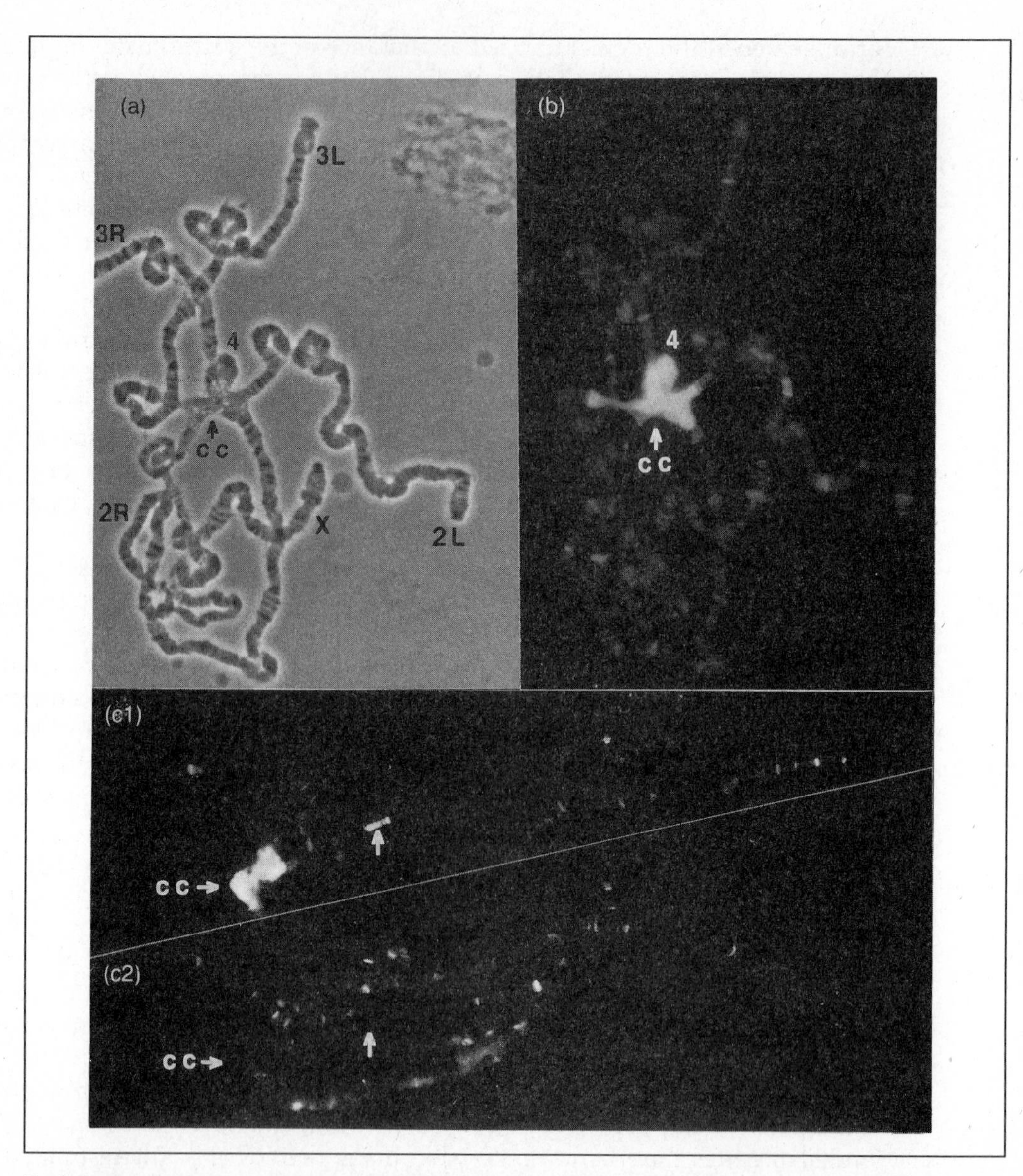

Fig. 1 The polytene chromosomes of *Drosophila melanogaster*: (a) phase contrast, and (b) immunofluorescence staining to localize heterochromatin protein 1 (HP1). Note that the chromocentre (cc) and fourth chromosome are intensely stained. (c) Double-immunofluorescence staining of salivary gland polytene chromosomes showing the different distributions of the *Polycomb* and HP1 proteins: (c1) staining with monoclonal mouse anti-HP1 antibodies and fluorescein-conjugated anti-mouse antibodies, (c2) staining with polyclonal rabbit anti-Pc antibodies and rhodamine-conjugated anti-rabbit antibodies. The arrow points to region 31, a site associated with HP1. Panels (a) and (b) provided by C. Craig; panels (c1) and (c2) provided by R. F. Clark.

stability is further underscored by the remarkable phenomenon of dominant *brown* (*bw*) variegation. Inactivation of the *bw* gene by mutation leads to flies that have brown rather than red eyes. Such mutations are generally recessive, but a striking exception is the *brown*^{Dominant} (*bw*^{D}) allele. This allele is capable of a mosaic, *trans*-inactivation of a wild-type *bw*$^{+}$ allele located at its normal position on the homologous chromosome, but has no effect on the expression of *bw*$^{+}$ alleles placed at other euchromatic sites (18, 19). *bw*^{D} is associated with insertion of a block of heterochromatin near or within the *brown* gene (20), and trans-dominant inactivation appears to depend on specific sequences near the *bw*$^{+}$ gene that may normally serve to promote pairing. *bw*^{D} variegation is suppressed by chromosomal rearrangements that move the *bw*^{D} allele further away from the pericentric heterochromatin, while it is enhanced in rearrangements which move it closer (21). Pairing effects might bring regions of DNA into the same spatial or temporal compartment of the nucleus for replication, helping to explain how the on/off decision is propagated through many rounds of mitotic cell division.

2.2 A mass-action assembly model

Any attempt to model the mechanism of PEV must explain how heterochromatin can exert a repressing effect on genes located many kilobases away. Genetic and cytological evidence (see below) indicates that heterochromatin is composed of a protein–DNA complex involving both histones and non-histone chromosomal proteins. The initial site of complex deposition may be specified by a DNA sequence or structure, but once initiated the complex is proposed to spread in a co-operative and sequence-independent fashion along the chromosome until some boundary or limit is encountered (22). In variegating chromosomal rearrangements, then, it is imagined that gene inactivation occurs because the spreading heterochromatin crosses the heterochromatin–euchromatin junction created by the rearrangement, invading the neighbouring euchromatin and establishing a mitotically stable heterochromatic domain (Fig. 2).

If the extent and stability of heterochromatin-mediated gene repression is, as proposed, a function of the extent and stability of a protein–DNA complex, it is reasonable to suppose that the levels of the various constituent proteins would influence the frequency with which gene silencing is observed in cases of PEV (23). If each heterochromatin protein is present in the cell in two forms, one as part of a silencing complex and the other free in the nucleus, then shifts in the intracellular concentration of any constituent protein would, by mass action, influence the extent of heterochromatin assembly and spreading (24). Chance differences in protein concentration between cells may account for the mosaic nature of PEV. More importantly, mutations negatively affecting the synthesis of any heterochromatin protein would reduce the extent of heterochromatin assembly, and thus the probability of silencing. Conversely, mutations reducing the synthesis of general euchromatic proteins required to maintain gene activity would enhance the probability of heterochromatin assembly in a competitive situation. In genetic

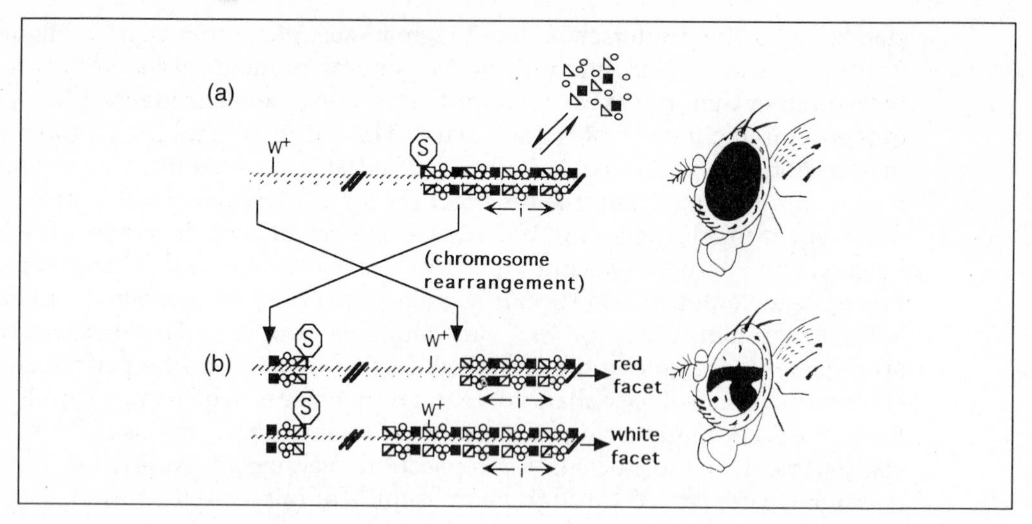

Fig. 2 A schematic illustration of *white* variegation in the X chromosome inversion *In(1)w*^{m4}. The *white* locus (w^+) is located in the distal euchromatin (thin line) of the wild-type X chromosome. It provides a function essential to the normal red pigmentation of the *Drosophila* eye. The inverted X chromosome, *In(1)w*^{m4}, was the result of chromosomal breaks which occurred adjacent to the *white* locus and within the pericentriomeric heterochromatin of the X chromosome (thick line). Thus, the *white* locus has come to lie 25 kb away from the heterochromatic breakpoint in this inversion. This abnormal juxtaposition gives rise to flies with mottled, or variegated, compound eyes composed of red (*white* gene is active) and white (*white* gene is inactive) eye facets. (a) In the normal chromosome, heterochromatin-specific proteins (represented by geometric symbols) assemble at initiation sites 'i' and co-operatively propagate a condensed heterochromatic structure in *cis* until a terminator or 'stop' signal is encountered. (b) A chromosome rearrangement which has one breakpoint between the 'i' and 'stop' signals permits the spread of heterochromatin into a normally euchromatic region of the chromosome. When the extent of spreading in a given cell includes the gene w^+, the gene is inactivated; when the spreading terminates before reaching gene w^+, the gene remains active in that cell. Adapted from (4).

parlance, the former class of mutations would behave as 'suppressors of variegation', while the latter would be 'enhancers of variegation'. Mutations that alter function or stability of the heterochromatin or the euchromatin structure as a whole could also appear as modifiers of PEV.

For structural components of heterochromatin, the mass-action assembly model implies that a gene may behave, when duplicated, as an enhancer of PEV and, when deleted, as a suppressor of PEV. Similarly, a structural component of euchromatin with antagonistic effects on heterochromatin spreading would behave, when duplicated, as a suppressor of PEV and, when deleted, as an enhancer of PEV. Display of both of these types of behaviours (termed 'antipodal effects' by Locke *et al.* (24)) is considered a signature for genes likely to encode structural components of heterochromatin and euchromatin.

2.3 Modifiers of heterochromatic silencing

The identification and characterization of modifiers of PEV in *Drosophila* has become an important and fruitful strategy in dissecting the composition and properties of heterochromatin. One of the earliest modifiers recognized was the Y chromosome, which is itself heterochromatic in somatic tissue; more Y chromosome material (e.g., a male with the sex chromosome constitution XYY instead of the normal XY) has the effect of suppressing PEV, whereas less Y chromosome material (e.g., an XO male) enhances PEV (6). Later investigations showed that X heterochromatin levels also can titrate PEV levels (25).

The effect of heterochromatin levels on PEV may be understood as a competition between the assembly of heterochromatin and the assembly of euchromatin, reflecting levels of structural proteins (23). The introduction of additional heterochromatin leads to a corresponding reduction in heterochromatin formation at a variegating breakpoint. This, in turn, leads to the inference that the amount of heterochromatin in a cell is primarily a function of the levels of heterochromatin structural proteins (23).

The histone genes in *Drosophila* are present in a tandem array of ~100 copies of a repeating unit which includes one copy each of the genes for H1, H3, H4, H2A, and H2B (26). Partial deletions of this cluster result in suppression of PEV (27), as does growth in the presence of butyrate, an inhibitor of histone deacetylation (28). These results implicate histone proteins themselves in the propagation and/or stabilization of heterochromatin and suggest that the deacetylated form is required for that purpose. However, butyrate has a number of effects. It has been pointed out (13) that histone gene deletions and butyrate feeding also cause delayed development, although their effects on PEV (i.e., suppression) are the opposite of the effect (i.e., enhancement) caused by low temperature-, pH-, and crowding-associated developmental delay (29). Recent immunolocalization studies using antibodies specific for H4 isoforms acetylated at each of four N-terminal lysine residues showed that the pericentric heterochromatin is relatively poor overall in acetylated H4, but relatively enriched in the H4 isoform(s) acetylated at $lysine_{12}$ (30). Thus, differential histone acetylation could well be involved in partitioning the genome into heterochromatin and euchromatin. However, a direct role for any histone isoform in heterochromatin-mediated gene repression is yet to be demonstrated. The genetic manipulation of histones is complicated in *Drosophila* (and most other eukaryotic organisms except *S. cerevisiae*) by the multiplicity of histone gene copies.

The genetic dissection of PEV has been intensively pursued through the identification, cloning, and characterization of dominant modifiers of PEV. Genetic screens (8, 24, 31, 32) have identified 40–50 modifiers on the two major autosomes of *D. melanogaster*. For some of these loci, comparisons of duplications and deficiencies overlapping the same interval show the antipodal effects anticipated for chromatin structural proteins (see above). Recently, several of these loci (listed in Table 1) have yielded to molecular cloning, and conceptual translation of the gene products and/or cytological localization studies have uncovered properties suggestive of structural chromosomal proteins or their modifiers. Some dosage-dependent modifiers of PEV also exhibit dosage-dependent effects on homeotic gene expression (33, 34), suggesting that the mechanisms of epigenetic silencing

Table 1 Summary of cloned loci from *Drosophila* encoding non-histone modifiers of PEV

Locus	Gene product	Antipodal effect	Reference
Su(var)205	Heterochromatin-associated protein (HP1)	Yes	35, 38
Su(var)3-7	Zinc-finger-containing protein	Yes	33
Modulo	DNA-binding protein	Unknown	91
Su(var)3-6	Protein phosphatase type I	No	92
E(var)3-93D	Protein with N-terminal homology to transcriptional activators	No	34

See (93) for more details.

by heterochromatin formation may overlap, at least in some respects, with mechanisms of developmentally regulated gene silencing (see below).

Probably the best-characterized heterochromatin-associated protein in *Drosophila* is heterochromatin protein 1 (HP1). This was initially identified by a set of monoclonal antibodies generated against a fraction of tightly bound non-histone chromosomal proteins isolated from *Drosophila* embryos (35). These antibodies localize to the pericentric heterochromatin, in a banded pattern on the fourth chromosome, at several telomeres, and at a small number of euchromatic sites in indirect immunofluorescent staining of formaldehyde-fixed polytene chromosome squashes (Fig. 1b) (35, 36). The cytological map position of the HP1 gene at 29A coincided with the genetic map position of *Su(var)205*, a dominant suppressor of PEV previously identified (31). Subsequent molecular cloning and sequencing has identified the mutational lesions in four alleles of this gene, including the original *Su(var)205* allele (37, 38; T. Hartnett and J. C. Eissenberg, unpublished data). All of these alleles of *Su(var)205* are both dominant suppressors of PEV and recessive lethals; both phenotypes are complemented by a heat-shock-driven HP1 cDNA-based transgene (38, 39).

Recent data on HP1 suggest that heterochromatin assembly and PEV may be dynamic and actively regulated during development. Transiently expressed HP1-β-galactosidase fusion protein can specifically associate with the pericentric heterochromatin, even when expression is delayed until the late third instar (~5–6 days of development (40)). Overexpression of a wild-type HP1 cDNA under a heat-shock-regulated promoter results in significant enhancement of PEV, even when this overexpression is delayed for up to seven days of development (39).

The physical association of HP1 protein with cytological heterochromatin, together with the genetic data showing that the gene encoding HP1 (named *Su(var)205*, after the first-published description of the locus) is a dosage-dependent modifier of PEV argue strongly for a structural role for HP1 in epigenetic repression by heterochromatin formation. This inference has gained substantial support from the results of a recent study using antibodies against HP1 to stain the polytene chromosomes of a variegating stock; where the variegating locus appears to be part of the condensed β-heterochromatin it is stained using HP1, while in those nuclei in which it is packaged in the banded, polytenized, euchromatic homologue it is unstained (10).

2.4 Organization of heterochromatin

What defines *Drosophila* heterochromatin at the DNA sequence level? What signals—sequences, or arrangements of repetitious sequences, perhaps—dictate the characteristic packaging, and/or drive gene silencing? Detailed mapping of the sequence organization of heterochromatin has been complicated both by the presence of repetitious DNA and by the paucity of unique markers. Structural analysis of three variegating breakpoints (22) found middle repetitive elements, structurally related to transposons, at the heterochromatic end of each breakpoint. Re-inversions of one of the variegating rearrangements restored wild-type gene expression, yet left the heterochromatic sequences in the immediate vicinity of the breakpoint intact (22), suggesting that the efficient propagation of heterochromatin-mediated silencing activity requires more than the sequences at the breakpoint itself. Interestingly, other studies of revertant chromosome rearrangements have shown that, if a strong enhancer of variegation is included in the genetic background of the revertant chromosome, some PEV is still detectable in many cases (41, 42). The amount of heterochromatin carried with each re-inversion was not measured; it could well be that only re-inversions that bring along large heterochromatic blocks, potentially including an 'initiator' sequence, retain silencing activity. Alternatively, a threshold quantity of repetitious sequence DNA could be required.

The sequence family best correlated with the distribution of HP1 is the Dr.D element (43), which is found in the β-heterochromatin and in blocks throughout the fourth chromosome of *D. melanogaster*. While such correlations are intriguing, no explicit test has yet been made of whether or not an ectopically located heterochromatin-associated sequence can result in (a) an ectopic site of HP1 binding, or (b) variegated silencing of a flanking marker. Small heterochromatin transpositions to new sites do stain autonomously with antibodies to HP1 (36), arguing that linear continuity with a centromere or the chromocentre is not a requirement for HP1 binding.

The ability to cause efficient transposition of P-element-derived visible markers to new chromosomal sites in the *Drosophila* genome has uncovered a wide variety of position effects (reviewed in 44). In instances where variegated expression of the transposon marker has been encountered, the insertion site has been found to be a telomere, the pericentric heterochromatin, or the fourth chromosome (45, 46; L. L. Wallrath and S. C. R. Elgin, unpublished data). Since the marker being used and the amount and sequence of flanking DNA are a constant, transposons will be especially effective probes for detecting structural and functional heterogeneity at these sites. For example, in contrast to what has been observed in yeast, variegating inserts in the telomeres do not respond to classical modifiers of PEV that do, on the other hand, affect silencing of inserts at the chromocentre (L. L. Wallrath and S. C. R. Elgin, unpublished data). It is likely that the protein composition and details of chromatin structure vary in different heterochromatin domains (47; L. L. Wallrath and S. C. R. Elgin, unpublished data).

3. Pattern formation and the mechanism of cellular memory

The process of pattern formation determines the structure and function that a cell and its descendants attain in the developed body. Mechanisms that generate differential patterns of gene expression form the basis for the creation of cellular diversity. In *Drosophila*, many of the factors that govern the process of development have been identified (48). Their spatial and temporal patterns of expression indicate that the basic body plan is laid down very early in development, beginning with maternal components that act during oogenesis to specify the major body axes in the egg. The subsequent regulatory cascade of the segmentation genes subdivides the zygote into the repetitive array of segmental structures that is characteristic of the fly body. This reiterated positional information is subsequently translated into distinct patterns of homeotic selector gene expression along the anterior–posterior body axis. The task of the homeotic genes is to provide each segment, through a combinatorial arrangement of products, with the information required for building segment-specific structures and appendages. Unlike the early maternal and segmentation factors, the homeotic genes are not only responsible for establishing positional values; the combination of their products is also required throughout development to maintain the determined state of a cell. Thus the homeotic genes act as key regulators throughout development.

3.1 The *Polycomb* group gene complex

Genetic analysis has led to the identification of a class of mutations, unlinked to the clustered homeotic genes, which none the less display complex homeotic phenotypes (49–52). The homeotic transformations seen are reminiscent of gain-of-function mutations of homeotic selector genes, in which a homeotic product is ectopically expressed in an incorrect segment. This misexpression results in a change of identity of an anterior segment into a more posterior one. The effect is most dramatically observed in the lethal mutant embryos of this class, in which all segments are transformed towards the identity of the eighth (most posterior) segment. By analysing *Polycomb* (*Pc*) mutations, it was realized that the phenotype was the result of complete derepression of the homeotic genes of the bithorax complex (BX-C) and the *Antennapedia* complex (ANT-C) (49). It was proposed that *Pc* acts as a segment-specific repressor of homeotic genes. So far, mutations in at least fifteen genes have been found to result in this particular phenotype; these genes are collectively called the *Polycomb* group (*Pc*-G) (51). By extrapolating data from genetic screens, it has been postulated that the class might comprise 30–40 members.

Once molecular markers for homeotic gene products became available, the genetic predictions could be substantiated by a visual assessment of the ectopic expression patterns seen in mutant embryos (53–55). However, the story obtained a twist when it was realized that, in the mutant embryos, the early pattern of homeotic products was indistinguishable from wild type. Only at germ band retraction did ectopic expression start to emerge. This result demonstrated that the *Pc*-G was not involved in the establishment of the homeotic pattern, but rather in its maintenance, which identified the genes of the *Pc*-G as an integral part of the cellular memory system. The products of the *Pc*-G would recognize an inactive state of a particular homeotic gene, as defined by the early initiation system, and would transmit this repressed state in a stable and heritable form through the many cell divisions occurring during the rest of development. Indeed, clonal analysis showed that the repressive role of *Pc*-G genes is required not only during differentiation of larval tissue, but also for development of adult structures (49, 50, 56).

Subsequent analysis of the regulatory functions of the genes of the *Pc*-G, either by a closer inspection of the mutant phenotypes or by molecular characterization, revealed that the homeotic genes are only a small subset of the target genes. Interestingly, many of the other genes that have been identified as targets turned out to encode factors important for regulating developmental processes. Thus, the primary task of the *Pc*-G seems to be the permanent inactivation of developmental regulators in defined domains of the body. We have learned a good deal about several of the proteins in this group.

Polycomb, a small gene extending over approximately 4 kb, encodes a 390 aa nuclear protein (57). Both the transcript and the *Polycomb* protein (Pc) are essentially homogeneously distributed in most tissues during development, although there is a preferential accumulation in tissues with high proliferative activity (58). In polytene chromosomes, the protein is found to be associated with the homeotic gene clusters of the ANT-C and BX-C (59). However, approximately 100 additional sites were identified, supporting the notion that the *Pc*-G is responsible for repressing a large set of genes. Interestingly, in this tissue other *Pc*-G loci were identified as targets of the *Pc* protein, suggesting a cross-regulatory activity within the class. In transgenic lines carrying an isolated copy of a *cis*-regulatory element of a homeotic gene, the *Pc* protein was found associated with the transgene at the site of integration (60). This reflects an obvious specificity in the association of the protein with chromosomes. However, neither known DNA-binding motifs nor direct *in vitro* DNA-binding activity of the protein has been identified, implying that the specificity is modulated through additional factors. A 48 aa interval has been identified near the amino terminus that shows homology to HP1 (35, 57) (Fig. 3). This conserved protein motif has been termed the 'chromo domain' (for chromatin organization modifier, see below).

In addition to producing the homeotic transformations typical of embryos defective for *Pc*-G genes, mutations in *polyhomeotic* (*ph*) cause extensive cell death of the

```
(a)   Dm HP1   EEYAVEKII  DRRVRK GKVEYYLKWKGYPE TENIWEPEN NLDCQDLI
      Dv HP1   EEYAVEKIL  DRRVRK GKVEYYLKWKGYAE TENIWEPEG NLDCQDLI
      Mm M31   EEYVVEKVL  DRRVVK GKVEYLLKWKGFSD EDNIWEPEE NLDCPDLI
      Mm M32   EEFVVEKVL  DRRVVN GKVEYFLKWKGFTD ADNIWEPEE NLDCPELI
      HsM1     EEYVVEKVL  DRRVVK GKVEYLLKWKGFSD EDNIWEPEE NLDCPDLI
      HsaHP1   EEYVVEKVL  DRRVVK GQVEYLLKWKGFSE EHNIWEPEK NLDCPELI
      Pchet1   EEYVVEKII  DKRTVN GVKQYFLKWKGYDE SENIWEPHE NLECPELI
      Pchet2   EEFIVEKIL  DKRTEPDGSVRYLLKWKGYGD EDNIWEPPE NMDCPDLL
               EEy VEKIL  DrRv   G v Y LKWKGy e   NIWEP   Nld qdLi
                 f          k t    k       f d             me pe l

(b)   Dm Pc    LVYAAEKIIQKRVKK GVVEYRVKWKGWNQRYNTWEPEVNILDRRLI
      MmM33    QVFAAECILSKRLRK GKLEYLVKWRGWSSKHNSWEPEENILDPRLL
                V AAE I  KR  K G  EY VKW GW     N WEPE NILD RL

(c)   Ce est   KLYTVESILEHR  KKKGKSEFYIKWLGYDH THNSWEPKE NIVDPTLI
      swi6     DEYVVEKVLKHRMARKGGGYEYLLKWEGYDDPSDNIWSSEADCSGCKQLI
      CHD-1    EADGDPNAGFERN KEPGDIQYLIKWKGW SHIHNIWETEETLKQQNVR
```

Fig. 3 Comparative anatomy of chromo domain sequences. (a) Chromo domain sequences from HP1-related proteins: Dm HP1: *Drosophila melanogaster* (35, 37); Dv HP1: *D. virilis* (94); Mm M31, Mm M32: mouse (79); HsM1: human (79); HsaHP1: human (80); Pchet 1, Pchet 2: mealybug (95). (b) Chromo domain sequences from *Polycomb* protein and a related protein from mouse: Dm Pc: *D. melanogaster Polycomb* protein (57); MmM33: mouse *Polycomb* protein homologue (96). (c) Chromo domain sequence from Ce est: nematode (97); Swi6; *S. pombe* (81); and a novel mouse DNA-binding protein, CHD-1 (78). Shown at the bottom of panels (a) and (b) are consensus amino acids; lower-case characters indicate positions at which one of two alternative amino acids are found for HP1.

central epidermis, an abnormal pattern of axonal pathways in the CNS, and abnormalities in the patterns of the segmentation genes (61, 62). In contrast to the homogeneous pattern of *Pc* expression, the early transcript distribution of *ph* shows a pattern that suggests regulation by the early segmentation genes (63). Based on this finding, it has been proposed that *ph* could be a member of the *Pc*-G which relays the information between the initiation and maintenance phases of homeotic expression. In later stages, however, the *ph* protein shows a homogeneous distribution similar to that of the *Pc* protein (64). The 169 kD *ph* protein contains a single zinc-finger motif, a serine–threonine-rich region, and glutamine repeats, suggestive of a potential DNA-binding protein (65).

The embryonic phenotype of *Posterior sex combs* (*Psc*) mutants shows only weak homeotic transformations (66). This is due to a strong maternal contribution. Extensive maternal deposition of gene products into the unfertilized egg characterizes the entire class and probably reflects the need for large quantities of these proteins during embryonic development. The developmental profile of *Psc* protein expression coincides with the previously described pattern of *Pc* protein, being essentially homogeneous in all tissues tested during most of the embryonic and larval stages (66). The sequence of the protein reveals a scattered arrangement of zinc fingers. Interestingly, a 200 aa stretch encompassing these zinc fingers has been conserved during evolution and is also found in the mouse Bmi-1 protein and a related Mel-18 protein (67, 68). The murine *bmi-1* gene had previously been identified as a proto-oncogene (69).

3.2 *Enhancer of zeste* and the antipodal *trithorax* group

The *Enhancer of zeste (E(z))* gene has a homeotic phenotype that classifies the gene as a member of the *Pc*-G (52). Molecular analysis predicted a 760 aa protein with a cysteine-rich cluster but lacking any known functional motifs (70). In the carboxy terminus, a region of 116 amino acids shows a strong similarity to the *trithorax* gene product, a member of the *trithorax* group (*trx*-G) of genes. In contrast to the *Pc*-G, this class is considered to be important for maintaining the active state of homeotic gene expression during development (71). The molecular mechanism by which the products of the *trx*-G promote activation of transcription is not known. However, sequence homology between the *brahma* protein, a *trx*-G member, and the yeast Swi2/Snf2 factor, suggests that the process has been conserved (72). The Swi/Snf proteins of yeast are thought to facilitate transcriptional activation by altering chromatin structure and thus accessibility for transcription factors (see Chapter 5). The common protein motif shared by the *E(z)* and *trx* products might indicate that the proteins interact with a common target. Additional factors would subsequently determine the positive or negative state of the transcriptional regulation (70).

3.3 The chromo domain: a molecular link between PEV and homeotic gene silencing

There is growing evidence for a direct mechanistic connection between the modifiers of PEV and the *Pc*-G. Some of the loci originally identified as members of the *Pc*-G have also been found to act as modifiers of PEV (73). Conversely, certain modifiers of PEV cause weak homeotic transformations (34). Based on these findings, it has been suggested that the *Pc*-G uses a mechanism analogous to heterochromatin formation to transmit stably and heritably the inactive state of transcription (57, 74).

Cytological observations also support the analogy between silent homeotic chromatin and heterochromatin: the cytological region 89E, location of the BX-C, is a relatively densely staining constriction in salivary gland polytene chromosomes (where these genes are not expressed), but not in fat body chromosomes (where these genes are expressed) (75–76). Similarly, the *Ultrabithorax* sequences at 89E are late replicating in salivary glands but not in fat body (76). As noted above, dense staining, underreplication in polytene tissue, and late replication are all properties of heterochromatin as well.

The functional significance of the chromo domain homology remains unclear. HP1 and Polycomb protein have generally non-overlapping distributions on polytene chromosomes (36, 59) (Fig. 1c). The chromo domain is dispensable for nuclear localization and heterochromatin binding of HP1 (40), but appears to be essential for gene silencing in that a missense mutation in this region is a suppressor of variegation (T. Hartnett and J. C. Eissenberg, unpublished data). Interestingly, two *Pc* alleles have been isolated with point mutations in the chromo domain, and these mutations prevent the protein from binding to its target sequences in polytene chromosomes (77). The recent cloning of a mouse cDNA encoding a chromo domain-containing protein that also has a motif highly related to a helicase domain found in a number of transcriptional activators (78) seems paradoxical if the chromo domain is a mediator of gene silencing. However, this is reminiscent of the yeast Rap1 protein, which has clearly been shown to play a role both in activating and in repressing complexes (see Chapter 7). At present, it seems most likely that the chromo domain is involved in mediating a class of protein–protein interactions at the chromosomal level.

Reduced-stringency Southern blot hybridization using the chromo domain sequence as a probe indicates that this motif is found in many phyla (79); cDNA clones encoding HP1-like proteins from a wide variety of organisms show considerable evolutionary conservation both in the N-terminal chromo domain and in a C-terminal domain. Fig. 3 shows the degree of structural conservation found in chromo domains of HP1-like and otherwise unrelated proteins from diverse sources. In most cases, however, a functional role for these HP1-related proteins has not yet been established. A human HP1-like protein is associated with a subnuclear structure termed PIKA (polymorphic interphase karyosomal association (80)), which might suggest an involvement of chromo domain-containing proteins in nuclear compartmentalization of heterochromatin-like domains. In fission yeast, a gene originally identified as a represssor of the silent mating-type loci (*SWI6* of *S. pombe*) was found to encode a protein with a chromo domain (81). As this type of regulation is also considered to use an epigenetic silencing mechanism (see Chapter 7), this finding further underscores the important role chromo domain-containing proteins seem to play in chromatin organization.

3.4 *Pc*-G multimeric protein complexes

Double *Pc*-G mutant combinations result in stronger homeotic transformations than expected from the sum of the single phenotypic effects (51). During embryonic development, the *engrailed* gene becomes extensively derepressed in double *Pc*-G mutants, but only weakly derepressed in a singly mutant background

(82). In addition, the phenotypic consequences vary with different doses of the target genes. These synergistic effects suggest that the *Pc*-G proteins co-operate at a molecular level. These findings are reminiscent of the co-operative effects that have been proposed for the components of heterochromatin.

There is indeed accumulating evidence that the genetically defined *Pc*-G includes members that can interact in multimeric protein complexes. Immunoprecipitation experiments using embryonic nuclear extracts have shown that *Pc* protein is associated with 10–15 different proteins (64). Among these, the *ph* and *Psc* proteins were found to be co-precipitated with *Pc* protein (64; A. Franke, M. van Lohuizen, A. Berns, and R. Paro, unpublished results). This biochemical evidence of complex formation has been further substantiated by results showing co-localization of the proteins on the polytene chromosomes (64–66, 83) (Fig. 4).

Additional evidence for a functional role of some *Pc*-G proteins in multimeric complexes comes from the analysis of mutant proteins. A temperature-sensitive *E(z)* mutant, grown at a non-permissive temperature during the larval period, shows polytene chromosomes with a decondensed appearance and, in particular, a disruption of the binding of other *Pc*-G proteins such as *ph*, *Psc*, and *Pc* (83).

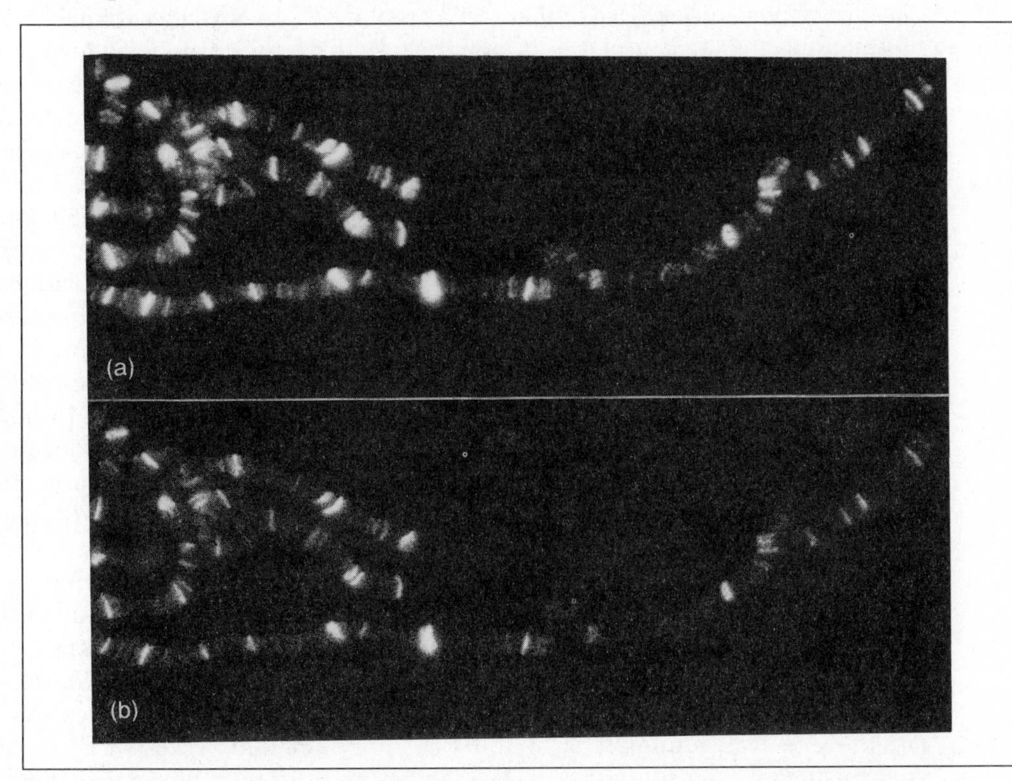

Fig. 4 Double-immunofluorescence of salivary gland polytene chromosomes showing the overlapping distributions of the *Polycomb* and *polyhomoeotic* proteins. (a) Staining with polyclonal mouse anti-Pc and DTAF-conjugated anti-mouse antibodies. (b) Staining with polyclonal rabbit anti-ph antibodies and rhodamine-conjugated anti-rabbit antibodies.

Mutations in the *Pc* chromo domain redistribute the *Pc* protein into a homogeneous pattern in the nucleus in comparison with the speckled localization in wild-type nuclei (77). Interestingly, in such mutants, the *ph* protein becomes subjected to the same kind of redistribution (A. Franke and R. Paro, unpublished results). Similar changes in distribution have been reported for mutations in some of the proteins involved in silencing in yeast (see Chapter 7). The aberrant function of one member of the complex seems to have a dramatic effect on the localization of the other components. The results indicate that these silencing systems function through the formation of multimeric complexes.

3.5 *Pc*-G response elements and DNA-binding specificity

A puzzling aspect of the *Pc*-G silencing is the fact that all members identified so far seem to be homogeneously expressed in all cells. A ubiquitous requirement for function is further supported by the widespread and pleiotropic mutant phenotypes. However, repressive patterns can vary substantially between cells, depending on the cell's position in the embryo. What tags a gene to be inactivated by the *Pc*-G remains unclear. That there is a definite DNA specificity imposed on the formation of *Pc*-G complexes is demonstrated by the protein patterns found in polytene chromosomes (see Fig. 4); however, in no case has any *in vitro* sequence specificity yet been demonstrated. Thus, alternative methods have been applied to identify the sequences through which the *Pc*-G acts. Reporter gene constructs with various *cis*-regulatory elements of potential *Pc*-G targets were found to be differentially repressed during embryonic and larval development (60, 84–86). In many cases, it was found that larger DNA fragments maintained the repressive effect better than smaller constructs, which often exhibited strong position effects. This suggested that the larger the size of the silencing complex established by the *Pc*-G proteins, the better the repression is sustained over developmental time. Those DNA elements that are subject to regulation by the *Pc*-G have been named PREs (*Pc*-G response elements; (85)). That a direct interaction of *Pc*-G proteins with the defined regulatory elements does exist can be demonstrated by immunostaining polytene chromosomes of transgenic lines (60, 65); at the integration site of the transgene containing a particular *cis*-regulatory element, a new *Pc*-G protein-binding site can be observed.

A further link between *Pc*-G-regulated silencing and heterochromatin-induced PEV was found in transgenic lines with a reporter gene construct carrying a PRE from the *polyhomeotic* regulatory region (87). This element induced mosaic expression not only of the *lacZ* reporter gene but also of a distantly located *white* gene, used as a marker in the P-element transformation vector. The mosaic expression of *white* in the eye was reminiscent of the PEV phenomenon, except that it could be modulated only by mutations in *Pc*-G genes and not by mutations in modifiers of PEV. This phenomenon has been termed DREV (developmental regulator effect variegation). The result demonstrates that *Pc*-G protein complexes, like heterochromatin, can have long-range silencing effects.

A high resolution map of such silencing complexes has been obtained in the BX-C region by applying a novel cross-linking strategy (88). Tissue culture cells are fixed *in vivo* with formaldehyde and chromatin is isolated and reduced in size by shearing. Because of the cross-linking, proteins as well as DNA fragments associated with the *Pc* protein could be purified by immunoprecipitation with anti-*Pc* antibodies. The recovered DNA fragments, enriched in PREs, were PCR amplified and used to probe a genomic walk of the BX-C. Interestingly, all sequences encompassing the *Ultrabithorax* (*Ubx*) and *abdominal-A (abd-A)* genes were represented in the enriched fraction, while the *Abdominal-B* (*Abd-B*) domain was apparently free of *Pc* protein (Fig. 5). Subsequent analysis revealed that, in these cells, *Ubx* and *abd-A* are silent, whilst the *Abd-B* gene is active. This result suggests that *Pc* and probably other associated *Pc*-G proteins can cover large chromosomal domains of repressed genes. In addition to the widespread distribution, certain regulatory fragments of the BX-C, were found to be particularly enriched for the *Pc* protein. Examples include the Mcp and Fab-7 elements, previously identified as potential border and PRE elements (89, 90; see Chapter 9). An attractive interpretation envisages these PREs as nucleation signals which induce packaging into silenced chromatin by self-assembly of *Pc*-G complexes.

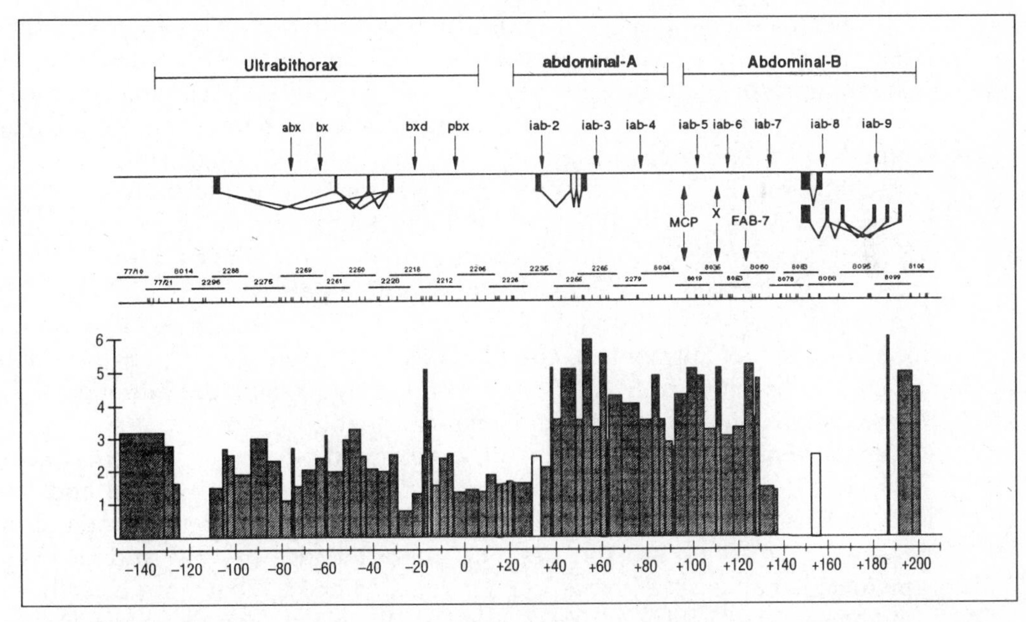

Fig. 5 *Polycomb* protein distribution in the BX-C of *Drosophila* SL-2 tissue culture cells. The organization of the complex is depicted in the upper part of the figure. The regions of the *Ubx*, *abd-A*, and *Abd-B* genes are drawn with their respective regulatory regions identified from genetic rearrangements. Transcription units are shown below with the corresponding exon structures as black boxes. The location of the presumptive boundary elements *Mcp* between *iab-4* and *iab-5*, *Fab-7* between *iab-6* and *iab-7*, and of a postulated element X between *iab-5* and *iab-6* are shown. Below, the line depicts the distribution of *Eco*RI sites in the BX-C of Canton-S wild-type flies, and the λ-clones covering the region. The bottom part of the figure shows the distribution of the *Pc* protein in the complex as deduced from the quantitation of the hybridization intensities of the immunoprecipitated chromatin fragments. From (88), reproduced by permission of Cell Press.

4. Final comments

4.1 Important questions

The data reviewed above highlight the many similarities between heterochromatic regions (particularly those created by rearrangements or insertions of markers that result in PEV) and developmental memory complexes. Both can result in a gene inactivation that is mitotically stable. Both result in dense staining, late replication, and possible underreplication in polytene tissue. Both appear to be products of multiprotein complexes, involving a protein that carries a chromo domain. All of these similarities suggest that, in both cases, formation of a relatively condensed chromatin structure has occurred. During the next few years, one can anticipate that chromatin structure mapping studies will be carried out to analyse the relationship between the nucleosome array and potential higher order organization at both heterochromatic sites and developmental memory complexes.

A major question to be resolved is the nature of the 'start' and 'stop' signals involved in the formation of such packaging domains. The size of the regions affected clearly makes propagation of a self-assembled structure attractive; in the case of PEV a gradient of repression, reminiscent of that seen in the yeast telomeric regions, is observed, albeit on a much larger scale. Experiments with transgenes have clearly indicated the presence of PRE (*Pc*-G response elements), although the minimal element has not been identified. The 'start' site for heterochromatin formation remains a mystery— although there are some clues that suggest that repeated DNA elements may be sufficient, acting through some sort of pairing mechanism (98). This idea, however, raises immediate questions as to how multi-copy genes, such as those encoding the histones, remain active.

That boundaries (stop sites) exist for these self-assembled chromatin structures is certainly implied by the discrete distribution patterns of the identified proteins involved, such as HP1 and *Pc* (see Figs 1, 4, 5). In fact, a possible boundary for the developmental memory complex has been identified; the *Fab-7* element can limit repression within *abd-A* (see Chapter 9). Again, the situation for heterochromatin is more problematic. Heterochromatin formation can spread over hundreds of kilobases of DNA during PEV, no doubt covering several special chromatin structure (scs) elements in the process. However, the structural and functional characteristics of the fourth chromosome suggest that discrete boundaries do exist.

Most mysterious are the defining characteristics of PEV and developmental memory complexes. How is memory maintained? What do we really mean by 'higher order structure'? How ordered are the alternative structures, or packaging modes, and how does this impinge on the gene activation process? The answers to these questions, and other questions not yet defined, await further experimentation.

4.2 Discussion

Several interesting questions were raised in discussions with other authors. One would like to have direct evidence of any influence of heterochromatin and/or Pc-

G proteins on large scale elements of nuclear structure and organization. There is little such evidence at present: however, it has long been recognized that heterochromatin is normally associated with the nuclear membrane in interphase nuclei. The fact that large scale chromosome rearrangements are well tolerated (excepting the genes at or very near the breakpoints themselves) argues against an essential role for fixed nuclear compartments. In specific *Drosophila* genotypes, however, trans-interactions between loci on homologous chromosomes have been detected (e.g., bw^D, transvection); such interactions are efficiently disrupted by chromosome rearrangements. The proteins governing such pairing-dependent trans-regulation are now being characterized, and heterochromatin-associated and/or Pc-G proteins will likely be among them.

As mentioned above, many investigators expect that there are DNA sequences that are both necessary and sufficient to trigger formation of heterochromatin. An explicit test for the ability of a particular sequence to behave as an 'initiator' of heterochromatin would be to fuse such a sequence to a reporter gene (e.g. *white*) and introduce the construct into the fly genome by germ line transformation, testing for sequences that impose variegated silencing on the reporter. Many of the satellite and middle repetitive DNAs found in heterochromatin have been cloned and could be tested; however, it may be that large tandem arrays of such sequences will be necessary, increasing the technical difficulty of such experiments.

Both biochemical and genetic strategies have been employed successfully to identify the protein components of heterochromatin and of the Pc-G complexes. In the future, immunological probes to identified proteins will increasingly be used to define interacting members of protein complexes (immunoprecipitation, colocalization in polytene chromosomes, etc.); screens for interacting proteins conducted in yeast appears to be another promising avenue. Genetic results indicate that these systems are complex (10–100 proteins), so much work remains to be done.

Acknowledgements

We thank our colleagues and fellow authors for comments on the manuscript. Work described herein from the Eissenberg lab has been supported by NIH grant GM40732; and Grant No.DB-55 from American Cancer Society; from the Elgin lab by NIH grants HD23844 and GM31532; from the Paro lab by a grant from the Deutsche Forschungsgemeinschaft.

References

1. Cartwright, I. L., Hertzberg, R. P., Devan, P. B., and Elgin, S. C. R. (1983) Recognition of the nucleosomal structure of chromatin by a cleavage reagent with low sequence preference: (methidium propyl-EDTA) iron (II). *Proc. Natl Acad. Sci. USA*, **80,** 3212.
2. Urieli-Shoval, S., Gruenbaum, Y., Sedat, J., and Razin, A. (1982) The absence of detectable methylated bases in *Drosophila melanogaster* DNA. *FEBS Lett.*, **146,** 148.

3. Heitz, E. (1928) Das heterochromatin der Moose. *Jb. Wiss. Bot.*, **69,** 728.
4. Eissenberg, J. C. (1989) Position effect variegation in *Drosophila*: towards a genetics of chromatin assembly. *Bioessays*, **11,** 14.
5. Lyon, M. F. (1972) X-chromosome inactivation and developmental patterns in mammals. *Biol. Rev.*, **47,** 1.
6. Spofford, J. B. (1976) Position-effect variegation in *Drosophila*. In *The genetics and biology of Drosophila, Vol. 1c.* Ashburner, M. and Novitski, E. (ed.). Academic Press, New York, p. 955.
7. Henikoff, S. (1981) Position-effect variegation and chromosome structure of a heat shock puff in *Drosophila. Chromosoma*, **83,** 381.
8. Reuter, G., Werner, W., and Hoffman, H. J. (1982) Mutants affecting position effect heterochromatinization in *Drosophila melanogaster. Chromosoma*, **85,** 539.
9. Kornher, J. S. and Kauffman, S. A. (1986) Variegated expression of the *Sgs-4* locus in *Drosophila melanogaster. Chromosoma*, **94,** 205.
10. Belyaeva, E. S., Demakova, O. V., Umbetova, G. H., and Zhimulev, I. F. (1993) Cytogenetic and molecular aspects of position-effect variegation in *Drosophila melanogaster. Chromosoma*, **102,** 583.
11. Laird, C. D., Chooi, W. Y., Cohen, E. H., Dickson, E., Hutchinson, N., and Turner, S. H. (1973) Organization and transcription of DNA in chromosomes and mitochondria of *Drosophila. Cold Spring Harbor Symp. Quant. Biol.*, **38,** 311.
12. Karpen, G. H. and Spradling, A. C. (1990) Reduced DNA polytenization of a minichromosome region undergoing position-effect variegation in *Drosophila. Cell*, **63,** 97.
13. Rushlow, C. A., Bender, W., and Chovnick, A. (1984) Studies on the mechanism of heterochromatin position effect at the *rosy* locus of *Drosophila melanogaster. Genetics*, **108,** 603.
14. Hilliker, A. J., Appels, R., and Schalet, A. (1980) The genetic analysis of *Drosophila melanogaster* heterochromatin. *Cell*, **21,** 607.
15. Wakimoto, B. T. and Hearn, M. G. (1990) The effects of chromosome rearrangements on the expression of heterochromatic genes in chromosme 2L of *Drosophila melanogaster. Genetics*, **125,** 141.
16. Eberl, D. F., Duyf, B. J., and Hilliker, A. J. (1993) The role of heterochromatin in the expression of a heterochromatic gene, the *rolled* locus of *Drosophila melanogaster. Genetics*, **134,** 227.
17. Stern, C. and Kodani, M. (1955) Studies on the position effect at the *cubitus interruptus* locus of *Drosophila melanogaster. Genetics*, **40,** 343.
18. Henikoff, S. and Dreesen, T. D. (1989) Trans-inactivation of the *Drosophila brown* gene: evidence for transcriptional repression and somatic pairing dependence. *Proc. Natl Acad. Sci. USA*, **86,** 6704.
19. Dreesen, T. D., Henikoff, S., and Loughney, K. (1991) A pairing-sensitive element that mediates *trans*-inactivation is associated with the *Drosophila brown* gene. *Genes and Dev.*, **5,** 331.
20. Slatis, H. M. (1955) A reconsideration of the *brown-dominant* position effect. *Genetics*, **40,** 246.
21. Talbert, P. B., LeCiel, C. D. S., and Henikoff, S. (1994) Modification of the *Drosophila* heterochromatic mutation *brown*^{Dominant} by linkage alterations. *Genetics*, **136,** 559.
22. Tartof, K. D., Hobbs, C., and Jones, J. (1984) A structural basis for variegating position effects. *Cell*, **37,** 869.

23. Zuckerkandl, E. (1974) Recherches sur les proprietes et l'activite biologique de la chromatine. *Biochimie*, **56,** 937.
24. Locke, J., Kotarski, M. A., and Tartof, K. D. (1988) Dosage dependent modifiers of position effect variegation in *Drosophila* and a mass action model that explains their effect. *Genetics*, **120,** 181.
25. Hilliker, A. J. and Appels, R. (1982) Pleiotropic effects associated with the deletion of heterochromatin surrounding rDNA on the X chromosome of *Drosophila. Chromosoma*, **86,** 469.
26. Lifton, R. P., Goldberg, M. L., Karp, R. W., and Hogness, D. S. (1978) The organization of the histone genes in *Drosophila melanogaster*: functional and evolutionary implications. *Cold Spring Harbor Symp. Quant. Biol.*, **42,** 1047.
27. Moore, G. D., Sinclair, D. A. R., and Grigliatti, T. A. (1983) Histone gene multiplication and position-effect variegation in *Drosophila melanogaster. Genetics*, **105,** 327.
28. Mottus, R., Reeves, R., and Grigliatti, T. A. (1980) Butyrate suppression of position-effect variegation in *Drosophila melanogaster. Mol. Gen. Genet.*, **178,** 465.
29. Michailidis, J., Graves, J. A. M., and Murray, N. D. (1989) Suppression of position-effect variegation in *Drosophila melanogaster* by fatty acids and dimethylsulphoxide: implications for the mechanism of position-effect variegation. *J. Genet.*, **68,** 1.
30. Turner, B. M., Birley, A. J., and Lavender, J. (1992) Histone H4 isoforms acetylated at specific lysine residues define individual chromosomes and chromatin domains in *Drosophila* polytene nuclei. *Cell*, **69,** 375.
31. Sinclair, D. A. R., Mottus, R. C., and Grigliatti, T. A. (1983) Genes which suppress position-effect variegation in *Drosophila melanogaster* are clustered. *Mol. Gen. Genet.*, **191,** 326.
32. Wustmann, G., Szidonya, J., Taubert, H., and Reuter, G. (1989) The genetics of position-effect variegation modifying loci in *Drosophila melanogaster. Mol. Gen. Genet.*, **217,** 520.
33. Reuter, G., Giarre, M., Farah, J., Gausz, J., Spierer, A., and Spierer, P. (1990) Dependence of position-effect variegation in *Drosophila* on dose of a gene encoding an unusual zinc-finger protein. *Nature*, **344,** 219.
34. Dorn, R., Krauss, V., Reuter, G., and Saumweber, H. (1993) The enhancer of position-effect variegation of *Drosophila, E(var)3-93D*, codes for a chromatin protein containing a conserved domain common to several transcriptional regulators. *Proc. Natl Acad. Sci. USA*, **90,** 11376.
35. James, T. C. and Elgin, S. C. R. (1986) Identification of a nonhistone chromosomal protein associated with heterochromatin in *Drosophila* and its gene. *Mol. Cell. Biol.*, **6,** 3862.
36. James, T. C., Eissenberg, J. C., Craig, C., Dietrich, V., Hobson, A., and Elgin, S. C. R. (1989) Distribution patterns of HP1, a heterochromatin-associated nonhistone chromosomal protein of *Drosophila. Eur. J. Cell Biol.*, **50,** 170.
37. Eissenberg, J. C., James, T. C., Foster-Hartnett, D. M., Hartnett, T., Ngan, V., and Elgin, S. C. R. (1990) Mutation in a heterochromatin-specific chromosomal protein is associated with suppression of position-effect variegation in *Drosophila melanogaster. Proc. Natl. Acad. Sci. USA*, **87,** 9923.
38. Eissenberg, J. C., Morris, G. D., Reuter, G., and Hartnett, T. (1992) The heterochromatin-associated protein HP1 is an essential protein in *Drosophila* with dosage-dependent effects on position-effect variegation. *Genetics*, **131,** 345.
39. Eissenberg, J. C. and Hartnett, T. (1993) A heat shock-activated cDNA rescues the

recessive lethality of mutations in the heterochromatin-associated protein HP1 of *Drosophila melanogaster*. *Mol. Genet.*, **240,** 333.
40. Powers, J. A. and Eissenberg, J. C. (1993) Overlapping domains of the heterochromatin-associated protein HP1 mediate nuclear localization and heterochromatin binding. *J. Cell Biol.*, **120,** 291.
41. Reuter, G., Wolff, I., and Friede, B. (1985) Functional properties of the heterochromatic sequences inducing w^{m4} position-effect variegation in *Drosophila melanogaster*. *Chromosoma*, **93,** 132.
42. Pokholkova, G. V., Makunin, I. V., Belyaeva, E. S., and Zhimulev, I. F. (1993) Observations on the induction of position effect variegation of euchromatic genes in *Drosophila melanogaster*. *Genetics*, **133,** 231.
43. Miklos, G. L. G., Yamamoto, M.-T., Davies, J., and Pirrotta, V. (1988) Microcloning reveals a high frequency of repetitive sequences characteristic of chromosome 4 and the β-heterochromatin of *Drosophila melanogaster*. *Proc. Natl Acad. Sci. USA*, **85,** 2051.
44. Wilson, C., Bellen, H. J., and Gehring, W. J. (1990) Position effects on eukaryotic gene expression. *Annu. Rev. Cell Biol.*, **6,** 679.
45. Hazelrigg, T., Levis, R., and Rubin, G. M. (1984) Transformation of *white* locus DNA in *Drosophila*: dosage compensation, *zeste* interaction, and position effects. *Cell*, **36,** 469.
46. Steller, H. and Pirrotta, V. (1985) Expression of the *Drosophila white* gene under the control of the *hsp70* heat shock promoter. *EMBO J.*, **4,** 3765.
47. Bishop, C. P. (1992) Evidence for intrinsic differences in the formation of chromatin domains in *Drosophila melanogaster*. *Genetics*, **132,** 1063.
48. Lawrence, P. A. (1992) *The making of a fly*. Blackwell Scientific Publications, Oxford.
49. Duncan, I. and Lewis, E. B. (1982) Genetic control of body segment differentiation in *Drosophila*. In *Developmental order: its origin and regulation*. Subtelny, S. and Green, P. B. (ed.). Alan Liss, New York, p. 533.
50. Struhl, G. (1981) A gene product required for the correct initiation of segmental determination in *Drosophila*. *Nature*, **293,** 36.
51. Jurgens, G. (1985) A group of genes controlling the spatial expression of the *bithorax* complex in *Drosophila*. *Nature*, **316,** 153.
52. Jones, R. S. and Gelbart, W. M. (1990) Genetic analysis of the *Enhancer of zeste* locus and its role in gene regulation in *Drosophila melanogaster*. *Dev. Biol.*, **118,** 442.
53. Struhl, G. and Akam, M. (1985) Altered distributions of *Ultrabithorax* transcripts in *extra sex combs* mutant embryos of *Drosophila*. *EMBO J.*, **4,** 3259.
54. McKeon, J. and Brock, H. W. (1991) Interactions of the *Polycomb* group genes with homeotic loci of *Drosophila*. *Wilhelm Roux's Arch. Dev. Biol.*, **199,** 387.
55. Simon, J., Chiang, A., and Bender, W. (1992) Ten different *Polycomb* group genes are required for spatial control of the *abdA* and *AbdB* homeotic products. *Development*, **114,** 493.
56. Phillips, M. D. and Shearn, A. (1990) Mutations in *polyhomeotic*, a *Drosophila* Polycomb-group gene, cause a wide range of maternal and zygotic phenotypes. *Genetics*, **125,** 91.
57. Paro, R. and Hogness, D. S. (1991) The Polycomb protein shares a homologous domain with a heterochromatin-associated protein of *Drosophila*. *Proc. Natl Acad. Sci. USA*, **88,** 263.
58. Paro, R. and Zink, B. (1992) The *Polycomb* gene is differentially regulated during oogenesis and embryogenesis of *Drosophila melanogaster*. *Mech. Dev.*, **40,** 37.
59. Zink, B. and Paro, R. (1989) *In vivo* binding pattern of a trans-regulator of homoeotic genes in *Drosophila melanogaster*. *Nature*, **337,** 468.

60. Zink, B., Engstrom, Y., Gehring, W. J., and Paro, R. (1991) Direct interaction of the *Polycomb* protein with *Antennapedia* regulatory sequences in polytene chromosomes of *Drosophila melanogaster*. *EMBO J.*, **10**, 153.
61. Smouse, D., Goodman, C., Mahowald, A., and Perrimon, N. (1988) *polyhomeotic*: a gene required for the embryonic development of axon pathways in the central nervous system of *Drosophila*. *Genes and Dev.*, **2**, 830.
62. Dura, J.-M. and Ingham, P. (1988) Tissue- and stage-specific control of homeotic and segmentation gene expression in *Drosophila* embryos by the *polyhomeotic* gene. *Development*, **103**, 733.
63. Deatrick, J. (1993) Localization *in situ* of *polyhomeotic* transcripts in *Drosophila melanogaster* reveals spatially restricted expression beginning at the blastoderm stage. *Dev. Genet.*, **13**, 326.
64. Franke, A., DeCamillis, M., Zink, D., Cheng, N., Brock, H. W., and Paro, R. (1992) *Polycomb* and *polyhomeotic* are constituents of a multimeric protein complex in chromatin of *Drosophila melanogaster*. *EMBO J.*, **11**, 2941.
65. DeCamillis, M., Cheng, N., Pierre, D., and Brock, H. W. (1992) The *polyhomeotic* gene of *Drosophila* encodes a chromatin protein that shares polytene chromosome-binding sites with *Polycomb*. *Genes and Dev.*, **6**, 223.
66. Martin, E. C. and Adler, P. N. (1993) The *Polycomb* group gene *Posterior sex combs* encodes a chromosomal protein. *Development*, **117**, 641.
67. Brunk, B. P., Martin, E., and Adler, P. N. (1991) *Drosophila* genes *Posterior sex combs* and *Suppressor two of zeste* encode proteins with homology to the murine *bmi-1* oncogene. *Nature*, **353**, 351.
68. van Lohuizen, M., Frasch, M., Weintjens, E., and Berns, A. (1991) Sequence similarity between the mammalian *bmi-1* proto-oncogene and the *Drosophila* regulatory genes *Psc* and *Su(z)2*. *Nature*, **353**, 353.
69. van Lohuizen, M., Verbeek, S., Scheijen, B., Wientjens, E., van der Bulden, H., and Berns, A. (1991) Identification of cooperating oncogenes in Eμ-*myc* transgenic mice by provirus tagging. *Cell*, **65**, 737.
70. Jones, R. S. and Gelbart, W. M. (1993) The *Drosophila Polycomb*-group gene *Enhancer of zeste* contains a region with sequence similarity to *trithorax*. *Mol. Cell. Biol.*, **13**, 6357.
71. Kennison, J. A. (1993) Transcriptional activation of *Drosophila* homeotic genes from distant regulatory elements. *Trends Genet.*, **9**, 75.
72. Tamkun, J. W., Deuring, R., Scott, M. P., Kissinger, M., Pattatucci, A. M., Kaufman, T. C., and Kennison, J. A. (1992) *brahma*: a regulator of *Drosophila* homeotic genes structurally related to the yeast transcriptional activator SNF2/SWI2. *Cell*, **68**, 561.
73. Grigliatti, T. (1991) Position-effect variegation—an assay for nonhistone chromosomal proteins and chromatin assembly and modifying factors. *Meth. Cell. Biol.*, **35**, 587.
74. Paro, R. (1990) Imprinting a determined state into the chromatin of *Drosophila*. *Trends Genet.*, **6**, 416.
75. Akam, M. E. (1983) The location of *Ultrabithorax* transcripts in *Drosophila* tissue sections. *EMBO J.*, **2**, 2075.
76. Laird, C., Jaffe, E., Karpen, G., Lamb, M., and Nelson, R. (1987) Fragile sites in human chromosomes as regions of late-replicating DNA. *Trends Genet.*, **3**, 274.
77. Messmer, S., Franke, A., and Paro, R. (1992) Analysis of the functional role of the *Polycomb* chromodomain in *Drosophila melanogaster*. *Genes and Dev.*, **6**, 1241.
78. Delmas, V., Stokes, D. G., and Perry, R. P. (1993) A mammalian DNA-binding protein

that contains a chromodomain and an SNF2/SWI2-like helicase domain. *Proc. Natl Acad. Sci. USA*, **90,** 2414.

79. Singh, P. B., Miller, J. R., Pearce, J., Kothary, R., Burton, R. D., Paro, R., James, T. C., and Gaunt, S. J. (1991) A sequence motif found in a *Drosophila* heterochromatin protein is conserved in animals and plants. *Nucleic Acids Res.*, **19,** 789.
80. Saunders, W. S., Chue, C., Goebl, M., Craig, C., Clark, R. F., Powers, J. A., Eissenberg, J. C., Elgin, S. C. R., Rothfield, N. F., and Earnshaw, W. C. (1992) Molecular cloning of a human homologue of *Drosophila* heterochromatin protein HP1 using anti-centromere autoantibodies with anti-chromo specificity. *J. Cell Sci.*, **104,** 573.
81. Lorentz, A., Osterman, K., Fleck, O., and Schmidt, H. (1994) Switching gene swi6, involved in the repression of silent mating-type loci in fission yeast, encodes a homologue of chromatin-associated proteins from *Drosophila* and mammals. *Gene*, **143,** 139.
82. Moazed, D. and O'Farrell, P. H. (1992) Maintenance of the *engrailed* expression pattern by *Polycomb* group genes in *Drosophila*. *Development*, **116,** 805.
83. Rastelli, L., Chan, C. S., and Pirrotta, V. (1993) Related chromosome binding sites for *zeste*, suppressors of *zeste* and *Polycomb* group proteins in *Drosophila* and their dependence on *Enhancer of zeste* function. *EMBO J.*, **12,** 1513.
84. Muller, J. and Bienz, M. (1991) Long range repression conferring boundaries of *Ultrabithorax* expression in the *Drosophila* embryo. *EMBO J.*, **10,** 3147.
85. Simon, J., Chiang, A., Bender, W., Shimell, M. J., and O'Connor, M. (1993) Elements of the *Drosophila* bithorax complex that mediate repression by *Polycomb* group products. *Dev. Biol.*, **158,** 138.
86. Busturia, M. and Bienz, M. (1993) Silencers in *Abdominal-B*, a homeotic *Drosophila* gene. *EMBO J.*, **12,** 1415.
87. Fauvarque, M.-O. and Dura, J.-M. (1993) *polyhomeotic* regulatory sequences induce developmental regulator-dependent variegation and targeted P-element insertion in *Drosophila*. *Genes and Dev.*, **7,** 1508.
88. Orlando, V. and Paro, R. (1993) Mapping *Polycomb*-repressed domains in the *bithorax* complex using *in vivo* formaldehyde cross-linked chromatin. *Cell*, **75,** 1187.
89. Gyurkovics, H., Gausz, J., Kummer, J., and Karch, F. (1990) A new homeotic mutation in the *Drosophila bithorax* complex removes a boundary separating two domains of regulation. *EMBO J.*, **9,** 2579.
90. Galloni, M., Gyurkovics, H., Schedl, P., and Karch, F. (1993) The bluetail transposon: evidence for independent *cis*-regulatory domains and domain boundaries in the *Bithorax* complex. *EMBO J.*, **12,** 1087.
91. Garzino, V., Pereira, A., Laurenti, P., Graba, Y., Levis, R. W., Le Parco, Y., and Pradel, J. (1992) Cell lineage-specific expression of *modulo*, a dose-dependent modifier of variegation in *Drosophila*. *EMBO J.*, **11,** 4471.
92. Baksa, K., Morawietz, H., Dombradi, V., Axton, M., Taubert, H., Szabo, G., Torok, I., Udvardy, A., Gyurkovics, H., Szoor, B., Glover, D., Reuter, G., and Gausz, J. (1993) Mutations in the protein phosphatase 1 gene at 87B can differentially affect suppression of position-effect variegation and mitosis in *Drosophila melanogaster*. *Genetics*, **135,** 117.
93. Reuter, G. and Spierer, P. (1992) Position effect variegation and chromatin proteins. *BioEssays*, **14,** 605.
94. Clark, R. F. and Elgin, S. C. R. (1992) Heterochromatin protein 1, a known suppressor of position-effect variegation, is highly conserved in *Drosophila*. *Nucleic Acids Res.*, **20,** 6067.

95. Epstein, H., James, T. C., and Singh, P. B. (1992) Cloning and expression of *Drosophila* HP1 homologs from a mealybug, *Planococcus citri. J. Cell Sci.*, **101,** 463.
96. Pearce, J. J. H., Singh, P. G., and Gaunt, S. J. (1992) The mouse has a Polycomb-like chromobox gene. *Development*, **114,** 921.
97. Waterston, R., Martin, C., Craxton, M., Huynh, C., Coulson, A., Hillier, L., Durbin, R., Green, P., Shownkeen, R., Halloran, N., Metzstein, M., Hawkins, T., Wilson, R., Berks, B., Du, Z., Thomas, K., Thierry-Mieg, J., and Sulston, J. (1992) A survey of expressed genes in *Caenorhabditis elegans. Nature Genet.*, **1,** 114.
98. Dorer, D. R. and Henikoff, S. (1994) Expansions of transgenic repeats cause heterochromatin formation and gene silencing in *Drosophila. Cell*, **77**, 993.

9 Domains and boundaries

PAUL SCHEDL and FRANK GROSVELD

1. Introduction

The very large genomes found in many higher eukaryotes must be condensed into a relatively small volume inside the nucleus. This is accomplished by packaging the DNA into a complex, multi-level nucleoprotein structure, chromatin. At the first level of organization the DNA is assembled into a beads-on-a-string nucleosome array, the 10 nm fibre visualized by electron microscopy (see Chapter 1). In the second level of organization, the 10 nm fibre is coiled into a 30 nm fibre. While this transition is thought to be facilitated by H1 binding (1), the precise organization of the nucleosomes is not completely resolved. In one widely accepted model (2, 3) they are coiled into a fairly homogeneous solenoid-like structure, six nucleosomes per turn. Recently, some electron microscopy studies have suggested that the principle building block of the 30 nm fibre is a di-nucleosomal unit that appears to zigzag across the fibre axis, giving a much more irregular and potentially dynamic structure (4). Whichever model is correct, the 10 nm and 30 nm fibres most certainly coexist in the interphase nucleus, with specific regions of the chromosome undergoing transitions from one state to the other, presumably as a consequence of transcriptional activity or DNA replication and recombination (see 5–9). Additional levels of organization above the 30 nm fibre are required to fit the DNA inside the interphase nucleus and to achieve the even higher levels of DNA compaction observed in mitotic chromosomes. This higher order folding of the 30 nm fibre is very poorly understood, and there is still a great deal of controversy about the principles governing the 3-D orgnization of chromatin inside the nucleus.

2. Cytology and chromatin domains

Several of the key ideas about higher order structure were first suggested by cytological studies on the lampbrush chromosomes of amphibian oocytes (10–12) and on the 'specialized' polytene chromosomes of insects (13, 14). The most important idea to emerge was that the chromosome is subdivided into a series of discrete and topologically independent domains. In lampbrush chromosomes the chromatin fibre is organized into a series of large loops emanating from the main axis of the chromosome. While individual fibres cannot be readily visualized in polytene chromosomes, the latter also appear to be organized into discrete domains, displayed in the characteristic and highly reproducible banding patterns (see Fig. 1, Chapter 8). That the 30 nm chromatin fibre is segregated into topologically independent domains is supported from more recent biochemical experiments (see 15 and Section 8 below). The second idea was that each topological domain is likely to be specified (directly or indirectly) by the underlying DNA sequence. Thus specific DNA sequences were postulated to be required for the formation of each loop in lampbrush chromosomes. The third idea was that each domain corresponds to a functionally autonomous genetic unit. In lampbrush chromosomes, transcriptionally active genes are located in the loops, while in polytene chromosomes the chromatin fibre in a band decondenses and forms a puff when high levels of transcription are induced.

Most of the current models for higher order chromatin structure follow the lead of these earlier cytological investigations, postulating that the chromosome is subdivided into a series of discrete and topologically independent domains. In one popular model the chromatin fibre is organized into a series of looped domains which are anchored either to the nuclear matrix (in interphase nuclei) or to the main axis of the chromosome (in mitotic chromosomes). Such models typically assume that there must be a special class of nucleoprotein structures—boundaries—which function to delimit the higher order domains. Such domain boundaries would be assembled at specific sites along the chromosome, and would be composed of a special set of proteins.

3. Are chromatin domains determined by regions of nuclear matrix association?

Because of their central importance in defining the higher order organization of the chromatin fibre, domain boundaries have been a subject of considerable interest. Domain boundaries are generally assumed to be extremely stable nucleoprotein complexes, firmly attached to some supranuclear structure, e.g., the 'nuclear matrix' or 'scaffold'. Consequently, many of the biochemical strategies to identify putative domain boundaries have employed quite harsh high salt extraction procedures to strip away the less firmly bound nuclear proteins such as the histones. DNA sequences (and proteins) corresponding to the putative 'domain boundaries' are then identified by fragmenting the DNA and partitioning it into soluble and insoluble fractions.

The results obtained from such biochemical experiments were generally contradictory and inconclusive until a LIS (lithium diiodosalicylate) extraction protocol was introduced (16). In this procedure the chromosomal DNA is first stripped of histones and many non-histone chromosomal proteins by LIS extraction; it is then digested with restriction enzymes (in a low salt buffer) and the resulting fragments are separated into soluble and insoluble fractions by centrifugation. A clear-cut specificity in DNA binding is observed with this procedure; certain sequences remain preferentially associated with the insoluble matrix fraction, whereas others are released into the soluble supernatant fraction. Matrix or scaffold attachment regions (MARs or SARs) have been mapped by this method in the 5.0 kb *Drosophila* histone gene repeat unit (near one end of the long H1–H3 non-transcribed spacer) and in the 87A7 heat shock locus (in the ~ 1.4 kb intergenic spacer separating the two divergently transcribed *hsp70* genes) (see Fig. 1 and ref. 16). Subsequent fine structure mapping localized the former MAR to two small regions of about 150–200 bp in length (17). These observations suggested a model in which the 100 or so histone repeat units are organized into a series of small looped domains with each repeat anchored to the matrix by a domain boundary located in the H1–H3 spacer (18). The partitioning of sequences from 87A7 into soluble and insoluble fractions does not depend upon transcriptional activity; precisely the same distribution of fragments was observed before and after heat induction of the *hsp70* genes.

Apparent confirmation of the results obtained from LIS extraction has come from a modified version of the high salt extraction procedure. In this procedure, a high salt matrix protein fraction (devoid of chromosomal DNA) is incubated (in low salt buffer) with exogenously added labelled restriction fragments (and competitor DNA); matrix association is then determined by pelleting the matrix and assaying the soluble and insoluble DNA (19). Significantly, the domain boundaries or MARs identified with this high salt procedure overlap with those identified by LIS extraction and are 'evolutionarily' conserved (e.g., the MAR in the *Drosophila* H1–H3 spacer binds to a mouse nuclear matrix fraction). The LIS procedure and the modified high salt extraction procedures have now been used to map MARs in and around many different genes in a wide range of organisms (20–22). MARs are typically detected in DNA segments that are AT rich, and contain sequences matching the consensus for topoisomerase II cleavage of naked DNA substrates (18, 23). MARs are often localized close to DNA sequences that have been shown to play a key role in transcriptional regulation (18, 21).

While the apparently very tight and highly specific binding of MAR DNA segments to the nuclear matrix fraction would appear to make these elements attractive candidates for the boundaries of chromatin domains, there are some potential problems with this experimental strategy. This approach requires that the boundaries be undisturbed by the procedures used to prepare the nuclear matrix and to assay for matrix association. Only in this case can one be certain that a chromatin domain boundary identified by virtue of its association with the nuclear matrix *in vitro* corresponds to a relevant complex *in vivo* that has been faithfully preserved in going from cell, to nuclei, to matrix. Hence, this experimental strategy is predicated on the assumption that MARs are extremely stable structures that are completely resistant to extraction procedures that disrupt most DNA–protein complexes. If the DNA–protein complexes which define boundaries *in vivo* are inherently unstable, or if the extraction and incubation procedures disrupt or alter these complexes, then the association of a particular naked DNA fragment with the nuclear matrix does not constitute good evidence that this fragment corresponds to a boundary *in vivo*. Rather, the observed matrix attachment may be simply the result of the adventitious reassociation of matrix proteins with newly accessible sequences once the chromatin organization is damaged or destroyed. Particularly troubling in this regard have been studies which demonstrated that exogenously added MAR DNA can readily displace endogenous MAR DNA from the matrix (24), a result that does not appear to be consistent with the key premise of this experimental approach.

4. Are chromatin domains determined by specialized chromatin structures?

Several other approaches and additional criteria have been used to look for putative domain boundaries. The scs-like elements in *Drosophila* were initially identified on the basis of cytological analysis of the heat shock puff at 87A7 and chromatin-mapping experiments (5). *In situ* hybridization with ^{3}H-labelled probes was used to localize sequences from the 87A7 region that were within or outside of the heat shock puff. The primary domain of decondensation encompassed a DNA segment of about 15 kb. After the two *hsp70* genes at 87A7 are induced by heat shock, this domain has an average ratio of compaction over B-form DNA approaching that expected for a beads-on-a-string nucleosome array. Probes outside of this primary decondensation domain appeared to bracket the puff, suggesting that they might be located in immediately adjacent domains. Indirect end-labelling experiments were then used to examine the chromatin organization of the DNA segments that spanned the proximal and distal edges of the puff. These experiments revealed two 'specialized chromatin structures', scs and scs′, located on the proximal and distal sides respectively of the 87A7 heat shock locus (see Fig. 1). Each element is defined by a pair of very strong nuclease-hypersensitive sites flanking a 250–350 bp nuclease-resistant sequence. The nuclease cleavage pattern around each element is altered by the transcriptional activation of the heat shock genes. This effect on the nucleoprotein structure of scs and scs′ appears to be indirect, as both elements are located several kb downstream from the DNA segments (the *hsp70* genes and immediate 3′ flanking spacers) whose chromatin organization is directly disrupted by RNA polymerase transcription (25). Both scs and scs′ are *in vivo* targets for topoisomerase II after heat shock induction, which is consistent with the possible role of these structures in facilitating the decondensation of the 87A7 chromatin fibre (6).

Chromatin structures similar to scs have now been found elsewhere in the *Drosophila* chromosome, and in some higher eukaryotes. An scs-like element has been identified in the chicken (and human) β-globin complex. This constitutive nuclease hypersensitive site is located at or near the 5′ edge of a chromatin domain that spans the globin locus; the domain shows enhanced DNase I sensitivity in chromatin digests and increased levels of histone acetylation in erythropoietic cells (26–29). In contrast to the *Drosophila* scs-like element, the β-globin element does not have a nuclease-resistant core flanked by hypersensitive sites, but a single hypersensitive site which is thought to define a site of action for topoisomerase II (30).

While these scs-like elements would appear to be plausible candidates for boundaries of chromatin domains, the two lines of evidence discussed above—their location near the edges of chromatin domains and their biochemical properties (e.g., nuclease hypersensitivity, sites of action for topoisomerase II)—are circumstantial. Many other 'regulatory elements' (e.g., promoters, enhancers, or even replication orgins) exhibit nuclease hypersensitivity in chromatin digests, and can define sites of action for topoisomerase II *in vivo*.

5. Functional assays for boundaries of chromatin domains

One of the key suggestions that emerged from the early cytological work on lampbrush and polytene chromosomes was the idea that each chromatin domain corresponds to a functionally autonomous genetic unit. One function of a domain boundary, then, would be to establish a unit of independent gene activity. When genes are reintroduced into the genomes of higher eukaryotes they are often not functionally autonomous, but are instead subject to local chromosomal position effects: at some insertion sites their expression is repressed, while at other sites it is enhanced. This suggests that it might be possible to test for boundary function *in vivo* using a 'position effect assay'. When DNA segments containing boundaries are placed on both sides of a reporter construct, they should be able to establish a domain of independent gene activity insulating the reporter against such local position effects.

A position effect assay has been used to test DNA segments containing scs and scs' from the *Drosophila* 87A7 locus, and HS5 from the chicken β-globin complex. The insulating activity of scs and scs' was tested with two different *white* reporter genes (31) (see Fig. 1). The first reporter, a *white* maxi-gene, contains the entire *white* transcription unit plus all of the regulatory sequences thought to be required for 'wild type' levels of expression. While this reporter is capable of giving a near wild-type eye colour, it is subject to negative position effects; and many transformants have a red-brown or even orange eye colour instead of bright red. When 'random' DNA fragments were placed on either side of the maxi-gene, no insulation was observed, and eye colour varied with the site of insertion. In contrast, when the maxi-gene was bordered by scs and scs', essentially the same (bright red) eye colour was observed at all euchromatic insertion sites, suggesting that these elements can insulate against negative position effects. The second reporter was a *white* mini-gene. This construct has the *white* promoter but lacks the upstream regulatory elements; consequently, transformants typically have only a light yellow eye colour. This reporter is subject to positive position effects; at many sites of insertion *white* expression is enhanced and the eye colour can be orange or even red. When scs or scs' were placed upstream of this *white* mini-gene they

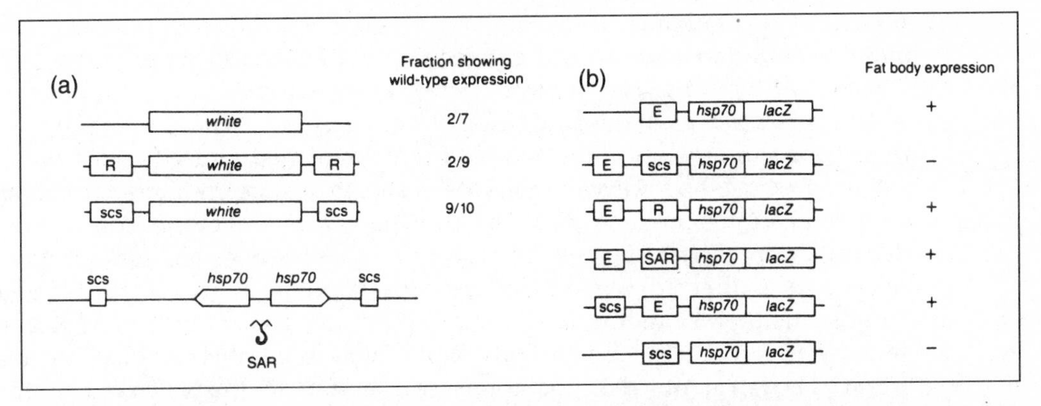

Fig. 1 Specialized chromatin structure (scs) elements flanking the *Drosophila* 87A7 *hsp70* genes can insulate a transduced *white* gene from position effects and can block enhancer activity when inserted between the enhancer and a target promoter. (a) Expression resulting from stably transduced copies of a full-length *white* gene without flanking DNA, with randomly chosen prokaryotic flanking sequences (R), and with flanking scs elements. The map immediately below indicates the normal position of scs and scs elements relative to the *hsp70* gene copies found at cytological position 87A7 (redrawn from 5). Location of a previously identified SAR associated with this locus (16) is shown below the map. (b) Expression of β-galactosidase in *Drosophila* fat body, driven by a yolk protein enhancer, is blocked by insertion of an scs element between the enhancer (E) and a target promoter, but not by insertion of a randomly chosen pBR322 plasmid DNA fragment (R), the SAR sequences from the 87A7 heat shock locus (SAR), or an scs element upstream.

were unable to block chromosomal position effects, and eye colour depended on the site of insertion. In contrast, when the mini-gene was bordered by scs and scs′, a light yellow eye colour phenotype was observed independent of the insertion site. These findings indicate that scs and scs′ can insulate against positive position effects and also demonstrate that the elements must flank the reporter in order to establish a domain of independent gene activity (31).

The *white* mini-gene was also used to test the insulating activity of the chicken β-globin 5′ element (HS5) in flies (26). The mini-gene was protected against positive position effects when it was bordered on both sides by two copies of the β-globin 5′ element. When the converse experiment was carried out, that is, a positive position effect assay in transgenic mice where a reporter gene was flanked by the *Drosophila* scs and scs′ elements, no significant insulating activity was observed. The scs-scs′ transgenics showed a similar number of sites of expression in developing mouse embryos as the construct which lacked the elements (D. Meijer and F. Grosveld, unpublished data).

A second assay for boundary function was suggested by the ability of enhancers and silencers to control the transcriptional activity of one or more promoters located many kilobases away. This raises the question of how these regulatory elements are directed to act on their appropriate target genes, but prevented from acting on other nearby genes. One factor which might help limit the scope of action of these regulatory elements would be the higher order organization of the chromosome. In this case, domain boundaries might function to restrict the activity of regulatory elements to the chromosomal domain in which they reside. This suggested an 'enhancer blocking assay' in which DNA fragments containing putative boundaries are placed between an enhancer and a promoter.

Enhancer blocking assays have been used to examine the 'boundary function' of several scs-like elements and a SAR from *Drosophila*. These putative boundaries were tested in a construct in which a female fat body specific *yp-1* yolk-protein enhancer was used to drive β-galactosidase expression from a heterologous *hsp70* promoter:*lacZ* fusion gene (see Fig. 1). When DNA fragments containing the scs elements were interposed between the *yp-1* enhancer and the *hsp70* promoter, they blocked *yp-1* directed β-galactosidase expression in adult female fat bodies. In contrast, neither a random DNA segment of similar size nor the SAR from 87A7 had any apparent effect on fat body expression. The enhancer-blocking activity of the scs-like elements did not appear to be due to the silencing of the *hsp70* promoter, as the promoter was still fully inducible by heat shock. Moreover, in order to block enhancer action, the scs element had to be interposed between the enhancer and the promoter; it had no effect on *yp-1*-directed fat body expression when placed upstream of the enhancer (32).

A similar strategy has been used to test the boundary function of the chicken β-globin 5′ constitutive hypersensitive element, HS5 (26). The construct used had a strong enhancer element (HS2) from the mouse β-globin locus control region (LCR) activating a human $^{A}\gamma$-globin promoter: *neomycin* resistance fusion gene ($^{A}\gamma$-*neo*). Placing one copy of the β-globin 5′ element between the HS2 enhancer and the $^{A}\gamma$-*neo* fusion gene suppressed the formation of neomycin-resistant colonies nearly ten-fold, while two copies of the element reduced colony formation nearly thirty-fold. The chicken β-globin 5′ element exhibits the same directionality as the scs elements. It must be interposed between the HS2 enhancer and the promoter in order to reduce colony formation. The chicken 5′ element appears to exert its effect by preventing the HS2 from altering the chromatin structure of the $^{A}\gamma$-globin promoter. A variety of earlier experiments suggested that one of the functions of the globin LCR is to facilitate the formation of a nuclease-hypersensitive region at the globin promoters. This hypersensitive region is not observed when the chicken 5′ element HS5 is placed between the HS2 enhancer and $^{A}\gamma$-globin promoter. The *Drosophila* scs elements showed no blocking activity in this assay system (G. Felsenfeld, personal communication).

The human equivalent of the chicken 5′ element has also been tested in an enhancer-blocking assay. Like the chicken element, the human 5′ element has little or no enhancer activity and cannot itself activate a human β-globin gene in transgenic mice (Raguz-Bolognesi, unpublished data.). To test for enhancer-blocking activity, a complete globin LCR (HS1–4) was placed upstream of a human β-globin gene flanked by one or two of the 5′ element (HS5); this construct was then introduced as single-copy inserts into transgenic mice (see Fig. 2). The β-globin gene was found to be active independently of whether or not the human 5′ element was placed between the gene and the LCR (G. Zaffarana, D. Meijer, and F. Grosveld, in preparation). It is unclear why different results were obtained with the chicken and human 5′ elements; the elements may not be functionally equiva-

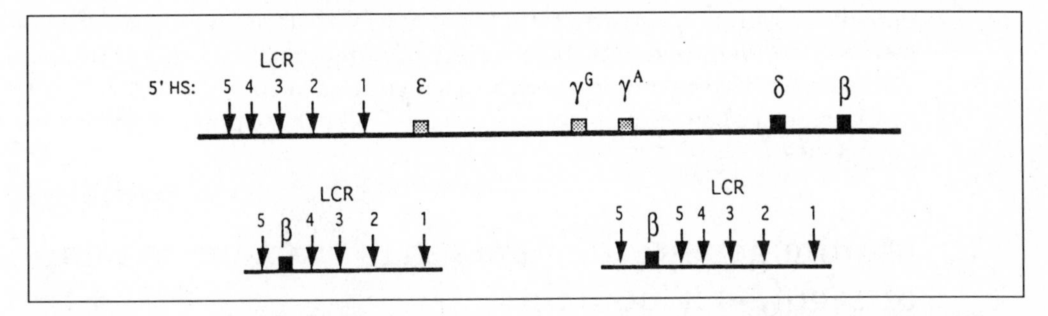

Fig. 2 Testing of a human 5′ element. The globin locus is shown on top. When a reporter β gene was placed between the LCR and the 5′ element HS5 (left) or between two 5′ elements (right), the expression remained at 100% per copy of the gene.

lent, or, differences in the assay systems used might be critical. The first study used drug selection in tissue culture cells, while the second examined β-globin expression in transgenic animals. Another factor that could be important is the strength or properties of the activating element; the first study used only the enhancer part of the mouse LCR while the second used the entire human LCR. That the properties of the enhancer might be relevant is suggested by experiments on scs and scs′ (J. Vazquez, unpublished data) and on *su(Hw)* (see below) which indicate that the efficacy of enhancer blocking depends upon features of the enhancer.

The biological activities of MARs from higher eukaryotes have also been examined using similar experimental approaches. The chicken lysozyme gene A elements provide a good example. These MARs are located near the edge of the lysozyme gene chromatin domain which exhibits enhanced sensitivity to DNase I (20, 33). In tissue culture cells the A elements are found to block enhancer action and are required for postion-independent gene expression (34). However, in a test in transgenic mice (33, 35), the A elements did not appear to be required for a high level of 'position-independent expression' in the appropriate cells; this activity was conferred by the combined action of the several enhancers which together constitute the locus control element (see below). On the other hand, the incidence of ectopic expression increased when the construct contained an incomplete lysozyme locus control element and lacked the A elements. Hence, the A elements would appear to 'insulate' the reporter construct against 'positive' position effects in this transgenic assay system.

While these experiments indicate that higher eukaryotes have genetic elements that can restrict the action of enhancers and silencers, and presumably delimit *regulatory* domains, it is not yet clear whether such elements actually correspond to the boundaries of *structural* domains. All of these functional assays for *chromatin* boundaries are based on the speculation that *chromatin* domains correspond to units of autonomous genetic function. Whether this speculation is correct has not yet been demonstrated. (What is still required is a reliable assay for *chromatin* domains as units of structure rather than as units of regulation.) Moreover as discussed below, there may be a variety of other genetic elements (as even simple competitive effects combined with distance) that in one context or another have insulating or enhancer-blocking activity much like these putative chromatin domain boundaries.

6. Mobile genetic elements that function as boundaries of regulatory domains

Many of the spontaneous mutations recovered in *Drosophila* prior to the spread of PM and IR dysgenesis were induced by the insertion of retrotransposons. Long before anyone realized this fact, Lewis (36) discovered that a group of 15 spontanteous mutations in some 11 different genes were suppressible by a second site mutation in the *su(Hw)* locus. Thirty years later Modolell *et al.* (37) demonstrated that all but two of these mutations had a *gypsy* retrotransposon inserted at the cytological position of the mutant locus. Many of these *gypsy*-containing loci have since been cloned and the location of the retrotransposon determined. Paradoxically it was found that the *gypsy* transposon can induce a mutation not only when it is inserted within a transcription unit, but also when it is located outside in the flanking non-transcribed sequences, often many kilobases away (38, 39).

Analysis of a series of partial and complete revertants of the *gypsy*-induced mutations revealed that the simple insertion of foreign DNA is not in itself usually responsible for the mutagenic effects of the *gypsy* transposon; rather there is a small ~500 bp DNA segment within the transposon that is (in most cases) required for inducing the mutant phenotype (40, 41). Deletions which remove all or part of this 500 bp DNA segment reduce or eliminate the phenotypic effects of the *gypsy* insertion. This short DNA segment causes the same phenotypic effects as an intact *gypsy* transposon when it is inserted at approximately the same position in the *yellow* gene as the transposon in the y^2 mutant allele (42). It is also this 500 bp DNA segment that is the target for the *su(Hw)* gene product. *su(Hw)* encodes a protein consisting of a central zinc-finger domain containing 12 'fingers', a C-terminal leucine zipper, and N- and C-terminal acidic domains (43, 44). The *su(Hw)* protein recognizes a 27 bp sequence consisting of an octomer motif and flanking A/T tracts (45, 46). There are 12 copies of this sequence in the ~500 bp *gypsy* DNA segment and, assuming that *su(Hw)* protein binds to its recognition sequence as a monomer, there could be as many as 12 protein moieties associated with *gypsy* transposons *in vivo*. Since the progressive deletion of these binding sites causes a corresponding reduction in the severity of the mutant phenotype, it is clear that the mutagenic effect is dependent on the number of bound protein molecules. The disposition of these protein molecules may also be important, as insertions which disrupt the reiterated array of *su(Hw)*-binding sites also ameliorate the phenotypic effects of *gypsy* insertions (47).

How do the reiterated *su(Hw)*-binding sites disrupt gene function? A variety of studies indicate that the reiterated *su(Hw)*-binding sites function as boundaries of regulatory domains in a manner quite similar to that observed for the scs-like chromatin structures. Insertion of *su(Hw)*-binding sites in the 5′ end of the *hsp70* gene, between the heat shock factor sites and the promoter, suppresses heat-inducible transcription (48). The *su(Hw)*-binding sites in the 500 bp *gypsy* DNA segment can prevent a variety of enhancers in the *yellow* gene from activating the *yellow* promoter (42). As observed for the scs-like elements, this enhancer-blocking activity depends upon the position of the reiterated *su(Hw)*-binding sites in between the enhancer and promoter. DNA segments containing reiterated *su(Hw)*-binding sites are capable of establishing a domain of independent gene activity, and can insulate a *white* reporter gene from chromosomal position effects (49). Finally, the idea that the reiterated *su(Hw)*-binding sites function as boundaries of regulatory domains is entirely consistent with the phenotypic effects of the many different *su(Hw)*-suppressible *gypsy*-transposon-induced mutations. In most cases it is possible to explain the mutant phenotype by an enhancer-blocking activity that prevents enhancer(s) in the mutant locus from directing the appropriate pattern(s) of expression.

Other than mediating the mutagenic effects of the *gypsy* transposon, the function of *su(Hw)* is poorly understood. The gene is expressed throughout development and *su(Hw)* protein can be detected in the nuclei of all tissues examined. In salivary gland polytene chromosomes, it has a rather ubiquitious distribution, though the levels of protein seems to be higher in some cytogenetic intervals than in others (44, 45). It is not clear whether the *D. melanogaster* genome has reiterated arrays of the 27 bp binding sequence (other than in the few *gypsy* transposons), and consequently the localization observed in polytene chromosomes may reflect the binding of the protein to DNA segments containing only one or two closely linked binding sites. In spite of the very general tissue and temporal distribution of *su(Hw)* protein, the only phenotype of null mutations is female sterility. *su(Hw)* mutations appear to first perturb oogenesis prior to the onset of vitellogenesis. Although the initial endomitotic replication of the nurse cell chromosomes in the very early egg chambers is apparently normal, the chromosomes do not properly decondense, and in stage 4 and later chambers they have an irregular bulbous appearance that is not evident in wild-type chambers of the same age (44). Thus *su(Hw)* protein may be required in the higher order organization of the chromatin fibre in nurse cell nuclei. Conceivably, the *su(Hw)* protein could have a similiar function in somatic tissues; however, the fact that it is dispensible in somatic cells would argue that there must be other proteins in these cells that are capable of carrying out this function.

7. Regulatory domains in the bithorax complex

After the gap and pair-rule genes subdivide the early *Drosophila* embryo into fourteen units (called parasegments (PS)) each parasegment acquires its specific identity through the action of the homeotic genes in the *Antennapedia* and bithorax complexes (ANT-C and BX-C respectively). The BX-C contains three genes, *Ubx*, *abd-A*, and *Abd-B*, which specify parasegmental identities in the posterior two-thirds of the animal from PS5 to PS14. In the adult fly these parasegments correspond to the posterior half of the thorax and all of the abdominal segments (for review see 50, 51). The *Ubx*, *abd-A*, and *Abd-B* genes encode homeodomain proteins; each is expressed in an intricate temporal and spatial pattern in an overlapping set of parasegments: *Ubx* is expressed from PS5 to 13, *abd-A* from PS7 to 12; and *Abd-B* from PS10 to 14. The proper elaboration of the different parasegment-specific expression patterns of each gene is required for the morphogenesis of segmentally distinct structures; mutations that alter *Ubx*, *abd-A*, or *Abd-B* activity can transform parasegment identity.

The intricate patterns of BX-C gene expression are generated by a complex ~300 kb *cis*-regulatory region. This very large *cis*-regulatory region is sub-divided into nine parasegment-specific units—*abx/bx*, *bxd/pbx*, *iab-2*, *iab-3*, *iab-4*, *iab-5*, *iab-6*, *iab-7*, and *iab-8* (see Fig. 5 in Chapter 8), which are arranged in the same order along the chromosome as the parasegments they affect are arranged along the body. Each of these parasegment specific units controls the expression of one of the three homeodomain transcription units, *Ubx*, *abd-A*, or *Abd-B*, in a specific parasegment. Thus, expression of the *Ubx* transcription unit (which is located on the centromere-proximal side of the complex) is controlled by the parasegment specific *cis*-regulatory units *abx/bx* and *bxd/pbx*, which direct *Ubx* transcription in PS5 and PS6 respectively. The *abd-A* gene is in the middle of the complex, and its expression is controlled by *cis*-regulatory units *iab-2*, *iab-3*, and *iab-4*, which direct *abd-A* expression in PS7, 8, and 9 respectively. The *Abd-B* gene, which is on the distal side of the bithorax complex, expresses transcripts from several different promoters (52). The centromere-proximal promoter is responsible for expressing the smallest class of *Abd-B* transcripts, class A. The activity of this promoter is controlled by the *cis*-regulatory units *iab-5*, *iab-6*, *iab-7*, and *iab-8*, which direct expression in PS10, PS11, PS12, and PS13 respectively (see Fig. 3). In PS14 at least three other *Abd-B* promoters appear to be activated; however, a PS 14-specific *cis*-regulatory unit has not yet been identified.

Loss-of-function mutations in one of these nine parasegment specific *cis*-regulatory units typically transform the corresponding parasegment into a copy of the parasegment immediately anterior to it. Consistent with this transformation in segmental identity, the normal temporal and spatial pattern of the relevant BX-C homeotic gene is replaced by an expression pattern which mimics that found in the parasegment immediately anterior to it (see 53–55). For example, a mutation that removes much of the *iab-7* *cis*-regulatory region ($iab\text{-}7^{Sz}$) results in the transformation of PS12 into PS11 (or, in the adult, A7 into A6) (55). In this case, instead of the normal *Abd-B* expression pattern in PS12, a PS11-like pattern that is generated by *iab-6*, is observed (see Fig. 3).

Besides these nine *cis*-regulatory units, genetic studies have identified two unusual *cis*-acting elements, *Mcp* and *Fab-7*, which are defined by small deletions on the distal side of the BX-C. *Mcp* is located between the *iab-4* and *iab-5* *cis*-regulatory

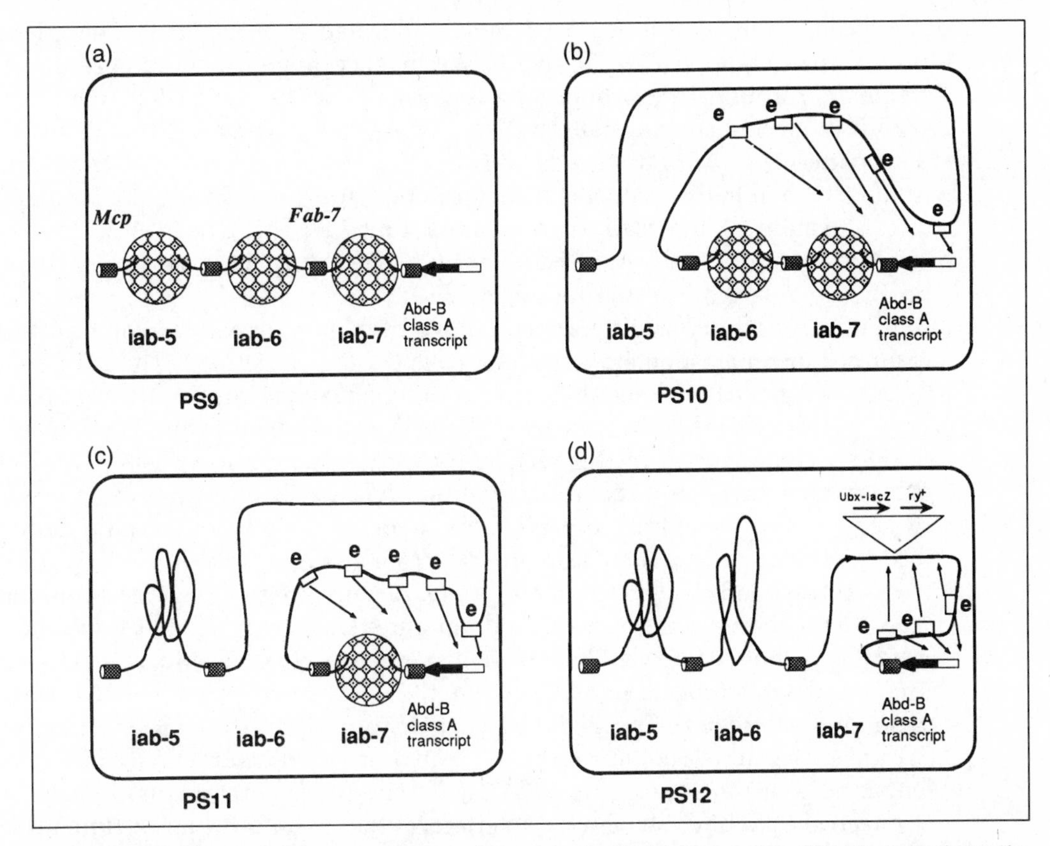

Fig. 3 Model for the functional organization of the *iab cis*-regulatory domains. This diagram shows the organization of the *iab cis*-regulatory regions *iab-5*, *iab-6*, and *iab-7*, into 'independent chromatin domains' when the bithorax complex functions in the 'maintenance mode' and is dependent upon the products of the *Polycomb* and *trithorax* group genes. (a) In PS9, all three *cis*-regulatory domains are in an inactive configuration, and are complexed with *Polycomb* group proteins. As a consequence, the *Abd-B* gene is off. (b) In PS10, the *iab-5 cis*-regulatory domain is in an active configuration, while both *iab-6* and *iab-7* are inactive. Regulatory elements (e) in the active *iab-5* domain interact with the *Abd-B* promoter, generating a PS10-specific expression pattern. (c) In PS11, the *iab-6 cis*-regulatory domain is in an active configuration, while *iab-7* is inactive. Elements in the *iab-6 cis*-regulatory domain interact with the *Abd-B* promoter to generate a PS11-specific expression pattern. (d) Finally in PS12, the *iab-7 cis*-regulatory domain is in an active configuration, and regulatory elements in this domain direct *Abd-B* expression in a PS12-specific expression pattern. This *iab-7 cis*-regulatory domain contains the *bluetail* transposon, which like the *Abd-B* gene is also activated by the *iab-7* regulatory elements. The *Abd-B* gene and promoter to the distal side of the complex are indicated by filled and open regions of the arrow respectively. Code: PS = parasegment; e = enhancer elements in different *cis*-regulatory domains. The putative boundaries *Mcp*, *Fab-7* (and '*Fab-6*') are indicated by the cylinders separating adjacent *cis*-regulatory domains.

units, while *Fab-7* is located between *iab-6* and *iab-7*. Mutations in *Mcp* and *Fab-7* differ from mutations in the nine *cis*-regulatory units in two critical respects. First, they are dominant gain-of-function mutations instead of recessive loss-of-function mutations. Second, they transform the affected parasegment into a copy of the parasegment immediately posterior to it instead of that immediately anterior. In an *Mcp* mutant, PS9 is replaced by a duplicated copy of PS10. This phenotypic transformation appears to be due to the inappropriate functioning of the *iab-5 cis*-regulatory unit in PS9, where *iab-4* is normally active. Consistent with this view, *Abd-B* (which is not normally transcribed in PS9) is expressed in PS9 in a PS10-like pattern in *Mcp* mutant embryos (53, 54). Mutations in *Fab-7* transform PS11 into PS12 (56). In this case, the transformation appears to be due to the inappropriate activation of the *iab-7 cis*-regulatory unit in PS11 where *Abd-B* is normally controlled by *iab-6*. Indeed, in *Fab-7* embryos the pattern of *Abd-B* protein expression in PS 11 is identical to that found in PS12 (55).

A number of different hypotheses have been advanced to explain the dominant gain-of-function phenotypes of the *Mcp* and *Fab-7* mutations. The simplest model is that these genetic elements correspond to parasegment-specific negative regulators, or silences—*Mcp* would be a PS9-specific silencer, while *Fab-7* would be a PS11-specific silencer. In this view, *Mcp* would prevent the positive regulators in *iab-5* (which are responsible for activating *Abd-B* in PS10) from turning on the *Abd-B* gene in PS9. Similarly, the silencers associated with *Fab-7* would prevent the PS12-specific positive regulators in *iab-7* from acting on *Abd-B* in PS11. If *Mcp* and *Fab-7* correspond to parasegment-specific negative regulators, then the dominant gain-of-function phenotypes of each mutant should be reverted by second site *cis*-mutations that inactivate the positive regulators in *iab-5* and *iab-7* respectively. Indeed, deletions or other rearrangements which disrupt *iab-5* revert *Mcp*, while deletions or other rearrangements that disrupt *iab-7* revert *Fab-7*. On the other hand, if *Mcp* and *Fcp-7* are negative regulators, then mutations which inactivate either *iab-4* or *iab-6* respectively (but leave the positive regulators in *iab-5* and *iab-7* intact) should have absolutely no effect on the dominant gain-of-function phenotypes of these mutants. However, this is not the case—mutations which disrupt *iab-4* can revert *Mcp*, apparently preventing the regulatory elements in *iab-5* from functioning inappropriately in PS9 (57). Similarly, mutations which disrupt *iab-6* can revert *Fab-7*, again apparently preventing regulatory elements in *iab-7* from functioning inappropriately in PS11 (56). Another hypothesis is that the *Mcp* and *Fab-7* DNA segments contain a complex array of binding sites not only for negative but also for positive regulatory factors. In this case, the removal of these key regulatory targets by the *Mcp* and *Fab-7* deletions would shift the balance between negative and positive regulatory signals, leading to the inappropriate activation of *Abd-B* in these mutants. However, even this more sophisticated model is difficult to reconcile with the reversion of the gain-of-function phenotypes by second site mutations in *either* the proximal or the distal *cis*-regulatory units that flank *Mcp* and *Fab-7*. In addition, such a model does not seem to be consistent with the phenotypic transformations of the *Mcp* and *Fab-7* mutations. Thus, it is not clear why eliminating key positive and negative regulatory targets would have no effect on the more posterior parasegment (PS10 for *Mcp* and PS12 for *Fab-7*) while it has such a profound effect in the more anterior parasegment (PS9 for *Mcp* and PS11 for *Fab-7*). In fact, if key regulatory targets are eliminated by the *Mcp* and *Fab-7* deletions, entirely novel patterns of expression might be expected in both PS9 and

PS10 for *Mcp* and PS11 and PS12 for *Fab-7*. Instead, the *Abd-B* expression pattern in these parasegment pairs closely mimics that normally observed in wild-type animals in the more posterior parasegment of each pair (PS10 for *Mcp* and PS12 for *Fab-7*).

One model which appears to account for the known genetic properties of *Mcp* and *Fab-7* is that these elements function as regulatory boundaries, insulating each BX-C *cis*-regulatory domain from the regulatory influences of adjacent *cis*-regulatory domains (see Fig. 3). In this model, the deletion of the *Mcp* regulatory boundary would permit interactions between regulatory elements in the adjacent *cis*-regulatory domains *iab-4* and *iab-5*, leading to the inappropriate activation of *iab-5* in PS9. Similarly, removal of the *Fab-7* boundary would allow interactions between regulatory elements in *iab-6* and *iab-7* resulting in the inappropriate activation of *iab-7* in PS11 (see Fig. 3). The idea that the gain-of-function phenotypes of *Mcp* and *Fab-7* are due to adventitious interactions between normally independent *cis*-regulatory domains would explain why *Mcp* and *Fab-7* can be reverted not only by mutations which inactivate *iab-5* and *iab-7*, respectively, but also by mutations that inactivate *iab-4* and *iab-6*.

Additional support for a boundary model comes from the *bluetail* transposon (55), a *lac-Z* reporter inserted on the distal side of *Fab-7*, just within the *iab-7 cis*-regulatory domain (see Fig. 3). This *bluetail* reporter is subject to control elements in the *iab-7 cis*-regulatory domain, and it is activated in PS12 and in more posterior parasegments. In PS12, where the *iab-7 cis*-regulatory domain normally functions, the spatial and temporal pattern of β-galactosidase expression closely resembles that observed for *Abd-B*. Moreover, this PS12-like expression pattern appears to be reiterated in the more posterior parasegments. While the *bluetail* transposon seems to be under the regulatory influences of *iab-7*, it appears to be insulated from the control elements in the nearby *iab-6* (PS11) and *iab-5* (PS10) *cis*-regulatory domains, and it is not activated in parasegments anterior to PS12 (see Fig. 3). It seems likely that the *Fab-7* boundary is responsible for this insulation, since the *bluetail* transposon responds to control elements in *iab-6* when the *Fab-7* boundary is deleted (J. Mihaly, J. Gausz, H. Gyurkovics, and F. Karch, unpublished data).

In this context, it is important to note that, although the *Fab-7* boundary prevents control elements in the more proximal *iab-5* and *iab-6 cis*-regulatory domains from acting on the *bluetail* reporter, these *cis*-regulatory domains are able to interact with the *Abd-B* promoter which is located on the other side of *Fab-7* distal to *iab-7* (see Fig. 3). Why the *Fab-7* boundary does not prevent these *cis*-regulatory domains from acting on *Abd-B* is not at all clear. It is possible that *Fab-7* (as well as *Mcp*) have properties that are different from other regulatory boundaries such as the scs-like elements or the reiterated *su(Hw)*-binding sites. On the other hand, it is curious that most of the *gypsy* transposon-induced mutations in the BX-C are in regulatory domains which control *Ubx* expression, while there are no known *gypsy*-induced mutations in the regulatory domains controlling *Abd-B* expression. This raises the possibility that the *Abd-B cis*-regulatory domains contain elements that facilitate long-distance interactions with the *Abd-B* promoter, and circumvent the effects of regulatory boundaries such as *Fab-7*.

As has been observed for the scs-like boundaries, both *Mcp* and *Fab-7* have unusual chromatin structures. In the case of *Mcp*, genetic analysis has defined a ~400 bp DNA segment that appears to be essential for its boundary function (58). This DNA segment spans a large chromatin-specific nuclease-hypersensitive region that is observed throughout embryogenesis and in tissue culture cells (where the BX-C is off). The sequences required for *Fab-7* boundary function also have an unusual chromatin structure (55). However, instead of one major hypersensitive region, the *Fab-7* DNA segment contains three major hypersensitive regions plus several minor sites. The more complex chromatin organization of *Fab-7* seems to be consistent with genetic analysis which indicates that the sequences required for full boundary function are spread over a larger DNA segment than is the case for *Mcp*.

8. Domains in higher eukaryotes

The enormous size of the genomes in higher eukaryotes presents organizational and computational problems of an order not faced by lower eukaryotes such as yeast or fruit flies. Not only does a much larger amount of DNA have to be packed into the nucleus and passed through the cell cycle in an orderly manner, but also a much larger amount of information has to be correctly recognized, processed, and ultimately 'remembered' once specific developmental pathways have been selected. For example, one consequence of these expanded genomes is that there is a very good chance that recognition sequences for cell-type-specific transcription factors will be present not only in regions of the genome which should be active, but also in regions of the genome which should be transcriptionally silent in a given cell type that contains these factors. One way to circumvent such difficulties would be through the use of domains, i.e., the organization of chromatin inside the nucleus to subdivide the genome into large functional units. The subdivision of the chromosome into domains would provide a means for keeping parts of the genome repressed in particular cell types. This would reduce the problems of information recognition, processing, and memory posed by the use of regulatory factors that are capable of recognizing only very short sequence motifs. The concept of the domain offers an additional layer of control to influence gene expression in a directed and non-random manner.

The notion that chromosomes of higher eukaryotes are subdivided into chromatin domains has been supported by a number of lines of evidence in addition to the cytological and genetic studies discussed above. Of particular importance has been the finding that active or potentially active genes are packaged into a chromatin structure that is typically much more sensitive to agents such as DNase I than is the chromatin of inactive genes (see 27, 28, 33, 59, 60). Enhanced DNase I sensitivity can extend for considerable distances both upstream and downstream of the transcript; this sensitive region is flanked by DNA segments whose DNase I sensitivity is much like that of inactive genes. Other features of chromatin structure also distinguish the active domain. These include a greater extent of core histone acetylation, reductions in the levels of linker histones such H1 and H5, enhanced salt solubility of oligonucleosomes, and a reduced ability to form stable pseudo-higher-order structures (28, 29). These findings suggest that active or potentially active genes are organized into discrete domains with a chromatin structure that differs with respect to both its biochemical properties and its 3-D folding from that of inactive domains. Together with cytological studies (7, 8) these results imply that active domains have a more open and accessible organization. Within this accessible region, transcription factors would be able to bind and express (or repress) target genes. In contrast, in a non-accessible domain these same factors would not be able to bind even if the domain contained target sites, and genes within that domain would be inactive.

Such a model requires mechanisms to initiate and subsequently maintain either the active or the inactive chromatin configuration of a domain. In considering how inactive domains might be established, it would be reasonable to suppose that the organization of DNA into chromatin would have generally repressive effects. Recall that deletion of histone H4 in *S. cerevisiae* results in the constitutive expression of a number of genes (61 and Chapter 7). There is also a good deal of evidence for repressive effects of nucleosomes positioned over regulatory regions (62–65). In addition, studies described in Chapters 7 and 8 have established that there are clusters of non-histone proteins, (such as the *Polycomb* group (Pc-G) proteins) that are critical for stable repression. These accessory proteins provide a memory mechanism whereby a regulatory decision (i.e., the off state of a gene) can be stably propagated over many cell generations. At present there is no clear role for the homologues of the *Pc*-G proteins in vertebrates. Another class of proteins which could have an important role in gene repression is those which are dependent on DNA methylation for binding (66 and Chapter 10). The use of DNA methylation to mark permanently domains of inactive chromatin in somatic tissues would support the argument that faithfully remembering developmental choices over successive cell generations, in some cases for many decades, is a problem of special importance in higher eukaryotes (see Chapter 10).

In order to maintain active gene domains there must be mechanisms to counterbalance the repressive effects of histones and accessory factors. Perhaps the best example of proteins in lower eukaryotes that function to counteract repressive chromatin states and activate gene expression are members of the *trithorax* group. This group includes the GAGA factor, which appears to play a critical role in the formation of nuclease-hypersensitive sites (67–70). Proteins in the *trithorax* group, like their *Polycomb* group counterparts, also function in maintaining a particular developmental decision; however, in this case it is propagating the active, rather than the repressed state, through successive cell divisions. Homologues of the *trithorax*-group proteins are also found in higher eukaryotes, although there is little information about their likely functions. As noted above, other factors such as histone acetylation (71, 72,) or the association of HMG proteins (73) may also be important in conferring and maintaining an active chromatin configuration.

9. Locus control regions

The biochemical properties that are used to define active or inactive regions of chromatin (such as DNase I sensitivity) are markers for chromatin domains whose activity state has already been established. How is the activity state of a chromatin domain initially selected and what are critical *cis*- and *trans*-acting elements involved in this process? Good candidates for such a role are the locus control regions (LCRs). LCRs were originally defined in functional studies on the human β-globin gene in transgenic mice. The β-globin gene is normally sensitive to chromosomal position effects; however, when a region from the 5′ end of the globin domain containing a series of DNase-I-hypersensitive sites (HS1–HS5) was coupled to the globin transgene, the expression level in red blood cells became dependent on the number of copies of the introduced transgene and was independent of the site of integration in the host genome (74). Many enhancers tested had failed to activate expression, or gave only low levels of expression in transgenic animals despite the fact that they were able to stimulate high levels of expression in tissue culture cell assays using either transient transfection or stable transformation. This difference probably reflects inadequacies of the tissue culture cell assay systems—in the transient transfection experiments the proper chromatin structure may not be formed, while for the stable transformation assays the selection procedure for transformants ensures that only those cells survive in which the incoming DNA has been integrated into areas of the genome already in an active configuration. This is not the case in a transgenic mouse experiment; by inference, an LCR must have the capacity not only to activate gene expression, but also to establish *de novo* an active chromatin domain independent of its site of integration.

The human β-globin locus is about 75 kb in length and contains five tissue-specific and developmentally regulated genes (see Fig. 2). The LCR contains five tissue-specific hypersensitive sites (HS1–5), of which HS1–4 have the capacity to activate globin gene transcription, while the most 5′ site, HS5, does not have this activating property, but is a candidate site to function as a boundary element (see section 5). HS1–4 each consists of a small region of about 300 bp containing a high number of transcription factor binding sites (75). Only the HS2 element has classical enhancer activity when tested in transient transfection assays (76); this activity is due to the presence of recognition sequences for the transcription factor NF-E2 (77). All four of the hypersensitive sites display LCR activity when tested in multiple copies in transgenic mice (78). Detailed analysis has shown that, at least for HS3, this activity is dependent on the precise configuration of specific factor-binding sites (79, 80). When HS2 and HS3 were each tested in a single copy in combination with a globin gene, only HS3 still functioned as an LCR in transgenic mice, whereas HS2 showed a very low and non-reproducible activity (Ellis *et al.*, in preparation). This indicates that LCR activity may be dependent on a redundancy of certain binding sites which is exclusively in the normal configuration of HS3. The smallest HS3 core (which lacks this redundancy) does not function as an LCR; a larger fragment is required (Ellis *et al.*, in preparation). Redundancy and/or a certain amount of spacing between HS3 and the promoter may be required, e.g., to allow proper looping. Alternatively, the larger fragment may bind additional factors that are not detected in the usual *in vitro* DNA/factor-binding experiments, or *in vivo* footprinting experiments. Note that an increased dosage of a transcription factor can overcome *SIR*-mediated silencing in yeast (81). PEV is also responsive to the dosage of some gene products (Chapter 8). LCRs have also been described for other genes (see 82). For the LCR associated with the chicken lysozyme gene it is clear that more than one HS is required for LCR activity (83).

The properties of the globin LCR indicate that a limited number of transcription factor binding sites are capable of changing inactive chromatin into an active state (or alternatively of preventing inactivation), thus changing the accessibility of neighbouring sequences to nuclear factors. The IgH μ LCR consists of a core enhancer which is flanked on either side by MARs (84). The intact μ LCR is able to induce DNase-I-hypersensitive sites and confer accessibility of a linked T7 promoter to the T7 RNA polymerase in isolated nuclei. In contrast, while the core enhancer by itself appears to be sufficient to confer T7 promoter accessibility, it does not induce formation of a detectable DNase I-hypersensitive site on its own. This analysis has recently been extended to an immunoglobulin transgene (85). Like the globin LCR, the intact μ LCR confers position-independent, cell-type-specific expression of the immunoglobulin reporter. However, when the flanking MARs are deleted, the level of tissue-specific expression is substantially reduced in most transgenic lines and the gene exhibits a position-dependent variability not evident with the intact μ LCR. Differences are also observed at the level of chromatin organization. The intact μ LCR induces the formation of an extended DNase-I-sensitive chromatin domain with hypersensitive sites in the core enhancer and at the μ promoter. When the MARs are deleted, the core enhancer is able to induce only local effects on chromatin structure—the enhancer becomes hypersensitive, but not the promoter, and an extended DNase-I-sensitive domain is not formed. This suggests that the MARs are required either to mediate communication between the core enhancer and the μ promoter, or to help propagate an enhancer-induced alteration in chromatin structure throughout the entire transgene domain. Somewhat similar results are observed for the globin LCR. Although a single copy of either HS2 and HS3 from the globin LCR is able to induce the formation of local hypersensitive sites, these LCR sub-elements differ in their ability to transmit this accessibility over the locus (Ellis *et al.*, submitted); a single copy of HS3 can also induce a hypersensitive site at the globin promoter, while a single copy of HS2 can not. Perhaps local hypersensitivity can be induced by several different types of elements, but these elements differ in their capacity to transmit this accessibility over distance to neighbouring sequences. The complete globin LCR is certainly capable of achieving the creation of such sites over more than a 100 kb (86); however it is not clear yet whether sub-elements from the globin LCR (or for that matter even the intact μ LCR) are capable of acting over such large distances. A particular combination of sub-elements may be required for such long-distance action.

10. Looping in domains

One of the salient features of the domain concept is that the chromatin fibre within the domain would be organized into some type of 3-D structure that permits interactions between different DNA–protein complexes located many kilobases apart. Support for a looping model rather than some type of linear DNA-scanning mechanism has been provided by different lines of evidence. The classical example is transvection in flies where enhancers (or silencers) on one chromosome interact with a promoter on the homologous chromosome, activating (or shutting down) gene expression. A linear scanning mechanism is not necessary for enhancer action; an enhancer is still capable of exerting its action when linked via a non-nucleic acid bridge to the promoter of a gene (87). In the β-globin locus there is competition between the genes for activation by the LCR (88, 89). A β-globin gene located close to the LCR appears to compete with a distal β-globin gene more effectively than when both genes are placed a long distance from the LCR (Fig. 4). This observation is most easily explained by a model proposing direct interaction between the genes and the LCR, rather than a spreading effect along the chromatin

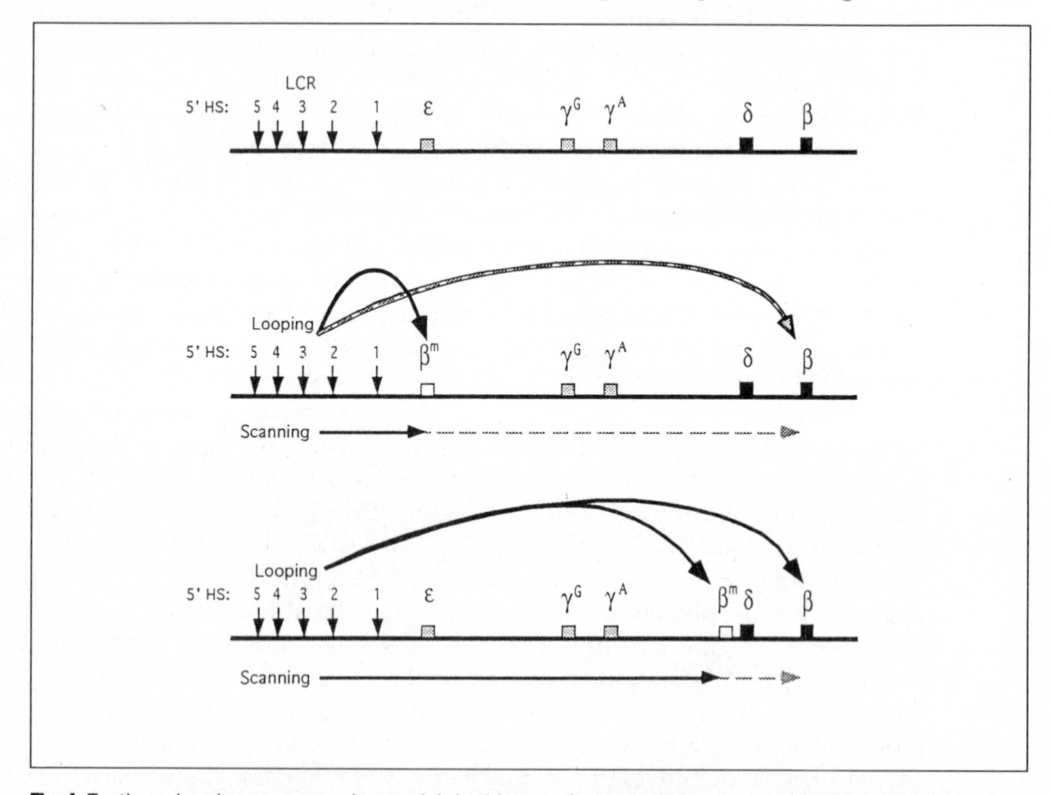

Fig. 4 Testing a looping vs. a scanning model. In this experiment a marker gene is inserted at the position of the embryonic gene or just 5′ to the adult genes, resulting in a difference in relative distance between the β^m, the β gene, and the LCR. The concomitant change in the ratio of β^m to β expression favours a looping vs. a scanning-type mechanism. From Dillon, unpublished data.

fibre (Dillon *et al.*, in preparation). Experiments of this nature do not, of course, address the issue of whether a domain is *first* activated by a looping- or by a scanning-type mechanism such as a replication event, or the initiation of a transcript that extends from one end of the locus to the other.

11. Final comments

Although the chromatin domain model for the organization of chromosomes in multicellular eukaryotes has considerable appeal, it remains to be proved that a domain constitutes both a unit of structure and a unit of gene expression. The latter possibility clearly has had a strong influence on our thinking about the mechanisms which might be involved in eukaryotic gene regulation and also about what features of chromatin organization might contribute to the elaboration of complex developmental pathways in multicellular organisms. The most important aspects of such studies in the future will be the design of novel assays to define domains. Clearly there are sequences (LCRs) which positively activate gene expression irrespective of their position of integration in the genome and which can generate a domain that has an altered chromatin configuration. Here the open questions concern the mechanisms involved in initiating an active chromatin configuration and subsequently maintaining this active state over multiple cell generations. It is also important to understand how this chromatin configuration can be propagated across an entire domain, and in this case what constitutes the limits or boundaries of the activated domain. From this perspective, finding the sequences that correspond to the domain boundaries presents a much more difficult problem, particularly when 'negative' properties are measured. This is additionally complicated because the insertion of multiple gene units and creation of deletions and other rearrangements are very common occurrences in transfections or transgenic animal experiments. The possibility of carrying out homologous recombination directly in the genome or via yeast artificial chromosome transgenics will be of particular value because these techniques will allow the analysis of mutations in the context of the complete functional locus.

Acknowledgements

We would like to thank members of our labs, and our colleagues in the scientific community who helped us in preparing this manuscript and provided us with unpublished data.

References

1. Zlatanova, J. and van Holde, K. (1992) Histone H1 and transcription: still an enigma? *J. Cell Science*, **103**, 889.
2. Widom, J. (1989) Toward a unified model of chromatin folding. *Annu. Rev. Biophys. Chem.*, **18**, 365.

3. Felsenfeld, G. and McGhee, J. D. (1986) Structure of the 30 nm fibre. *Cell*, **44,** 375.
4. Swedlow, J. R., Agard, D. A., and Sedat, J. W. (1993) Chromosome structure inside the nucleus. *Cur. Opinion Cell Biol.*, **5,** 412.
5. Udvardy, A., Maine, E., and Schedl, P. (1985) The 87A7 chromomere: identification of novel chromatin structures flanking the heat shock locus that may define the boundaries of higher order domains. *J. Mol. Biol.*, **185,** 341.
6. Udvardy, A. and Schedl, P. (1993) The dynamics of chromatin condensation—redistribution of topoisomerase II in the 87A7 heat shock locus during induction and recovery. *Mol. Cell. Biol.*, **13,** 7522.
7. Björkroth, B., Ericsson, C., Lamb, M. M., and Daneholt, B. (1988) Structure of the chromatin axis during transcription. *Chromosoma*, **96,** 333.
8. Ericsson, C., Mehlin, H., Björkroth, B, Lamb, M. M., and Daneholt, B. (1989) The ultrastructure of upstream and downstream regions of an active Balbiani ring gene. *Cell*, **56,** 631.
9. Jupe, E. R., Sinden, R. R., and Cartwright, I. L. (1993) Stably maintained microdomain of localized unrestrained supercoiling at a *Drosophila* heat shock gene locus. *EMBO J.*, **12,** 1067.
10. Gall, J. (1956) On the submicroscopic structure of chromsomes. In *Mutation*, Brookhaven Symp. Biol, **8,** 17.
11. Callan, H. G. and Lloyd, L. (1960) The lampbrush chromosomes of the crested newt *Triturus cristatus. Philos. Trans. R. Soc. Lond. B. Biol. Med.*, **243,** 135.
12. Gall, J. and Callan, H. G. (1962) ^{3}H-uridine incorporation into lampbrush chromosomes. *Proc. Natl Acad. Sci. USA*, **82,** 562.
13. Alfert, M. (1956) Composition and structure of giant chromosomes. *Int. Rev. Cytol.*, **3,** 131.
14. Ritossa, F. (1962) New puff pattern induced by temperature shock and DMP in *Drosophila. Experientia*, **18,** 571.
15. Benyajatti, C. and Worcel, A. (1976) Isolation, characterization and structure of the folded interphase genome of *Drosophila melanogaster. Cell*, **9,** 393.
16. Mirkovitch, J., Mirault, M.-E., and Laemmli, U. K. (1984) Organization of the higher order chromatin loop: specific attachment sites on nuclear scaffold. *Cell*, **39,** 223.
17. Gasser, S. M. and Laemmli, U. K. (1986) The organization of chromatin loops. Characterization of a scaffold attachment site. *EMBO J.*, **5,** 511.
18. Gasser, S. M. and Laemmli, U. K. (1987) A glimpse at chromosomal order. *Trends Genet.*, **3,** 16.
19. Cockerill, P. N. and Garrard, W. T. (1986) Chromosomal loop anchorage of the kappa immunoglobulin gene occurs next to the enhancer in a region containing topoisomerase II sites. *Cell*, **44,** 273.
20. Phi-Van, L. and Stratling, W. H. (1988) The matrix attachment regions of the chicken lysozyme gene co-map with the boundaries of the chromatin domain. *EMBO J.*, **7,** 655.
21. Gasser, S. M. and Laemmli, U. K. (1986) Cohabitation of scaffold binding regions with upstream/enhancer elements of three developmentally regulated genes of *D. melanogaster. Cell*, **46,** 521.
22. Cockerill, P. N. (1990) Nuclear matrix attachment occurs in several regions of the Igh Locus. *Nucleic Acids Res.*, **18,** 2643.
23. Adachi, Y., Kas, E., and Laemmli, U. K. (1989) Preferential, cooperative binding of DNA topoisomerase II to scaffold associated regions. *EMBO J.*, **8,** 3997.

24. Izaurralde, E., Mirkovitch, J., and Laemmli, U. K. (1988) Interaction of DNA with nuclear scaffolds *in vitro*. *J. Mol. Biol.*, **2002,** 111.
25. Udvardy, A. and Schedl, P. (1984) Chromatin organization of the 87A7 heat shock locus of *Drosophila melanogaster*. *J. Mol. Biol.*, **172,** 385.
26. Chung, J. H., Whiteley, M., and Felsenfeld, G. (1993) A 5′ element of the chicken β-globin domain serves as an insulator in human erythroid cells and protects against position effect in *Drosophila*. *Cell*, **74,** 505.
27. Stalder, J., Larsen, A., Engel, J. D., Dolan, M., Groudine, M., and Weintraub, H. (1980) Tissue specific DNA cleavages in the globin chromatin domain introduced by DNase I. *Cell*, **20,** 451.
28. Verreault, A. and Thomas, J. O. (1994) Chromatin structure of the β-globin chromosomal domain in adult chicken erythrocytes. *Cold Spring Harbor Symp. Quant. Biol.*, **LVIII,** 15.
29. Hebbes, T. R., Clayton, A. L., Thorne, A., and Crane-Robinson, C. (1992) Core histone hyperacetylation co-maps with generalized DNase I sensitivity in the chicken β-globin chromosomal domain. *EMBO J.*, **13,** 1823.
30. Reitman, M. and Felsenfeld, G. (1990) Developmental regulation of topoisomerase II sites and DNAase I hypersensitive sites in the chicken β-globin locus. *Mol. Cell. Biol.*, **10,** 2774.
31. Kellum, R. and Schedl, P. (1991) A position effect assay for boundaries of higher order chromatin domains. *Cell*, **64,** 941.
32. Kellum, R. and Schedl, P. (1992) A group of scs elements function as domain boundaries in an enhancer-blocking assay. *Mol. Cell. Biol.*, **12,** 2424.
33. Sippel, A., Schafer, G., Faust, N., Saueressig, H., Hecht, A., and Bonifer, C. (1994) Chromatin domains constitute regulatory units for the control of eukaryotic genes. *Cold Spring Harbor Symp Quant. Biol.*, **LVIII,** 37.
34. Steif, A., Winter, D. M., Stratling, W. H., and Sippel, A. E. (1989) A nuclear DNA attachment element mediates elevated and position-independent gene activity. *Nature*, **341,** 343.
35. Bonifer, C., Yanoutos, N., Kriger, G., Grosveld, F., and Sippel, A. (1994) Dissection of the LCR function located on the chicken lysozyme domain in transgenic mice. Submitted.
36. Lewis, E. (1949) *Drosophila Information Service*, **23,** 59.
37. Modolell, J., Bender, W., and Meselson, M. (1983) *D.melanogaster* mutations suppressible by the *suppressor of Hairy-wing* are insertions of a 7.3 kb mobile element. *Proc. Natl Acad. Sci. USA*, **80,** 1678
38. Peifer, M. and Bender, W. (1986) The anterobithorax and bithorax mutations of the bithorax complex. *EMBO J.*, **9,** 2293.
39. Jack, J., Dorsett, D., Delotto, Y., and Liu, S. (1991) Expression of the *cut* locus in the *Drosophila* wing margin is required for cell type specification and is regulated by a distal enhancer. *Development*, **113,** 735.
40. Peifer, M. and Bender, W. (1988) Sequences of the *gypsy* transposon of *Drosophila* neccessary for its effects on adjacent genes. *Proc. Natl Acad. Sci. USA*, **85,** 9650.
41. Smith, P. A. and Corces, V. G. (1992) The *suppressor of Hairywing* binding region is required for *gypsy* mutagenesis. *Mol and Gen. Genet.*, **233,** 65.
42. Geyer, P. and Corces, V. (1992) DNA position-specific repression of transcription by a *Drosophila* zinc finger protein. *Genes and Dev.*, **6,** 1865.

43. Parkhurst, S. M., Harrison, D. A., Remington, M. P., Spana, C., Kelley, R. L., Coyne, R. S., and Corces, V. G. (1988) The *Drosophila su(Hw)* gene which controls the phenotypic effect of the *gypsy* transposable element, encodes a putative DNA binding protein. *Genes and Dev.*, **2**, 1205.
44. Harrison, D. A., Gdula, D. A., Coyne, R. S., and Corces, V. G. (1993) A leucine zipper domain of the *suppressor of Hairy-wing* protein mediates its repressive effect on enhancer function. *Genes and Dev.*, **7**, 1966.
45. Spana, C., Harrison, D. A., and Corces, V. G. (1988) The *Drosophila melanogaster suppressor of Hairy-wing* protein binds to specific sequences of the *gypsy* retrotransposon. *Genes and Dev.*, **2**, 1414.
46. Spana, C. and Corces, V. G. (1990) DNA bending is a determinant of binding specificity for a *Drosophila* zinc finger protein. *Genes and Dev.*, **4**, 1505.
47. Flavell, A. J., Alphey, L. S., Ross, S. J., and Leigh-Brown, A. J. (1990) Complete reversions of a *gypsy* retrotransposon-induced *cut* locus mutation in *Drosophila melanogaster* involving *jockey* transposon insertions and flanking *gypsy* sequence deletions. *Mol. Gen Genet.*, **220**, 181.
48. Holdridge, C. and Dorsett D. (1991) Repression of *hsp70* heat shock gene transcription by the suppressor of hair-wing protein of *Drosophila melanogaster. Mol. Cell. Biol.*, **11**, 1894.
49. Roseman, R. R., Pirotta, V., and Geyer, P. K. (1993) A *su(Hw)* protein insulates expression of the *Drosophila melanogaster white* gene from chromosomal position effects. *EMBO J.*, **12**, 435.
50. Duncan, I. (1987) The bithorax complex. *Annu. Rev. Genet.*, **21**, 285.
51. Mahaffey, J. and Kaufman, T. (1988) The homeotic genes of the *Antennapedia* complex and bithorax complex of *Drosophila*. In *Developmental genetics of higher organisms*. G. M. Malacinsky (ed.). MacMillan, New York, p. 239.
52. Zavortink, M. and Sakonju, S. (1989) The morphogenetic and regulatory functions of the Drosophila *Abdominal-B* gene are encoded in overlapping RNAs transcribed from separate promoters. *Genes and Dev.*, **3**, 1969.
53. Celniker, S., Sharma, S., Keelan, D., and Lewis, E. (1990) The molecular genetics of the bithorax complex of *Drosophila: cis*-regulation in the *Abdominal-B* domain. *EMBO J.*, **9**, 4277.
54. Sanchez-Herrero, E. (1991) Control of the expression of the bithorax complex *abdominal-A* and *Abdominal-B* by *cis*-regulatory regions in *Drosophila* embryos. *Development*, **111**, 437.
55. Galloni, M., Gyurkovics, H., Schedl, P., and Karch, F. (1993) The *bluetail* transposon: evidence for independent *cis*-regulatory domains and domain boundaries in the bithorax complex. *EMBO J.*, **12**, 1087.
56. Gyurkovics, H., Gausz, J., Kummer, J., and Karch, F. (1990) A new homeotic mutation in the *Drosophila* bithorax complex removes a boundary separating two domains of regulation. *EMBO J.*, **9**, 2579.
57. Karch, F., Weiffenbach, B., Peifer, M., Bender, W., Duncan, I., Celniker, S., Crosby, M., and Lewis, E. B. (1985) The abdominal region of the bithorax complex. *Cell*, **43**, 81.
58. Karch, F., Galloni, M., Sipos, L., Gausz, J., Gyurkovics, H., and Schedl, P. (1994) *Mcp* and *Fab-7*: molecular analysis of putative boundaries of *cis*-regulatory domains in the bithorax complex of *Drosophila melanogaster*. *Nucleic Acid. Res.*, in press.
59. Weintraub, H. and Groudine, M. (1976) Chromosomal subunits in active genes have an altered conformation. *Science*, **193**, 848.

60. Wood, W. I. and Felsenfeld, G. (1982) Chromatin structure of the chicken β-globin gene region: sensitivity to DNase I, micrococcal nuclease, and DNase II. *J. Biol. Chem.*, **257,** 7730
61. Kim, U.-J., Han, M., Kayne, P., and Grunstein, M. (1988) Effects of histone H4 depletion on the cell cycle and transcription of *Saccharomyces cerevisiae. EMBO J.*, **7,** 2211.
62. Schmid, A., Fascher, K., and Horz, W. (1992) Nucleosome disruption at the yeast *PHO5* promoter upon *PHO5* induction occurs in the absence of DNA replication. *Cell*, **71,** 853.
63. Shimizu, M., Roth, S. Y., Szent-Gyorgi, C., and Simpson, R. T. (1991) Nucleosomes are positioned with base pair precision adjacent to the α_2 operator in *Saccharomyces cerevisiae. EMBO J.*, **10,** 3033.
64. Archer, T. K., Cordingley, M. G., Woliord, R. G., and Hager, G. L. (1991) Transcription factor access is mediated by accurately positioned nucleosomes on the mouse mammary tumor virus promoter. *Mol. Cell Biol.*, **11,** 688.
65. Benezra, R., Cantor, C. R., and Axel, R. (1986) Nucleosomes are phased along the mouse β-major globin gene in erythroid and non-erythoid cells. *Cell*, **44,** 697.
66. Boyes, J. and Bird, A. (1991) DNA methylation inhibits transcription indirectly via a methyl-CpG binding protein. *Cell*, **64,** 1123.
67. Lu, Q., Wallrath, L. L., Granok. H., and Elgin, S. C. R. (1993) (CT)n·(GA)n repeats and heat shock elements have distinct roles in chromatin structure and transcriptional activation of the *Drosophila hsp26* gene. *J. Mol. Biol.*, **13,** 2802.
68. Glaser, R. L., Thomas, G., Siegfried, E. S., Elgin, S. C., and Lis, J. T. (1990) Optimal heat-induced expression of the *Drosophila hsp26* gene requires a promoter sequence containing (CT)n·(GA)n repeats. *J. Mol. Biol.*, **211,** 751.
69. Tsukiyama, T., Becker, P. B., and Wu, C. (1994) ATP dependent nucleosome disruption at a heat shock promoter mediated by binding of GAGA transcription factor. *Nature*, **367,** 525.
70. Soeller, W., Oh, C. E., and Kornberg, T. B. (1993) Isolation of cDNAs encoding the *Drosophila* GAGA transcription factor. *Mol. Cell. Biol.*, **13,** 7961.
71. Turner, B. M. (1993) Decoding the nucleosome. *Cell*, **75,** 5.
72. Lee, D. Y., Hayes, J. J., Pruss, D., and Wolffe, A. P. (1993) A positive role for histone acetylation in transcription factor access to nucleosomal DNA. *Cell*, **72,** 73.
73. Brotherton, T. W., Reneker, J., and Ginder, G. D. (1990) Binding of HMG 17 to mononucleosomes of the avian β-glohin gene cluster in erythroid and non-erythroid cells. *Nucleic Acids Res.*, **18,** 2011.
74. Grosveld, F., van Assendelft, G. B., Greaves, D. R., and Kollias, G. (1987) Position independent, high level expression of the human β-globin gene in transgenic mice. *Cell*, **51,** 975.
75. Grosveld, F., Antoniou, M., Berry, M., De Boer, E., Dillon, N., Ellis, J., Fraser, P., Hurst, J., Imam, A., Meijer, D., Philipsen, S., Pruzina, S., Strouboulis, J., and Whyatt, D. (1994) Regulation of human globin gene switching. *Cold Spring Harbor Symp. Quant. Biol.*, **LVIII,** 7.
76. Tuan, D., Solomon, W., London, I., and Lee, D. (1989) An erythroid specific developmental stage independent enhancer far upstream of the human β-like globin genes. *Proc. Natl Acad. Sci. USA*, **86,** 6384.
77. Nay, P., Sorrentino, B., Lowvery, C., and Nienhur's, A. (1990) Inducibility of the HSII enhancer depends on binding of an erythroid specific nuclear protein. *Nucleic Acids Res.*, **18,** 6011.
78. Fraser, P., Hurst, J., Collins, P., and Grosveld, F. (1990) DNase I hypersensitive sites

1, 2, and 3 of the human β-globin DCR directs position independent expression. *Nucleic Acids Res.*, **18,** 3503.
79. Talbot, D., Philipsen, S., Fraser, F., and Grosveld, F. (1990) Detailed analysis of the site 3 region of the human β-globin dominant control region. *EMBO J.*, **7,** 2169.
80. Philipsen, S., Pruzina, S., and Grosveld, F. (1993) The minimal requirements for activity in transgenic mice of hypersensitive site 3 of the β-globin locus control region. *EMBO J.*, **3,** 1077.
81. Renauld, H., Aparicio, O., Zierath, P., Billington, B., Chablani, S., and Gottschling, D. (1993) Silent domains are assembled continuously from the telomere and are defined by promoter distance and strength and by SIR3 dosage. *Genes and Dev.*, **7,** 1133.
82. Dillon, N. and Grosveld, F. (1993) Transcriptional regulation of multigene loci: multilevel control. *Trends Genet.*, **9,** 134.
83. Bonifer, C., Yannoutos, B., Kruger, G., Grosveld, F., and Sippel, A. (1994) Dissection of the LCR functions located in the chicken lysozyme domain in transgenic mice. Submitted.
84. Jenuwein, T., Forrester, W. C., Qiu, R.-G., and Grosschedl, R. (1993) The immunoglobulin μ enhancer core establishes local factor access in nuclear chromatin independent of transcriptional stimulation. *Genes and Dev.*, **7,** 2016.
85. Forrester, W. D., Jenuwein, T., Van Genderen, C., and Grosschedl, R. (1994) Enhancer-mediated activation of immunoglobulin μ gene transcription during lymphoid development is dependent upon matrix attachment regions. *Science*, in press.
86. Forrester, W., Epner, E., Driscoll, C., Enver, T., Brice, M., Papayannopoulou, T., and Groudine, M. (1990) A deletion of the human β-globin locus activation region causes a major alteration in chromatin structure and replication across the entire β-globin locus. *Genes and Dev.*, **4,** 1637.
87. Muller, M., Fogo, J., and Schaffner, W. (1989) An enhancer stimulates transcription in *trans* when attached to the promoter via a protein bridge. *Cell*, **58,** 767.
88. Hanscombe, O., Whyatt, D., Fraser, P., Yannoutsos, N., Greaves, D., Dillon, N., and Grosveld, F. (1991) Importance of globin gene order for correct developmental expression. *Genes and Dev.*, **51,** 387.
89. Peterson, K., Clegg, C., Huxley, C., Josephson, B., Mangers, M., Furukana, T., and Stamatoyaunnopoulos, G. (1993) Transgenic mice containing a 240 kb YAC carrying the human β-globlin locus display proper developmental control of human globin genes. *Proc. Natl Acad. Sci. USA*, **90,** 7593.

10 Epigenetic regulation in mammals

SHIRLEY M. TILGHMAN and HUNTINGTON F. WILLARD

1. Introduction

The term 'epigenetic' when applied to mammalian genes refers to different stable states of phenotypic expression usually thought to be due to differential effects of chromosome or chromatin packaging, and determined or influenced by control mechanisms which can be manifested over exceptionally long distances. Both the recognition of epigenetic states and the design of tractable experimental approaches to analyse them require an appreciation for mammalian gene expression in its dynamic setting. One must consider those aspects of chromatin assembly, structure, and replication (see Chapters 1–3) that determine whether, when, and at what level genes are expressed, the series of dynamic events that occur at each cell division, when and under what circumstances epigenetic states are stably transmitted through meiosis or mitosis, and when and under what circumstances epigenetic states are reversed in the course of normal development or as a result of mutation or chromosome rearrangement.

Here we consider examples of epigenetic regulation in mammals—genomic imprinting and X chromosome inactivation—that show differential gene expression of the two copies of a gene or chromosome in the same nucleus, one copy being transcribed normally while the homologous copy is silenced. While the basis for discrimination between the copies is still uncertain, both long- and short-range features of mammalian chromatin and chromosome organization are implicated. We review some of the experimental evidence for such epigenetic phenomena in mammals and the models proposed to explain gene silencing in the context of specific 'case studies' of imprinted or inactivated loci in humans and/or mice.

2. Models to explain epigenetics in mammals: the requirement for a 'mark'

The reductionist view, taken in large part from pioneering experiments in prokaryotes, sees the problem of regulated gene expression in terms of the DNA-binding proteins and their ability to recognize specific target sites. Thus whether or not a gene is expressed is governed by whether the appropriate transcription factors are present in the cell, and the level of expression is determined by the concentration of the factors, their affinities for the binding site, and the strength of the signal that binding transmits to the transcription machinery.

Epigenetic phenomena such as imprinting and X chromosome inactivation cannot be fully explained by this simple view. To achieve allele-specific expression of a gene, the cell must be able to discriminate between two possibly identical copies, and express one while keeping the other silent (Fig. 1). This requires some form of a 'mark' to act as the discriminator.

The mark must fulfil several important criteria. First, it must be capable of being propagated through many cell divisions. Consider the markings of pigment in the coat of the tabby cat. These patches are the result of the random choice of which X chromosome to inactivate in an early melanoblast stem cell; that choice is clonally propagated in that cell's progeny (1). Second, the mark must be erasable in germ cells, to be re-set in the next generation. For example, an imprinted gene which is inherited in a silent state from a father will be re-activated when it is passed to the next generation through a female.

Several mechanisms have been suggested to achieve these non-Mendelian patterns of gene expression and to account for the pleiotropic effects of epigenetic regulation (see below). Use of a reversible covalent modification of DNA, such as DNA methylation (where the cytosine residue in a CpG dinucleotide is methylated at the N_7 position of the pyrimidine ring) is illustrated in Fig. 2a. Many genes have been shown to be more heavily methylated when they are in a silent configuration than when they are active. For example, the DNA encoding the mouse immunoglobulin genes is undermethylated in expressing B-lymphocytes relative to that in a non-expressing tissue such as liver (2, 3). It has been demonstrated that CpG

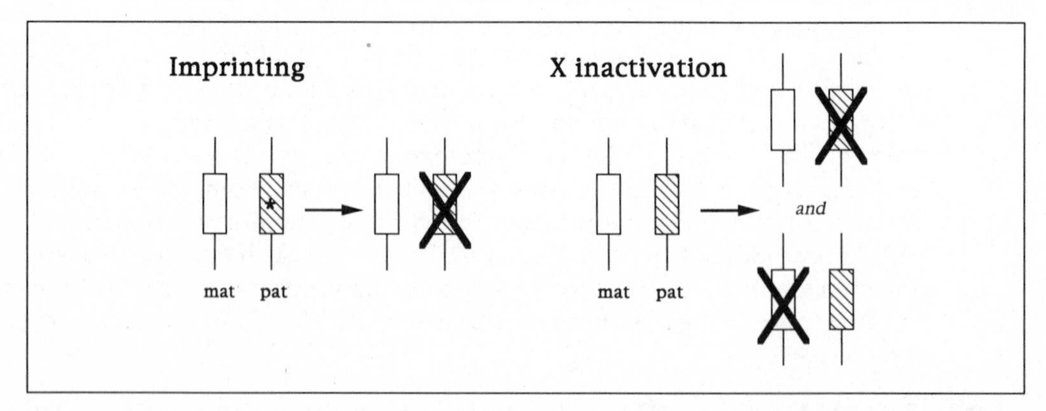

Fig. 1 Comparison of genomic imprinting and X chromosome inactivation. Both imprinting and X inactivation involve the differential expression of the maternally (mat) and paternally derived (pat) genes. In imprinting, one of the copies contains a mark (*) (shown here for the paternally derived copy) that determines that it will *not* be expressed in the embryo (denoted by the large black 'X'). In X inactivation, the choice of which copy will be expressed is random; some cells inactivate the paternally derived gene, while others inactivate the maternally derived gene. The embryo is, therefore, a mosaic of clonal patches where either the maternally or paternally derived X is inactive.

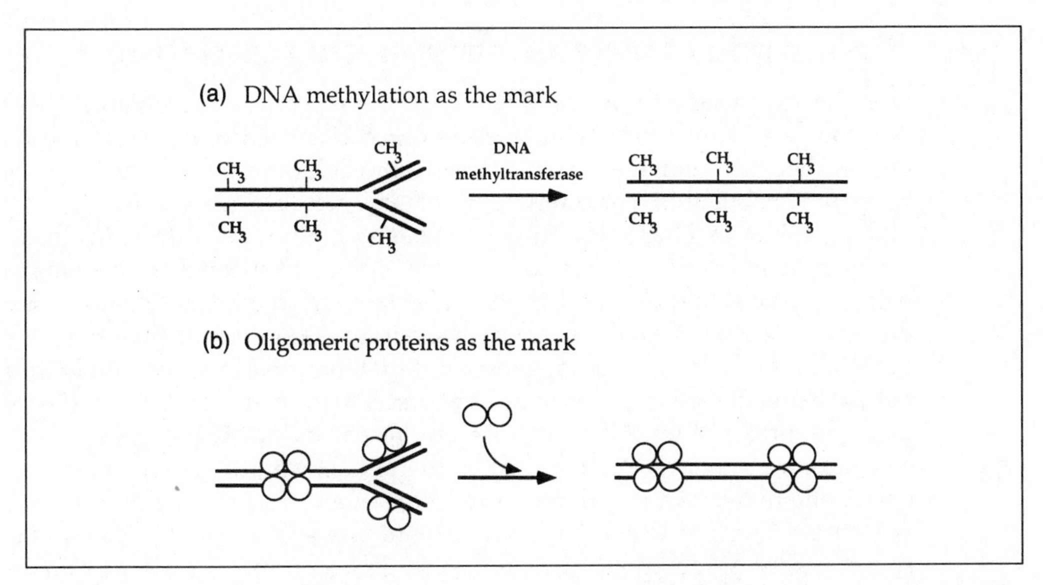

Fig. 2 Candidates for the epigenetic mark. (a) DNA at the replication fork of a methylated segment of DNA will be hemi-methylated. The maintenance methylase, DNA methyltransferase, will methylate the nascent strands to reconstitute the fully methylated DNA. (b) A hypothetical tetrameric protein is equally distributed at the replication fork. Co-operative binding of subunits ensures that the tetramer is re-established after replication only on the one allele that had been previously bound by protein.

methylation can interfere with the binding of specific transcription factors whose recognition sequence contains the dinucleotide, and that this interference can have an effect on gene expression, as measured by DNA transfection (4). What has been difficult to establish in most instances, however, is whether the methylation is sufficient to bring about the repression of gene expression in non-expressing tissues, or occurs as a by-product of some other mechanism.

Methylation is a particularly attractive means to achieve epigenetic differences because it can readily fulfil the criteria listed above. It is capable of propagation through many cell divisions, by virtue of the action of DNA methyltransferase, an enzyme whose substrate is hemi-methylated DNA (5–7) (Fig. 2a). Conversely, DNA methylation differences can be erased by the action of a DNA demethylase, although none have been characterized to date.

Another potential way to mark a gene differentially is through non-covalent modification, for example through differential protein binding to specific sites. To maintain the difference, however, one has to propose two requirements: first that the protein be an oligomer which binds co-operatively to DNA, and second that the protein subunits segregate equally to both daughter chromatids at replication forks (Fig. 2b). In this instance, the absence of bound protein at the replication fork of the previously unmarked chromosome will effectively ensure that no new protein can be bound, thereby maintaining the difference.

3. Pleiotropic effects of epigenetic regulation

Differential expression of homologous copies of genetic material in the same cell is accompanied by a variety of phenotypic, cellular, chromosomal, and molecular events that distinguish the two copies from each other. The two major classes of epigenetic regulation considered in this chapter are similar in many respects, although their mechanistic and genetic basis is different, at least in detail.

Genomic imprinting refers to the observation, from nuclear transplant experiments in mouse zygotes and from analysis of chromosomal deletions and partial disomies in both mice and humans, that certain regions of the genome are differentially 'marked' during gametogenesis and that maternally or paternally transmitted autosomal genes are not necessarily expressed equivalently (Fig. 1) (8–10). These parent-of-origin phenotypic effects are restricted to single genes or to small chromosomal regions, in contrast with X inactivation, in which the effects of epigenetic regulation are much more long range.

X chromosome inactivation is the *cis*-limited inactivation of genes on one of the two X chromosomes present in somatic cells of normal female mammals (1, 11). Like imprinting, it results in the silencing of one copy of the genetic material in a given cell. However, with the exception of extra-embryonic tissues in which the paternally derived X chromosome is always inactive, X inactivation differs from imprinting in being random with respect to parent-of-origin (1, 11) (Fig. 1). Thus, the epigenetic changes that differentiate between the maternally and paternally derived copies of the X are introduced anew in the embryo, rather than being transmitted directly through the germline.

Both imprinting and X inactivation correlate (locally in the case of imprinted genes and chromosomally in the case of the inactivated X chromosome) with changes in the timing of DNA replication, with changes in chromatin structure, and with DNA modification. It remains unclear which if any of these effects is responsible primarily for establishing the different epigenetic states and which are secondary phenomena associated with differences between active and inactive genes.

3.1 DNA replication

That genetically inert DNA replicates late in S-phase of the cell cycle has long been a feature of models relating chromatin structure to gene function (12). Experiments using tritiated thymidine to distinguish late- and early-replicating DNA established a correlation between the presence of inactive X chromosomes and late replication (13). Higher resolution optical assays using substitution of 5-bromodeoxyuridine (BrdU) into chromosomal DNA documented substantial differences in the regional order of replication on the active and inactive X chromosomes (14). The resolution of these assays remains quite poor; cytogenetically small regions on metaphase chromosomes contain thousands of kilobases of DNA, consisting of many replication origins that appear to intitiate DNA synthesis in relative synchrony within a defined period of S-phase.

More recently, molecular assays have been developed for use on mammalian interphase cells, based either on the method of fluorescence *in situ* hybridization to distinguish replicated from unreplicated DNA by the presence of hybridization signal on each of the replicated chromatids (15) or on the physical separation of BrdU-containing DNA following flow cytometry (16, 17). These assays have been used on a limited number of loci thus far, but have confirmed the later replication of specific gene sequences on the inactive X chromosome (16) and have demonstrated the late replication of imprinted autosomal genes (18, 19).

3.2 Chromatin effects

Differences between the chromatin of active and inactive regions have been apparent since the demonstration of the Barr body—the facultative heterochromatic state of the inactive mammalian X chromosome in interphase nuclei (20). The molecular basis for heterochromatin packaging of this type is unclear, since (unlike constitutive heterochromatin consisting of repetitive DNA, for example) the primary DNA sequences of the active, euchromatic X chromosome are identical to those of the inactive, heterochromatic X chromosome in the same nucleus. The heterochromatic X chromosome contains underacetylated histone H4, as determined by indirect immunofluorescence using anti-acetylated H4 antibodies (21; see Chapter 7). However, since histone acetylation correlates with gene activity generally (22), this is likely to be a consequence of X inactivation rather than a specific determinant of the epigenetic, inactive state. Histone acetylation has not yet been evaluated for imprinted genes; however, it would be surprising if there were not a systematic difference.

Chromatin structure can be assessed by a variety of nuclease accessibility assays. Actively transcribed regions of chromatin are generally more sensitive to nuclease digestion than are non-transcribed regions (23). DNase I hypersensitive sites have been detected both at several X-linked genes and at imprinted genes (see below). This relative sensitivity, as well as restriction endonuclease accessibility, presumably reflects a specific 'open' chromatin conformation at promoters and/or enhancers for actively transcribed genes.

3.3 DNA methylation

Differential DNA methylation has been correlated with differential gene expression for many years and, on that basis, has been considered as a leading candidate for the mark, both in imprinting and in X inactivation. Hypermethylation has been documented at the CpG islands associated with or near the promoters of both inactivated X-linked genes and imprinted autosomal genes.

Three general approaches have been used to study the role of methylation in epigenetic regulation. First, the use of methylation-sensitive restriction endonucleases has permitted direct assessment of the methylation status of a given CpG dinucleotide in genomic DNA (24). However, such assays are limited to the relatively small proportion of CpGs that are contained within appropriate recognition sites. Pfeifer *et al.* (25) have recently described an elegant (albeit technically challenging) method to modify cytosines *in vivo* only if they are unmethylated, which allows assessment of all cytosines by genomic DNA sequencing. Application of this method should lead to definitive identification of methylated regions. Second, a direct connection between DNA methylation and X inactivation was demonstrated by reactivation of genes from the inactive X using DNA-demethylating agents such as 5-azacytidine (26). Such experiments have implicated DNA methylation in the perpetuation of the inactive state in somatic cells, but are unable to address a possible role for DNA methylation as the mark responsible for initiation of DNA silencing in either imprinting or X inactivation. Third, mice deficient in DNA methyltransferase activity (27) have been used to establish a role for DNA methylation in genomic imprinting. Further application of this approach to both X inactivation and imprinting is eagerly anticipated since, in principle, such experiments should directly answer whether methylation is both necessary and sufficient for epigenetic regulation or whether it is a consequence of other primary effects responsible for gene inactivation.

4. Case studies of epigenetics in mammals

4.1 Parental imprinting

The functional non-equivalence of the maternal and paternal genomes in mammals was first suspected from the failure to observe parthenogenesis in mammals (28, 29). Parthenogenotes are derived from unfertilized eggs whose haploid nucleus has been induced to duplicate by physical or chemical means; their genomes are entirely maternally derived. Nuclear transplantation studies have confirmed that both maternal and paternal genomes are required for development (30, 31). In these experiments one of the two pronuclei was removed from a fertilized egg, and replaced with a pronucleus from a second (Fig. 3). When the reconstituted embryo contained pronuclei derived from both a male and a female, normal development ensued. However, when both pronuclei were derived from either males or females, the resulting embryo failed to thrive. In the female-derived embryo (gynogenote), the failure was largely in extra-embryonic tissues. With androgenotes (male derived), it was the embryo proper which was underdeveloped. While these experiments and others (32, 33), established the requirement for both parental pronuclei and led directly to the notion of genes inherited in differentially active states, they could not make any predictions about the nature or the number of such genes.

Nine imprinted genes have been identified to date in both mice and humans (Table 1). Here, we compare two clusters of these genes and consider whether there are common lessons to be learned from them regarding the mechanism of imprinting.

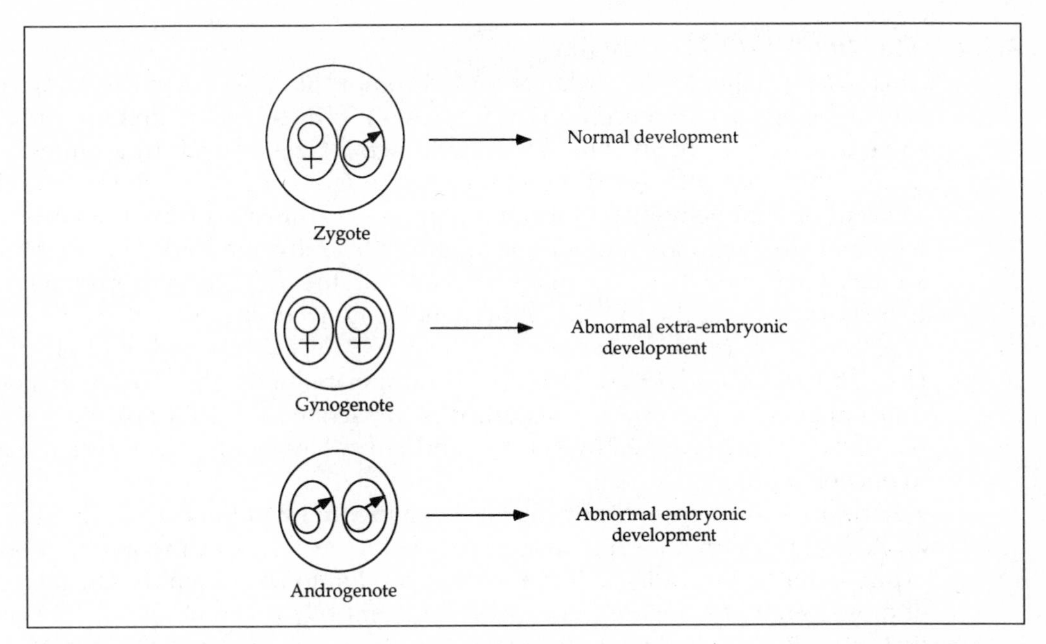

Fig. 3 Viability of zygotes created by nuclear transplantation. The genetic make-up and developmental fate of these zygotes are indicated. Adapted from (30, 31).

Table 1 Mammalian imprinted genes

Gene	Chromosome	Expression	Reference
Igf2	Mouse distal 7	Paternal	(35)
	Human 11p15.5	Paternal	(54, 53)
Igf2r	Mouse proximal 17	Maternal	(73)
	Human	Maternal[a]	(74, 75)
H19	Mouse distal 7	Maternal	(36)
	Human 11p15.5	Maternal	(56)
Ins-2	Mouse distal 7	Paternal[b]	(39)
Ins-1	Mouse 15	Paternal[b]	(39)
Snrpn	Mouse central 7	Paternal	(64)
	Human 15q11–q15	Paternal	(65, 66)
U2afbp-rs	Mouse proximal 11	Paternal	(108)
WT-1	Human 11p13	?[c]	(109)
Xist	Mouse X	Paternal[d]	(92, 93)

[a] Maternal expression of the *Igf2r* gene in humans has been observed in only a subset of individuals examined.
[b] The *ins-1* and *-2* genes are imprinted in extra-embryonic tissues of the mouse embryo, but not in pancreas. No information is available for the human gene.
[c] Monoallelic expression of the *WT-1* gene has been observed in fetal brain and placenta in a subset of individuals, and bialleleic expression has been seen in kidney.
[d] The *Xist* gene is paternally expressed only in the extra-embryonic tissues of the mouse embryo. In all other tissues, it is expressed from one of the parental alleles with random choice (Fig. 6). The paternal-specific expression of *XIST* in human embryos has not been tested.

4.1.1 The *Ins-2/Igf2/H19* cluster

This cluster maps to the distal end of chromosome 7 in mice; its counterpart in humans maps to chromosome 11p15.5 (Fig. 4). It includes the first imprinted gene to be discovered, the insulin-like growth factor 2 gene (*Igf2*). In a mutant mouse strain which has a targeted disruption in the *Igf2* gene, the heterozygote has 60 per cent normal body weight if the mutation was inherited from the father, but is a normal size when the mutation is inherited from the mothers (34, 35). Molecular studies confirmed that the maternal copy of the *Igf2* gene is normally silent, thereby explaining the unusual inheritance pattern of the phenotype.

A second imprinted gene in this region is *H19*; its expression is entirely maternal (36). Thus *Igf2* and *H19* are imprinted in opposite directions. The *H19* gene is an unusual gene of no known function. It is transcribed by RNA polymerase II, and the transcript is processed by splicing and polyadenylation, yet it does not appear to encode a protein (37, 38).

Recently a third imprinted gene has been identified in the cluster, insulin-2 (Ins-2). Although normally expressed in pancreatic islet cells in the adult, where it is expressed from both alleles, it is also expressed in the extra-embryonic membranes of mouse embryos, where the expression is entirely paternal (39). *Ins-2* is not the first example of a conditionally imprinted gene. *Igf2* is also expressed from both chromosomes in the choroid plexus and leptomeninges, two non-neuronal tissues in the brain (35).

The three genes lie in the same transcriptional orientation within a 120 kb domain of DNA in mice, as well as in humans (40). The gene order, in a 5′ to 3′

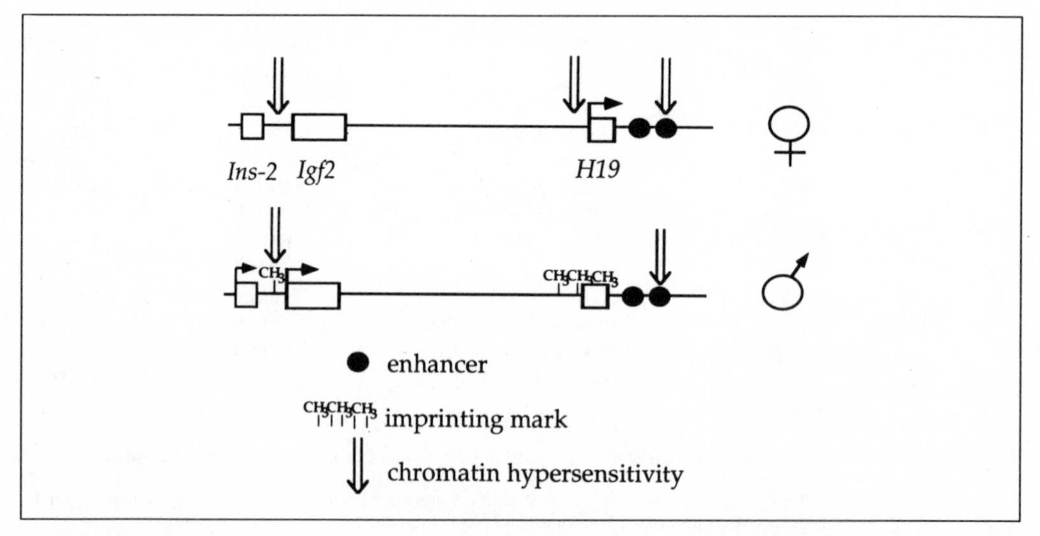

Fig. 4 Epigenetic modifications at the *Ins-2/Igf2/H19* locus. The *Ins-2*, *Igf2*, and *H19* genes are depicted with open boxes, and their transcriptional orientation and chromosomal transcription site indicated by the horizontal arrows. Two enhancers which lie 3′ of the *H19* gene are indicated by the closed circles. The positions of allele-specific methylation are indicated by the CH_3 symbols. The positions of DNase I and restriction enzyme hypersensitivity are indicated by the vertical arrows. Adapted from (43–6).

direction, is *Ins-2/Igf2/H19*. The physical linkage of these genes suggested that their imprinting was coupled in some manner (analogous perhaps to the chromosomal nature of X inactivation; see below), although their expression from different chromosomes eliminated the most straightforward models for *cis*-acting regulators of imprinting. At least two of the genes, *Igf2* and *H19*, are expressed in a very similar, if not identical pattern during development (41, 42). Both genes are activated at the time of implantation in extra-embryonic tissues and by day 7.5 in the mouse embryo proper. Expression of both genes persists in a wide array of tissues of endodermal and mesodermal origin, with no detectable expression in the central and peripheral nervous systems.

It has been established that DNA methylation plays a central role in the imprinting of *Igf2* and *H19*. In both mice and humans, the silent paternal *H19* gene is heavily methylated throughout a 7–9 kb region beginning 4–6 kb upstream of the promoter and continuing through the gene itself (43–45). Surprisingly, the only allele-specific methylation of the *Igf2* gene is found on the *active* maternal allele, upstream of its most proximal promoter (46). If this is also a mark, it must act positively, possibly preventing a repressor from binding or facilitating binding of an activator (Fig. 4).

Whilst the presence of parental-specific domains of DNA methylation within these imprinted genes is suggestive of a mark, it does not prove that methylation *per se* or these methyl groups in particular, functioned in that capacity. That issue was resolved when Li *et al.* (27, 47) generated a mouse containing a targeted mutation in the DNA methyltransferase gene, the gene responsible for maintaining the presence of methyl groups in DNA (Fig. 2). In homozygous mutant embryos, transcription of the normally silent paternal *H19* gene is activated at the same time that all *Igf2* expression is lost. This important result is the first to implicate DNA methylation directly in the mechanism of parental imprinting and confirms the fact that the direction of the mark is different for the two genes.

Is DNA methylation sufficient to fully explain imprinting? Are the methylation patterns inherited from the appropriate gamete, in this case from sperm, and do they survive the genome-wide demethylation event which occurs between the 8- and 32-cell stage of embryogenesis in mice (48, 49)? The first requirement is clearly met, as at least a subset of the methylation domain at *H19* and *Igf2* is present in sperm DNA. Whether these persist in early development, however, is still being investigated. At least one site in the *Igf2* promoter remains methylated throughout embryogenesis on some chromosomes, whereas none of the three methylation sites examined at *H19* do so (45). Although further work is necessary to assay all the relevant sites at both genes, the initial results suggest that the extensive methylation domains in differentiated somatic tissue are largely acquired after the early stages of embryogenesis. As to whether there are key 'nucleation' methylation sites which are established and maintained in the required manner, this remains to be determined.

The chromatin structure of this region has been examined for other signs of epigenetic markings (43, 44, 46). The promoter of the active *H19* gene is in an open conformation, as judged by access of a restriction enzyme or DNase I in chromatin. The silent paternal copy of the gene, however, is resistant to both kinds of endonucleolytic attack. In contrast, both promoters of the *Igf2* gene are in an active conformation, indicating that the maintenance of imprinting of these two genes may be achieved in different ways (see Fig. 4).

Another surprise came when the enhancers which lie 3′ to the *H19* gene were examined. These enhancers were the only regulatory elements in the entire region which had been identified, and they were considered specific to *H19* (50). However, they exhibited an open conformation on both alleles. As enhancers generally adopt open conformations only when they are engaged in transcription (51), one concludes that the *H19* enhancers are actively involved in transcription on both alleles.

The above observations led Bartolomei *et al.* (43, 52) to propose that the reciprocal imprinting of *Igf2* and *H19* is based on a competition between the genes for common regulatory elements such as the '*H19*' enhancers. On the paternal chromosome the inhibition of *H19* transcription by the allele-specific DNA methylation domain allows the enhancers to engage in *Igf2* transcription. On the apparently unmarked maternal chromosome, either the relative strength of the accessible *H19* promoter, its proximity to the enhancers, and/or the absence of methylation at the *Igf2* locus biases the competition in *H19*'s direction.

Another reason to consider the *Igf2*/*H19* locus as a single domain is the finding that it replicates as a unit, with the paternal copy replicating earlier than the maternal (18). This property of early paternal replication is shared with two other imprinted loci, the Prader–Willi region in human chromosome 15 and mouse chromosome 7, and the *Tme* region on mouse chromosome 17 (18, 19). Intriguingly, the imprinted genes examined have been found to be embedded within domains of up to 1–2 Mb showing asynchronous replication between the paternal and maternal copies. Thus imprinting may be regulated to some degree at the level of chromosomal domains that are themselves much larger than the individual imprinted loci contained within them (53). Viewed in this context, the difference between X chromosome inactivation, apparent at a chromosome-wide level, and autosomal imprinting, operating at the level of individual loci or small groups of loci, may simply be one of degree.

Imprinted genes are functionally hemizygous, as only one copy is expressed (Fig. 1). Therefore, mutations in imprinted genes will appear to be inherited in an autosomal dominant fashion in appropriate families, with transmission through only one parent. In addition, somatic mutations in imprinted genes will occur preferentially on one parental allele, as mutations in the imprinted gene will have no effect phenotypically. Likewise, when loss of heterozygosity (LOH) or trisomies occur, there should be a preferential retention or duplication of one of the two parental chromosomes. There are potentially two quite different phenotypes associated with mutations of imprinted genes: mutations that lead to loss of gene function, and mutations that lead to overexpression of the imprinted gene, possibly by inactivating the imprinting mechanism itself. In either case, there will be parental bias, a tell-tale sign of a mutation in an imprinted gene.

In humans, the imprinted *IGF2* and *H19* genes map to the proximal tip of chromosome 11, at band 15.5 (54–57), a region which has been implicated in at least two different human genetic diseases that display the genetic inheritance patterns outlined above. Beckwith–Wiedemann syndrome (BWS) is characterized by prenatal overgrowth leading to gigantism and macroglossia (an enlarged tongue), hemi-hypertrophy, and neonatal hypoglycaemia, along with a predisposition toward a variety of childhood tumours, such as hepatocellular carcinoma, rhabdomyosarcomas, and Wilm's tumour of the kidney (reviewed in 58). In its genetically transmitted form, BWS is an autosomal dominant disease, in which female carriers are more likely than male carriers to have affected children. The carrier females often display balanced germline translocations of 11p15, all of paternal origin. In sporadic cases, most exhibit paternal isodisomy of 11p15.5, while germline paternal duplications of 11p15 have been reported in some patients.

The *IGF2* gene is an excellent candidate for mutations in BWS, as a gain-of-function of *IGF2* could well explain the enhanced growth, macroglossia, and neonatal hypoglycaemia. The duplication of the paternal allele that occurs with paternal disomies and duplications would result in a doubling of IGF2 production (see Fig. 4). The disproportionate maternal inheritance of Beckwith–Wiedemann in the genetically transmitted disease could result from the inheritance of a mutation which derepresses the *IGF2* gene through perturbations in the imprinting mechanism. The identification of the translocation breakpoints in BWS patients might provide important clues as to the size of the chromosomal domain required for the appropriate imprinting of *IGF2*.

Several groups have now examined the expression of *IGF2* in BWS patients, and found biallelic expression in some but not all instances (55, 59). With respect to *H19*, the fact that trisomies as well as isodisomies are observed in BWS patients may argue that the loss of *H19*, which would only occur in the latter group, is not the major contributor to the phenotype.

Approximately 15 per cent of BWS patients develop childhood tumours of the liver, kidney, and muscle. In addition, one or more genes at 11p15.5 have been implicated in genetic susceptibility to childhood tumours such as Wilm's tumour, hepatocellular carcinoma, and rhabdomyosarcoma, leading to the proposal that a tumour-suppressor gene resides at this locus. When these tumours lose heterozygosity at 11p15.5, it is always the maternal allele which is lost. These tumours often display increased levels of *IGF2* transcripts, consistent with a gain-of-function of the growth factor, and two recent studies have shown that, in a subset of Wilm's tumours examined, biallelic expression of *IGF2*, *H19*, or both was observed (60, 61). The tissues most commonly susceptible to transformation in these syndromes are those which express *IGF2* and *H19*.

It is not clear as yet whether the gain of *IGF2* function is sufficient to explain the high incidence of tumours, or whether loss of a maternally expressed tumour-suppressor gene is also involved. Trisomic BWS patients are less likely to develop tumours than are those with no chromosome abnormalities (62). It is possible that the *H19* gene functions as a tumour suppressor, whose loss in paternal isodisomic tissue leads to tumour growth. The human *H19* gene can act as a tumour-suppressor gene in tissue culture cells (63). However, the possibility remains that another paternally imprinted gene with tumour-suppressor activity resides in the region.

4.1.2 The *Snrpn/Znf127* cluster

The second region that contains a cluster of imprinted genes maps to the central region of mouse chromosome 7 and to human chromosome 15q11–q13. It bears some striking similarities as well as differences to the *Ins-2/Igf2/H19* gene cluster. It is significantly larger, extending over several megabases of DNA, and contains at least three imprinted genes (Fig. 5). *Snrpn,* which encodes the small nuclear ribonucleoprotein polypeptide N (64), is expressed most abundantly in the brain of the mouse, and has been proposed to be involved in brain-specific splicing. Its expression is entirely paternal in both humans and mice (65, 66). Although the role of DNA methylation in establishing or maintaining the imprint on *Snrpn* has not been tested directly, the active paternal allele is selectively methylated within the fifth intron of the human *SNRPN* gene (66). This is very similar to what is observed for the methylation of the active copy of another imprinted gene, *Igf2r,* encoding the insulin-like growth factor receptor in mice (67).

Methylation differences in this region have been detected over a megabase of DNA from *SNRPN.* By screening a large contig of the region for parental-specific methylation differences in genomic DNA, a gene encoding a zinc-finger-containing protein (*ZNF127*) was identified (68). Although its pattern of expression has not been determined, a patchwork of maternal-specific and paternal-specific DNA methylation is present in the region. Between *SNRPN* and *ZNF127* lies a third DNA marker, *PW71,* which also exhibits maternal-specific methylation, although no transcript has yet been mapped to it (69) (Fig. 5).

The *SNRPN* gene maps to a region that has been implicated in a human genetic

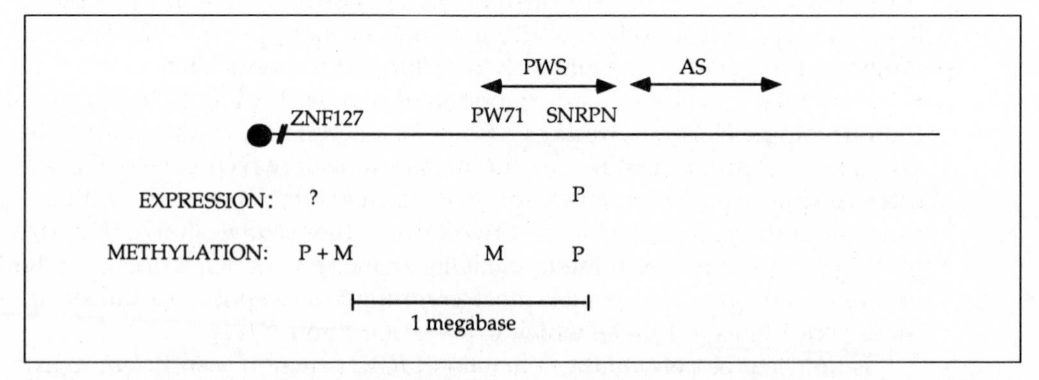

Fig. 5 The structure of the Prader–Willi and Angelman syndromes region. The positions of three markers of human chromosome 15q which have been implicated in Prader–Willi syndrome are indicated, along with the approximate chromosomal regions which define mutations which are unique to Prader–Willi and Angelman syndromes (double arrows). P and M indicate paternal and maternal expression and/or DNA methylation at the three loci. Adapted from (53).

disorder, the Prader–Willi syndrome (PWS) (70). PWS is characterized by neonatal hypotonia, severe obesity, hypogonadism, mild dysmorphism, and mild mental retardation. Immediately distal to the Prader–Willi region on human chromosome 15q is the map position of a genetically separable disorder, Angelman syndrome (AS) (70). Affected children display hyperactivity, ataxia, seizures, and severe mental retardation. These relatively rare syndromes have served as models for mutations in imprinted genes, as PWS is inherited as deletions entirely from fathers and AS is inherited as deletions entirely from mothers. Uniparental disomies have also been observed in both disorders, with maternal disomies occurring in PWS and paternal disomies in AS. These genetic observations have led to the hypothesis that the region contains at least two genes which are imprinted reciprocally, akin to *Igf2* and *H19* (70). Based on the nature of the mutations which are observed in PWS patients, it has been argued that the phenotype requires the loss of multiple gene products mapping to 15q11–q13, while AS behaves as a single gene disorder.

Insights into the regulation of this domain-wide parental methylation has come from studies of PWS and AS patients who have no deletions or uniparental disomy, that is, who have both maternal and paternal chromosomes intact (66, 71, 72). Surprisingly, disruptions in the normal methylation imprints at *ZNF127* and *PW71* are observed, in that they display uniparental patterns, with the PWS syndrome patients having only the maternal pattern and the AS syndrome patients having only the paternal pattern. This is so despite the fact that neither locus is disrupted in the patients. These studies suggest that mutations near the *SNRPN* gene are able to affect the methylation status of genes at a significant distance and implicate a mutation in the 'control region' for imprinting (53). The fact that AS patients as well as PWS patients exhibit these changes in genes that map away from the AS critical region also implies that the regulation of imprinting in these two regions has a common denominator.

4.1.3 Imprinting: summary and conclusions

The likely existence of reciprocally imprinted genes at both the *Ins-2/Igf2/H19* and *Snrpn* regions inevitably raises the question of whether all imprinted regions are similarly organized and, if so, whether the imprinting of these genes is mechanistically linked. The only other imprinted region that has been studied in some detail is the *Tme* locus on mouse chromosome 17. Only one imprinted gene, encoding the insulin-like growth factor 2 receptor gene, has been identified in this locus, while three other tightly linked genes have been shown to be biallelically expressed (73). In humans this gene is polymorphic; some individuals exhibit imprinting while others do not (74, 75). Thus its imprinting may not be controlled in the same way as the more tightly regulated loci.

Resolving the identity of the control elements responsible for imprinting will require two approaches: the use of transgenic mice to map the elements and the generation of germline mutations which disrupt them. The first approach has been successful for the *H19* gene, in that transgenic mice carrying a 14 kb transgene with

4 kb of 5′ and 8 kb of 3′ flanking sequence, along with an internally deleted structural gene, is correctly imprinted in heterologous chromosomal positions (43). As expected, the paternally inherited copy of the transgene is heavily methylated, whereas the maternally inherited copy is unmethylated. The same has not proved to be the case for *Igf2* transgenes, however, configured as 5′ flanking sequence driving β-galactosidase reporter expression (76). One possible conclusion is that the signals for imprinting for both genes lie proximal to the *H19* gene, although alternatives, including *Igf2*-specific imprinting elements not contained within the transgene, are possible.

4.2 X chromosome inactivation

In many respects, X chromosome inactivation is very similar to genomic imprinting as a form of epigenetic regulation, resulting in the silencing of one of the two homologous chromosomes in female somatic cells. At the level of individual genes, the mechanisms thought to be involved in the suppression of activity are virtually identical to those involved in genomic imprinting. There are two key differences, however, elucidation of which would go a long way towards explaining the elements of both imprinting and X inactivation.

First, X inactivation extends to much of an entire chromosome in a poorly understood *cis*-limited fashion (12). The effects of X inactivation are long range, far in excess of even the large replication domains that distinguish imprinted loci. It is unknown whether the chromosomal nature of X inactivation reflects an unusual mechanism that extends nearly the full length of the chromosome, or whether the chromosome consists of a series of much smaller, individual domains, each of which may be regulated precisely as in imprinting domains. Evidence in favour of the latter possibility has derived from experiments to address the distribution of genes that are subject to X inactivation and those that continue to be expressed from the otherwise inactive X, i.e., those said to 'escape' inactivation. To date, approximately forty human X-linked genes have been examined in a rodent/human somatic cell hybrid system in which the active and inactive human X chromosomes can be separated from each other and assayed independently (77). The available data, while limited, suggest that genes are clustered into groups that are subject to inactivation and other groups that escape inactivation (78). A detailed analysis of any given domain, including identification of many genes and systematic evaluation of X inactivation, DNA replication, DNA methylation, and associated chromatin alterations, is thus far lacking, however, and will be required to test this model definitively. Comparison of the mouse and human X chromosomes will be helpful since, with one exception (79, 80), all genes on the mouse X have been shown to be subject to X inactivation. Differences between homologous domains in the two species will be revealing.

A second major difference between mammalian X inactivation and genomic imprinting is the stability of the mark. In the mouse, it is apparent that the earliest form of X inactivation, as seen in the extra-embryonic trophectoderm and primitive endoderm lineages, is imprinted and is restricted to the paternally derived X chromosome (81). However, X inactivation in embryonic lineages, as occurs later during gastrulation, is random (1), affecting either the maternally or paternally derived X chromosome in any given cell (Fig. 1), reflecting the loss or instability of the original paternal imprint.

4.2.1 XIST, the X inactivation centre, and imprinting

Studies of abnormal X chromosomes in humans (82–84) and in mouse (85) have demonstrated that X inactivation involves the action of a locus, the X inactivation centre (*XIC* in humans, *Xic* in mouse), required in *cis* for X inactivation to occur (78). X inactivation can be conceptually divided into three stages: initiation in early development, during which the number of copies of *XIC* is 'counted' as a critical step in choosing one X to remain active; promulgation or spread of the inactivation signal in *cis* along the X chromosome(s) chosen to be inactive; and clonal maintenance of the inactive state through subsequent cell divisions. In principle, the *XIC* could be involved in any of these steps.

The *XIST (Xist)* gene has been cloned in both human and mouse; it is a candidate for the *XIC* since it maps to the region of the *XIC* in both species and since it is expressed exclusively from inactive X chromosomes (86–88). The product of the *XIST* locus is a long, untranslated RNA transcript that remains in the nucleus (89, 90) and is associated with the inactive X chromosome and with the Barr body in interphase nuclei (90). The organization and sequence of *XIST* are quite highly conserved in many mammalian species despite the lack of a significant open reading frame (78, 86–91).

Developmental studies have strongly supported a role for *Xist* in the initiation of X inactivation in mouse, as *Xist* expression is first detectable as early as the four-cell stage (92), prior to X inactivation. This expression is imprinted; only the paternally derived *Xist* allele is expressed in these early stages (92, 93) (Fig. 6). Any mark preventing expression on the maternal X must be erased subsequently in the embryo proper, since expression of the maternal *Xist* gene can be detected just prior to gastrulation (93). The imprinted *Xist* expression (restricted to the paternal X) is downregulated at the compacting morula stage, prior to random X chromosome choice as a result of an X chromosome counting mechanism (after which either the maternal or paternal *Xist* allele can be expressed in a given cell) (92). The erasure of the imprint is a result of the genome-wide demethylation that occurs during pre-implantation development between the eight-cell and blastocyst stages (21, 26).

According to one model of X inactivation, *XIST* RNA could be the *cis*-acting agent of X inactivation, either by itself or in concert with other factors. Thus, only the X chromosome expressing *XIST* (the paternal X in extra-embryonic lineages and either the maternal or paternal X in the embryo) would be capable of being silenced (Fig. 6). In humans, however, continued *XIST* expression is apparently not strictly required for maintenance of the inactive state in cultured cell lines (94, 95). This observation leaves open the question of why *XIST* is constitutively

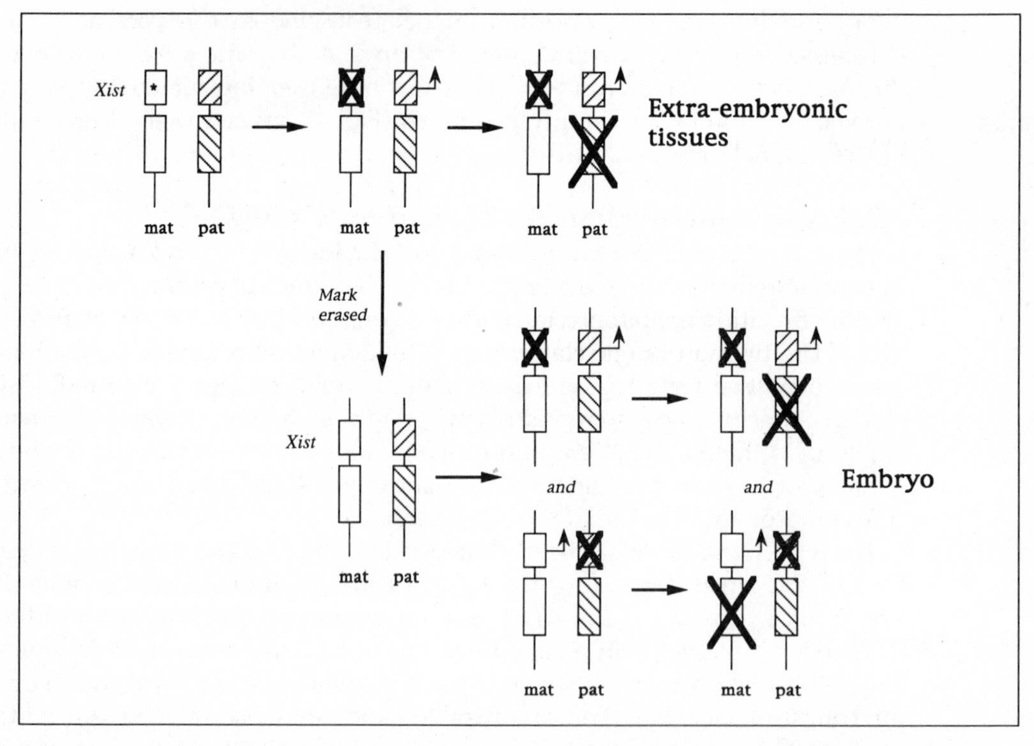

Fig. 6 Involvement of *Xist* and imprinting in X inactivation. *Xist* expression in early development is imprinted, due to a hypothesized mark (*) on the maternally derived *Xist* gene. In the early cleavage stages, the paternal *Xist* allele is expressed and the maternal *Xist* allele is silenced. This precedes non-random inactivation of the paternal X chromosome in extra-embryonic tissues. In the embryo itself, the mark is lost or erased, since *Xist* expression and X inactivation are random.

expressed from inactive Xs and what role such continued expression might play. There may be redundant levels of epigenetic regulation on the inactive X, such that removal of any one (i.e. deletion of the *XIST* gene (94, 95)) is not alone sufficient to reactivate the chromosome.

4.2.2 Role of DNA methylation

The role of DNA methylation in X inactivation can be considered from three perspectives—imprinting, initiation of X inactivation, and maintenance of the inactive state. There is abundant evidence that DNA methylation is involved in the maintenance of X inactivation (12), since reversal of methylation causes reactivation of previously inactivated genes (26). Further, analysis of the promoter region of two X-linked genes (*PGK1* and *HPRT*) using the genomic sequencing/methylation analysis method (30) has shown an extensive correlation between methylation/inactivation and demethylation/reactivation (96–98). Similarly, there are data that implicate methylation of the *XIST* promoter in the silencing of *XIST* on active X chromosomes in human, mouse, and cow (91, 99). These data clearly support a role for DNA methylation in the stability and somatic perpetuation of the inactive state of individual genes.

The data in early development are less clear, however. It is uncertain in most instances whether DNA methylation precedes or follows X inactivation (100, 101). In early pre-implantation embryos of mice, in which both X chromosomes are active, there is no detectable methylation at several genes at least up to the blastocyst stage. After this, different genes adopt different methylation profiles. Methylation at the *Pgk1* gene appears to precede that at the *G6pd* locus, but there is complete methylation at both by day 5.5 (101). Data on the *hprt* gene indicated that methylation is not detectable until day 8.5, well after X inactivation is initiated (100). The timing of methylation might be a function of the distance of individual genes from the *Xist* gene and the *Xic* (101). This issue notwithstanding, it is clear that gene inactivation can occur (at least for some genes) in the absence of methylation, since inactive X genes are not methylated (at least at the sites examined) in female primordial germ cells (101) or in embryonal carcinoma cells with an inactive X (102).

All of these data imply that, while DNA methylation at the promoter of individual genes is associated with X inactivation, it is probably as a consequence of a primary event elsewhere on the chromosome, presumably at the *Xic*. It remains an open question whether specific methylation events at the *Xist* gene (as seen, for example, in embryonal stem cells prior to X inactivation (99)) might represent this primary event. None the less, this conclusion and these studies can always be criticized for considering only a limited number of genes or CpG sites on the X chromosome. A more direct approach, in which DNA methylation is interfered with, either globally or at specific sites, is desperately needed.

A role for 'pre-emptive' DNA methylation in imprinting of the *Xist* gene has been demonstrated (99) and is largely consistent with the work summarized above implicating DNA methylation as the mark in autosomal imprinted genes. In extra-embryonic lineages, the paternally derived copy of *Xist* is fully unmethylated, whereas the maternally derived copy is fully methylated (99), as is the inactive copy of *Xist* in male cells (91, 99). In the male germline, developmentally regulated demethylation of *Xist* occurs at the onset of meiosis. This demethylation presumably renders the *Xist* promoter competent for transcription, both in male germ cells and in female embryos following fertilization. Viewed from the opposite perspective, it is the methylated state of the maternal *Xist* allele which represents the mark, since it is this imprinting event in female meiosis that appears to be responsible for silencing the maternal copy of *Xist* (93, 99) (Fig. 6). The differential methylation of the maternal and paternal copies of *Xist* apparently underlies the imprinted expression of the paternal *Xist* allele in early development and non-random inactivation of the paternal X chromosome.

4.2.3 Role of chromatin structure

Alterations in chromatin structure have been implicated both in the inactivation of specific genes and in the chromosomal aspects of X inactivation. Studies on the human *PGK1* gene (*in vivo* footprinting assay) have demonstrated that the active X promoter is associated with several transcription factors and is accessible to restriction enzyme cleavage (96, 97), whereas the inactive X *PGK1* promoter is completely blocked by positioned particles (presumably nucleosomes). In contrast to data demonstrating the presence of active X-specific DNase-I-hypersensitive sites at the promoters of several X-linked genes (e.g., 103), both the active X and inactive X *PGK1* genes were equally sensitive to general DNase I digestion (97). The data on *PGK1* (and similar data on the X-linked *HPRT* gene (104)) suggest that chromatin structure and methylation differences between copies of genes on the active or inactive X may be restricted to promoter regions (96, 97).

This restricted view of chromatin differences is difficult to reconcile with the cytological presence of heterochromatin (represented by the Barr body) or with the chromosome-wide histone H4 underacetylation observed on the inactive X (21), both of which suggest a more global effect. The Barr body is associated with much of the DNA of the inactive X chromosome and with the *XIST* RNA product (90). Exactly which DNA sequences are included or excluded from the Barr body (and which DNA sequences are associated with the *XIST* RNA) should be amenable to experimentation, using fluorescence hybridization *in situ*. It would be very illuminating to examine parts of the chromosome known to contain genes that escape X inactivation, for example. Many questions remain: What proteins are associated with the Barr body? What determines its specificity for the inactive X chromosome? How is it associated with the nuclear membrane? What happens to the Barr body and to the *XIST* RNA during mitosis and during the interphase–metaphase reorganization of chromatin? The composition and function of the Barr body are essentially unknown; this problem is badly in need of novel analytical approaches.

5. Final comments

The understanding of epigenetic regulation in mammalian cells has advanced significantly in the past few years, based on some of the experimental work discussed in this chapter. In addition to the details of individual case studies, however, a number of broader, global questions remain to be answered.

5.1 Discussion

Several interesting questions are raised by comparison of the mammalian system to the systems discussed in Chapters 7 and 8. Many organisms, such as *Drosophila*, manage life without DNA methylation. Is methylation really essential in mammals? Yes: while mouse cells lacking the major methyltransferase activity are viable, the homozygous mutation is lethal in the developing embryo, suggesting a crucial role for methylation during development and differentiation (27). It is not clear what defines the need for methylation; one wonders if there is some functionally analogous but otherwise different process in *Drosophila*. Are there modifiers of X-activation in the same way that there are modifiers (suppressors and enhancers) of PEV in *Drosophila*? Certainly the loss of methyltransferase activity acts as a suppressor; however, there has not been any systematic screen for such mutations.

5.2 Important questions

A critical issue in the mammalian system is to determine what event(s) are primary in both genomic imprinting and X inactivation—methylation? altered replication? chromosomal control elements? Much of the current research in this area is necessarily descriptive. No matter how well done, such studies are doomed to be correlative only. As geneticists, we are inclined to state the obvious: we need mutants! The recently described Prader–Willi syndrome patients with methylation abnormalities (69, 71, 72) may be very important in this respect. Similarly, mutants at the *Xce* locus in mouse (105), closely linked to and perhaps related to the *Xic* and to the *Xist* gene, should be very valuable for sorting out the relative roles and mechanisms of imprinting and *Xic* counting in the early stages of X inactivation. Human mutants in X inactivation would be extraordinarily useful and interesting, although it is currently difficult to know exactly how to detect such mutants. Rare familial cases of non-random X inactivation may be one possibility.

What is the genetic and molecular basis of the X inactivation centre? Are there 'imprinting centres' that are related mechanistically or structurally? The parallels between these two types of epigenetic regulation are striking, and it would be surprising perhaps if there were not chromosomal control elements shared between the two. How do such chromosomal control elements influence DNA methylation or replication (106, 107)? The circumstantial case for similarities between the *XIST* and *H19* genes, with their non-coding RNAs, is intriguing.

What establishes the number and extent of epigenetic domains? Replication data on imprinted domains and on late replication in X inactivation suggest chromosome models that invoke higher order chromatin effects and organization in the control of epigenetics (12, 53). Transgenic models in which one can manipulate different large chromosomal regions would appear to be a very powerful system to search for potential boundary elements both in imprinting and in X inactivation and to evaluate chromosomal, regional, or genic models of epigenetic control.

Lastly, why do we have epigenetic regulation? How did this unusual form of genetic control evolve? What are the consequences to the organism or to the cell in the absence of epigenetic regulation? To what extent and for what genes is double-dose expression from the X chromosome or from imprinted regions harmful? The development of new experimental approaches should at last provide a basis for testing critical hypotheses to address such questions and to integrate complementary perspectives on the chromosomal and epigenetic control of gene expression.

References

1. Lyon, M. (1961) Gene action in the X chromosome of the mouse (*Mus musculus* L.). *Nature*, **190**, 372.
2. Blackman, M. A. and Koshland, M. E. (1985) Specific 5′ and 3′ regions of the μ-chain gene are undermethylated at distinct stages of B-cell differentiation. *Proc. Natl Acad. Sci. USA*, **82**, 3809.
3. Kelley, D. E., Pollok, B. A., Atchison, M. L., and Perry, R. P. (1988) The coupling between enhancer activity and hypomethylation of κ immunoglobulin genes is developmentally regulated. *Mol. Cell. Biol.*, **8**, 930.
4. Busslinger, M., Hurst, J., and Flavell, R. A. (1983) DNA methylation and the regulation of globin gene expression. *Cell*, **34**, 197.
5. Bestor, T. H. and Ingram, V. M. (1983) Two DNA methyltransferases from murine erythroleukemia cells: purification, sequence specificity and mode of interaction with DNA. *Proc. Natl Acad. Sci. USA*, **82**, 2674.
6. Gruenbaum, Y., Cedar, H., and Razin, A. (1982) Substrate and sequence specificity of a eukaryotic DNA methylase. *Nature*, **295**, 620.
7. Riggs, A. D. (1975) X inactivation, differentiation, and DNA methylation. *Cytogenet. Cell Genet.*, **14**, 9.
8. Surani, M. A. (1993) Silence of the genes. *Nature*, **366**, 302.
9. Nicholls, R. D. (1993) Genomic imprinting and candidate genes in the Prader–Willi and Angelman syndromes. *Curr. Opinion Genet. Dev.*, **3**, 445.
10. Stewart, C. L. (1993) Genomic imprinting in the regulation of mammalian development. *Adv. Dev. Biol.*, **2**, 73.
11. Brown, C. J. and Willard, H. F. (1993) Molecular and genetic studies of human X chromosome inactivation. *Adv. Dev. Biol.*, **2**, 37.
12. Riggs, A. D. and Pfeifer, G. P. (1992) X chromosome inactivation and cell memory. *Trends Genet.*, **8**, 169.
13. Morishima, A., Grumbach, M., and Taylor, J. H. (1962) Asynchronous duplication of human chromosomes and the origin of the sex chromatin. *Proc. Natl Acad. Sci. USA*, **48**, 756.
14. Willard, H. F. and Latt, S. A. (1976) Analysis of DNA replication in human X chromosomes by fluorescence microscopy. *Am. J. Hum. Genet.*, **28**, 213.
15. Selig, S., Okamura, K., Ward, D. C., and Cedar, H. (1992) Delineation of DNA replication time zones by fluorescence *in situ* hybridization. *EMBO J.*, **11**, 1217.
16. Hansen, R. S., Canfield, T. K., Lamb, M. M., Gartler, S. M., and Laird, C. D. (1993) Association of fragile X syndrome with delayed replication of the FMR1 gene. *Cell*, **73**, 1403.
17. Furst, A., Brown, E. H., Braunstein, J. D., and Schildkraut, C. L. (1981) Alpha globin sequences are located in a region of early replicating DNA in murine erythroleukemia cells. *Proc. Natl Acad. Sci. USA*, **78**, 1023.
18. Kitsberg, D., Selig, S., Brandeis, M., Simon, I., Keshet, I., Driscoll, D. J., Nicholls, R. D., and Cedar, H. (1993) Allele-specific replication timing of imprinted gene regions. *Nature*, **364**, 459.
19. Knoll, J. H., Cheng, S. D., and Lalande, M. (1994) Allele specificity of DNA replication timing in the Angelman/Prader–Willi syndrome imprinted chromosomal region. *Nature Genet.*, **6**, 41.
20. Barr, M. L. and Bertram, E. G. (1949) A morphological distinction between neurons

of the male and female, and the behavior of the nucleolar satellite during accelerated nucleoprotein synthesis. *Nature*, **163,** 676.
21. Jeppesen, P. and Turner, B. M. (1993) The inactive X chromosome in female mammals is distinguished by a lack of histone H4 acetylation, a cytogenetic marker for gene expression. *Cell*, **74,** 281.
22. Turner, B. (1990) Histone acetylation and control of gene expression. *J. Cell Sci.*, **99,** 13.
23. Elgin S. (1981) DNase I-hypersensitive sites of chromatin. *Cell*, **27,** 413.
24. Bird, A (1992) The essentials of DNA methylation. *Cell*, **70,** 5.
25. Pfeifer, G., Steigerwald, S., Mueller, P., Wold, B., and Riggs, A. D. (1989) Genomic sequencing and methylation analysis by ligation mediated PCR. *Science*, **246,** 810.
26. Mohandas, T., Sparkes, R., and Shapiro, L. (1981) Reactivation of an inactive human X chromosome: evidence for X inactivation by DNA methylation. *Science*, **211,** 393.
27. Li, E., Bestor, T. H. and Jaenisch, R. (1992) Targeted mutation of the DNA methyltransferase gene results in embryonic lethality. *Cell*, **69,** 915.
28. Graham, C. F. (1974) The production of parthenogenetic mammalian embryos and their use in biological research. *Biol. Rev.*, **49,** 399.
29. Kaufman, M. H. (1983) *Early mammalian development: parthenogenetic studies*. Cambridge University Press, Cambridge/New York.
30. McGrath, J. and Solter, D. (1983) Nuclear transplantation in mouse embryos. *J. Exp. Zool.*, **228,** 355.
31. McGrath, J. and Solter, D. (1984) Completion of mouse embryogenesis requires both the maternal and paternal genomes. *Cell*, **37,** 179.
32. McGrath, D. and Solter, D. (1986) Nucleocytoplasmic interactions in the mouse embryo. *J. Embryol. Exp. Morphol.*, **97,** 277.
33. Surani, M. A. H., Barton, S. C., and Norris, M. L. (1986) Nuclear transplantation in the mouse: heritable differences between parental genomes after activation of the embryonic genome. *Cell*, **45,** 127.
34. DeChiara, T. M., Efstratiadis, A., and Robertson, E. J. (1990) A growth-deficiency phenotype in heterozygous mice carrying an insulin-like growth factor II gene disrupted by targeting. *Nature*, **345,** 78.
35. DeChiara, T. M., Robertson, E. J., and Efstratiadis, A. (1991) Parental imprinting of the mouse insulin-like growth factor II gene. *Cell*, **64,** 849.
36. Bartolomei, M. S., Zemel, S., and Tilghman, S. M. (1991) Parental imprinting of the mouse H19 gene. *Nature*, **351,** 153.
37. Pachnis, V., Brannan, C. I., and Tilghman, S. M. (1988) The structure and expression of a novel gene activated in early mouse embryogenesis. *EMBO J.*, **7,** 673.
38. Brannan, C. I., Dees, E. C., Ingram, R. S., and Tilghman, S. M. (1990) The product of the H19 gene may function as an RNA. *Mol. Cell. Biol.*, **10,** 28.
39. Giddings, S. J., King, C. D., Harman, K. W., Flood, J. F., and Carnaghi, L. R. (1994) Allele specific inactivation of insulin 1 and 2, in the mouse yolk sac, indicates imprinting. *Nature Genet.*, **6,** 310.
40. Zemel, S., Bartolomei, M. S., and Tilghman, S. M. (1992) Physical linkage of two mammalian imprinted genes. *Nature Genet.*, **2,** 61.
41. Lee, J. E., Pintar, J., and Efstratiadis, A. (1990) Pattern of the insulin-like growth factor II gene expression during early mouse embryogenesis. *Development*, **110,** 151.
42. Poirier, F., Chan, C.-T. J., Timmons, P. M., Robertson, E. J., Evans, M. J., and Rigby, P. W. J. (1991) The murine H19 gene is activated during embryonic stem cell differen-

tiation *in vitro* and at the time of implantation in the developing embryo. *Development*, **113,** 1105.

43. Bartolomei, M. S., Webber, A. L., Brunkow, M. E., and Tilghman, S. M. (1993) Epigenetic mechanisms underlying the imprinting of the mouse H19 gene. *Genes and Dev.*, **7,** 1663.
44. Ferguson-Smith, A. C., Sasaki, H., Cattanach, B. M., and Surani, M. A. (1993) Parental-origin-specific epigenetic modifications of the mouse H19 gene. *Nature*, **362,** 751.
45. Brandeis, M., Kafri, T., Ariel, M., Chaillet, J. R., McCarrey, J., Razin, A., and Cedar, H. (1993) The ontogeny of allele-specific methylation associated with imprinted genes in the mouse. *EMBO J.*, **12,** 3669.
46. Sasaki, H., Jones, P. A., Chaillet, J. R., Ferguson-Smith, A. C., Barton, S., Reik, W., and Surani, M. A. (1992) Parental imprinting: potentially active chromatin of the repressed maternal allele of the mouse insulin-like growth factor (*Igt2*) gene. *Genes and Dev.*, **6,** 1843.
47. Li, E., Beard, C., and Jaenisch, R. (1993) The role of DNA methylation in genomic imprinting. *Nature*, **366,** 362.
48. Monk, M., Boubelik, M., and Lehnert, S. (1987) Temporal and regional changes in DNA methylation in the embryonic, extraembryonic and germ cell lineages during mouse embryo development. *Development*, **99,** 371.
49. Kafri, T., Ariel, M., Brandeis, M., Shemer, R., Urven, L., McCarrey, J., Cedar, H., and Razin, A. (1992) Developmental pattern of gene-specific DNA methylation in the mouse embryo and germ line. *Genes and Dev.*, **6,** 705.
50. Yoo-Warren, H., Pachnis, V., Ingram, R. S., and Tilghman, S. M. (1988) Two regulatory domains flank the mouse H19 gene. *Mol. Cell. Biol.*, **8,** 4707.
51. Reitman, M., Lee, E., Westphal, H., and Felsenfeld, G. (1993) An enhancer/locus control region is not sufficient to open chromatin. *Mol. Cell. Biol.*, **13,** 3990.
52. Bartolomei, M. and Tilghman, S. M. (1992) Parental imprinting of mouse chromosome 7. *Semin. Dev. Biol.*, **3,** 107.
53. Nicholls, R. D. (1994) New insights reveal complex mechanisms involved in genomic imprinting. *Am. J. Hum. Genet.*, **54,** 733.
54. Giannoukakis, N., Deal, C., Paquette, J., Goodyer, C. G., and Polychronakos, C. (1993) Parental genomic imprinting of the human IGF2 gene. *Nature Genet.*, **4,** 98.
55. Ohlsson, R., Nystrom, A., Pfeifer-Ohlsson, S., Tohonen, V., Hedborg, F., Schofield, P., Flam, F., and Ekstrom, T. J. (1993) IGF2 is parentally imprinted during human embryogenesis and in the Beckwith–Wiedemann syndrome. *Nature Genet.*, **4,** 94.
56. Zhang, Y. and Tycko, B. (1992) Monoallelic expression of the human H19 gene. *Nature Genet.*, **1,** 40.
57. Zhang, Y., Shields, T., Crenshaw, T., Hao, Y., Moulton, T., and Tycko, B. (1993) Imprinting of human H19: allele-specific CpG methylation, loss of the active allele in Wilm's tumor, and potential for somatic allele switching. *Am. J. Hum. Genet.*, **53,** 113.
58. Junien, C. (1992) Beckwith–Wiedemann syndrome, tumourigenesis and imprinting. *Curr. Opinion Genet. Dev.*, **2,** 431.
59. Weksberg, R., Shen, D. R., Fei, Y. L., Song, Q. L., and Squire, J. (1993) Disruption of insulin-like growth factor 2 imprinting in Beckwith–Wiedemann syndrome. *Nature Genet.*, **5,** 143.
60. Rainier, S., Johnson, L. A., Dobry, C. J., Ping, A. J., Grundy, P. E., and Feinberg, A. P. (1993) Relaxation of imprinted genes in human cancer. *Nature*, **362,** 747.

61. Ogawa, O., Eccles, M. R., Szeto, J., McNoe, L. A., Yun, K., Maw, M. A., Smith, P. J., and Reeve, A. E. (1993) Relaxation of insulin-like growth factor II gene imprinting implicated in Wilm's tumour. *Nature,* **362,** 749.
62. Little, M., Van Heyningen, V., and Hastie, N. (1991) Dads and disomy and disease. *Nature,* **351,** 609.
63. Hao, Y., Crenshaw, T., Moulton, T., Newcomb, E., and Tycko, B. (1993) Tumour-suppressor activity of H19 RNA. *Nature,* **365,** 764.
64. Leff, S. E., Brannan, C. I., Reed, M. L., Ozcelik, T., Francke, U., Copeland, N. G., and Jenkins, N. A. (1992) Maternal imprinting of the mouse *Snrpn* gene and conserved linkage homology with the human Prader–Willi syndrome region. *Nature Genet.,* **2,** 259.
65. Reed, M. L. and Leff, S. E. (1994) Maternal imprinting of human SNRPN, a gene deleted in Prader–Willi syndrome. *Nature Genet.,* **6,** 163.
66. Glenn, C. C., Porter, K. A., Jong, M. T., Nicholls, R. D., and Driscoll, D. J. (1993) Functional imprinting and epigenetic modification of the human SNRPN gene. *Hum. Mol. Genet.,* **2,** 2001.
67. Stoger, R., Kubicka, P., Liu, C.-G., Kafri, T., Razin, A., Cedar, H., and Barlow, D. P. (1993) Maternal-specific methylation of the imprinted mouse *Igf2r* locus identifies the expressed locus as carrying the imprinting signal. *Cell,* **73,** 61.
68. Driscoll, D. J., Waters, M. F., Williams, C. A., Zori, R. T., Glenn, C. C., Avidano, K. M., and Nicholls, R. D. (1992) A DNA methylation imprint, determined by the sex of the parent, distinguishes the Angelman and Prader–Willi Syndromes. *Genomics,* **13,** 917.
69. Glenn, C. C., Nicholls, R. D., Robinson, W. P., Saitoh, S., Niikawa, N., Schinzel, A., Horsthemke, B., and Driscoll, D. J. (1993) Modification of 15q11-q13 DNA methylation imprints in unique Angelman and Prader–Willi patients. *Hum. Mol. Genet.,* **2,** 1377.
70. Knoll, J. H. M., Nicholls, R. D., Magenis, R. E., Graham, J. M. J., Lalande, M., and Latt, S. A. (1989) Angelman and Prader–Willi syndromes share a common chromosome 15 deletion but differ in parental origin of the deletion. *Am. J. Med. Genet.,* **32,** 285.
71. Reis, A., Dittrich, B., Greger, V., Buiting, K., Lalande, M., Gillessen-Kaesbach, G., Anvret, M., and Horsthemke, B. (1994) Imprinting mutations suggested by abnormal DNA methylation patterns in familial Angelman and Prader–Willi syndromes. *Am. J. Hum. Genet.,* **54,** 741.
72. Buiting, K., Dittrich, B., Robinson, W. P., Guitart, M., Abeliovich, D., Lerer, I., and Horsthemke, B. (1994) Detection of aberrant DNA methylation in unique Prader–Willi syndrome patients and its diagnostic implications. *Hum. Mol. Genet.,* **3,** 893.
73. Barlow, D. P., Stoger, R., Herrmann, B. G., Saito, K., and Schweifer, N. (1991) The mouse insulin-like growth factor type-2 receptor is imprinted and closely linked to the *Tme* locus. *Nature,* **349,** 84.
74. Xu, Y., Goodyer, C. G., Deal, C., and Polychronakos, C. (1993) Functional polymorphism in the parental imprinting of the human IGF2R gene. *Biochem. Biophys. Res. Commun.,* **197,** 747.
75. Kalscheuer, V. M., Mariman, E. C., Schepens, M. T., Rehder, H., and Ropers, H. H. (1993) The insulin-like growth factor type-2 receptor gene is imprinted in the mouse but not in humans. *Nature Genet.,* **5,** 74.
76. Lee, J. E., Tantravahi, U., Boyle, A. L., and Efstratiadis, A. (1993) Parental imprintng of an *Igf-2* transgene. *Mol. Reprod. Dev.,* **35,** 382.

77. Brown, C. J., Flenniken, A., Williams, B. R. G., and Willard, H. F. (1990) X chromosome inactivation of the human TIMP gene. *Nucleic Acids Res.*, **18,** 4191.
78. Willard, H. F., Brown, C. J., Carrel, L., Hendrich, B., and Miller, A. P. (1993) Epigenetic and chromosomal control of gene expression: molecular and genetic analysis of X chromosome inactivation. *Cold Spring Harbor Symp. Quant. Biol.*, **58,** 315.
79. Wu, J., Ellison, J., Salido, E., Yen, P., Mohandas, T., and Shapiro, L. J. (1994) Isolation and characterization of XE169, a novel human gene that escapes inactivation. *Hum. Mol. Genet.*, **3,** 153.
80. Agulnik, A. I., Mitchell, M., Mattei, M., Borsani, G., Avner, P. A., Lerner, J. L., and Bishop, C. E. (1994) A novel X gene with a widely transcribed Y-linked homologue escapes X inactivation in mouse and human. *Hum. Mol. Genet.*, **3,** 879.
81. Takagi, N. and Sasaki, M. I. (1975) Preferential inactivation of the paternally-derived X chromosome in the extraembryonic membranes of the mouse. *Nature*, **256,** 640.
82. Therman, E., Sarto, G. E., Palmer, C., Kalio, H., and Denniston, C. (1979) Position of the human X inactivation center on Xq. *Hum. Genet.*, **50,** 59.
83. Brown, C. J., Lafreniere, R. G., Powers, V., Sebastio, G., Ballabio, A., Pettigrew, A., Ledbetter, D., Levy, E., Craig, I. W., and Willard, H. F. (1991) Localization of the X inactivation centre on the human X chromosome in Xq13. *Nature*, **349,** 82.
84. Leppig, K., Brown, C. J., Bressler, S., Gustashaw, K., Pagon, R., Willard, H. F., and Disteche, C. (1993) Mapping of the distal boundary of the X inactivation center in a rearranged X chromosome from a female expressing XIST. *Hum. Mol. Genet.*, **2,** 883.
85. Rastan, S. and Robertson, E. J. (1985) X chromosome deletions in embryo-derived cell lines associated with lack of X chromosome inactivation. *J. Embryol. Exp. Morph.*, **90,** 379.
86. Brown, C. J., Ballabio, A., Rupert, J., Lafreniere, R., Grompe, M., Tonlorenzi, R., and Willard, H. F. (1991) A gene from the region of the human X inactivation centre is expressed exclusively from the inactive X chromosome. *Nature*, **349,** 38.
87. Borsani, G., Tonlorenzi, R., Simmler, M., Dandalo, L., Arnaud, D., Capre, V., Grompe, M., Pizzati, A., Muzay, D., Lawrence, C., Willard, H., Avner, P., and Ballabio, A. (1991) Characterization of a murine gene expressed from the inactive X chromosome. *Nature*, **351,** 325.
88. Brockdorff, N., Ashworth, A., Kay, G. F., Cooper, P., Smith, S., McCabe, V., Norris, D., Penny, G., Patel, D., and Rastan, S. (1991) Conservation of position and exclusive expression of mouse Xist from the inactive X chromosome. *Nature*, **351,** 329.
89. Brockdorff, N., Ashworth, A., Kay, G., McCabe, V., Norris, D., Cooper, D., Swift, S., and Rastan, S. (1992) The product of the mouse Xist gene is a 15 kb inactive X-specific transcript containing no conserved ORF and located in the nucleus. *Cell*, **71,** 515.
90. Brown, C. J., Hendrich, B., Rupert, J., Lafreniere, R., Xing, Y., Lawrence, J., and Willard, H. F. (1992) The human XIST gene: analysis of a 17 kb inactive X-specific RNA that contains conserved repeats and is highly localized within the nucleus. *Cell*, **71,** 527.
91. Hendrich, B. D., Brown, C., and Willard, H. F. (1993) Evolutionary conservation of possible functional domains of the human and murine XIST genes. *Hum. Mol. Genet.*, **2,** 663.
92. Kay, G. F., Barton, S. C., Surani, M. A., and Rastan, S. (1994) Imprinting and X chromosome counting mechanisms determine Xist expression in early mouse development. *Cell*, **77,** 639.

93. Kay, G. F., Penny, G. D., Patel, D., Ashworth, A., Brockdorff, N., and Rastan, S. (1993) Expression of *Xist* during mouse development suggests a role in the initiation of X chromosome inactivation. *Cell*, **72,** 171.
94. Brown, C. J. and Willard, H. F. (1994) The human X inactivation centre is not required for the maintenance of X chromosome inactivation. *Nature*, **368,** 154.
95. Rack, K., Chelly, J., Gibbons, R., Rider, S., Benjamin, D., Lafreniere, R., Oscier, D., Hendriks, R., Craig, I., Willard, H., Monaco, A., and Buckle, V. J. (1994) XIST is not required for the maintenance of X inactivation in acquired isodicentric X chromosomes. *Hum. Mol. Genet.*, **3,** 1053.
96. Pfeifer, G., Steigerwald, S., Hensen, R., Gartler, S., and Riggs, A. D. (1990) Polymerase chain reaction-aided genomic sequencing of an X chromosome-linked CpG island: methylation patterns suggest clonal inheritance, CpG site autonomy, and an explanation of activity state stability. *Proc. Natl Acad. Sci. USA*, **87,** 8252.
97. Pfeifer, G., Tanguay, R., Steigerwald, S., and Riggs, A. D. (1990) *In vivo* footprint and methylation analysis by PCR-aided genomic sequencing: comparison of active and inactive X chromosomal DNA at the CpG island and promoter of human PGK1. *Genes and Dev.*, **4,** 1277.
98. Hornstra, I. K. and Yang, T. P. (1994) High-resolution methylation analysis of the human HPRT gene 5′ region on the active and inactive X chromosomes: correlation with binding sites for transcription factors. *Mol. Cell. Biol.*, **14,** 1419.
99. Norris, D. P., Patel, D., Kay, G. F., Penny, G. D., Brockdorff, N., Sheardown, S., and Rastan, S. (1994) Evidence that random and imprinted Xist expression is controlled by preemptive methylation. *Cell*, **77,** 41.
100. Lock, L., Takagi, N., and Martin, G. (1987) Methylation of the hprt gene on the inactive X occurs after chromosome inactivation. *Cell*, **48,** 39.
101. Grant, M., Zuccotti, M., and Monk, M. (1992) Methylation of CpG sites of two X-linked genes coincides with X inactivation in the female mouse embryo but not in the germ line. *Nature Genet.*, **2,** 161.
102. Bartlett, M. H., Adra, C. N., Park, J., Chapman, V., and McBurney, M. W. (1991) DNA methylation of two X chromosome genes in female somatic and embryonal carcinoma cells. *Somat. Cell Mol. Genet.*, **17,** 35.
103. Wolf, S. and Migeon, B. R. (1985) Clusters of CpG dinucleotides implicated by nuclease hypersensitivity as control elements of housekeeping genes. *Nature*, **314,** 467.
104. Hornstra, I. K. and Yang, T. P. (1992) Multiple in vivo footprints are specific to the active allele of the X-linked human hypoxanthine phosphoribosyltransferase gene 5′ region: implications for X chromosome inactivation. *Mol. Cell. Biol.*, **12,** 5345.
105. Heard, A. and Avner, P. (1994). Role play in X inactivation. *Hum. Mol. Genet.*, **3,** in press.
106. Razin, A. and Cedar, H. (1994) DNA methylation and genomic imprinting. *Cell*, **77,** 473.
107. Bird, A. P. (1993) Imprints on islands. Recent results with parentally imprinted mouse genes suggest that the imprinting mechanism may involve de novo methylation of CpG islands. *Curr. Biol.*, **3,** 275.
108. Hayashizaki, Y., Shibata, H., Hirotsune, S., Sugino, H., Okazaki, Y., Sasaki, N., Hirose, K., Imoto, H., Okuizumi, H., Muramatsu, M., Komatsubara, H., Shiroishi, T., Moriwaki, K., Katsuki, M., Hatano, N., Sasaki, H., Ueda, T., Mise, N., Takagi,

N., Plass, C., and Chapman, V. M. (1994) Identification of an imprinted U2af binding protein related sequence on mouse chromosome 11 using the RLGS method. *Nature Genet.*, **6**, 33.

109. Jinno, Y., Yun, K., Nishiwaki, K., Kubota, T., Ogawa, O., Reeve, A. E., and Niikawa, N. (1994) Mosaic and polymorphic imprinting of the WT1 gene in humans. *Nature Genet.*, **6**, 305.

Index